Nuclear reactor systems

A technical, historical and dynamic approach

Bertrand Barré,
Pascal Anzieu, Richard Lenain, Jean-Baptiste Thomas

EDP Sciences
17, avenue du Hoggar
Parc d'activités de Courtabœuf, BP 112
91944 Les Ulis Cedex A, France

Cover illustration. From top to bottom and left to right,
- AREVA's European Pressurized Reactor (EPR): a Gen-III new build in Flamanville [© EDF médiathèque – Alexis Morin, Antoine Soubigou];
- the DRAGON High Temperature Reactor (1959–67): an OECD project [© DR];
- TRISO (Tristructural Isotropic) nuclear fuel: from the sixties to a promising future [© Wikimedia Commons, gen-iv.ne.doe.gov];
- the Gen-IV Sodium cooled Fast Reactor (SFR) schema [from "A Technology Roadmap for Generation IV Nuclear Energy Systems", US DOE Nuclear Energy Research Advisory Committee and the Generation IV International Forum, December 1, 2002].

Printed in France
ISBN : 978-2-7598-0669-0

This work is subject to copyright. All rights are reserved, whether the whole or part of the material is concerned, specifically the rights of translation, reprinting, re-use of illustrations, recitation, broadcasting, reproduction on microfilms or in other ways, and storage in data bank. Duplication of this publication or parts thereof is only permitted under the provisions of the French Copyright law of March 11, 1957. Violations fall under the prosecution act of the French Copyright law.

© EDP Sciences 2016

Introduction to the *Nuclear Engineering* book series

INSTN, the National Institute for Nuclear Science and Technology, is a higher education institution founded in 1956 as part of French Alternative Energies and Atomic Energy Commission (CEA). INSTN is specialized in nuclear education and training, and contributes to the human resources development required by nuclear research and industry, from operators to engineers, and researchers. INSTN's main objective is to contribute to disseminating CEA's expertise through specialized courses and continuing training, not only on a national scale, but across Europe and worldwide.

Bolstered by the CEA's efforts to build partnerships with universities and engineering schools, the INSTN has developed links with other higher education institutions, leading to the organisation of more than thirty jointly-sponsored Masters graduate diplomas. There are also courses covering disciplines in the health sector: nuclear medicine, radio-pharmacy and also a specific degree for hospital physicists.

Continuous education is another important sector of INSTN's activities that relies on the expertise developed within the CEA and by its industry partners.

INSTN's "Génie Atomique" known as "GA" course is a specialised course in nuclear engineering that can be considered as a master after the master course. The course was first taught in 1954 at the CEA Saclay research centre, where the first experimental piles were built, and since 1978 it has also been taught in CEA Cadarache research centre, where the fast neutron research reactors were developed. Starting from 1958, the "GA" course is taught at the School for the Military Applications of Atomic Energy (EAMEA), under the responsibility of the INSTN. Since its creation, the INSTN has graduated over 5000 engineers who did work in major companies or public-sector bodies in the French nuclear industry: CEA, EDF, AREVA, Marine Nationale (the French navy), IRSN (French TSO)... Many foreign students from a variety of countries have also studied for this diploma.

There are two categories of student: civilian and military. Civilian students will obtain jobs in the design, the construction or the operation of nuclear power plants or research establishments as well as in the fuel processing facilities. They can aim to become expert consultants, analysing nuclear risks or assessing environmental impact. The EAMEA provides education for officers assigned to French nuclear submarines or the aircraft carrier.

The teaching faculty comprises CEA research scientists, experts from the Nuclear Safety and Radiation Protection Institute (IRSN), and engineers working in industry (EDF, AREVA, etc.). The main subjects are: nuclear and neutron physics, thermal hydraulics, nuclear materials, mechanics, radiological protection, nuclear instrumentation, operation and safety of Pressurized Water Reactors (PWR), nuclear reactor systems, and the nuclear fuel cycle. These courses are taught over a seven-month period, followed by a final project that rounds out the student's training by applying it to an actual industrial situation. These projects take place in the CEA's research centres, companies in the nuclear industry (EDF, AREVA, etc.), and even abroad (USA, Japan, Canada, United Kingdom, etc.). A key feature of this programme is the emphasis on practical work carried out using the INSTN facilities (ISIS training reactor, PWR simulators, radiochemistry laboratories, etc.).

Even now that the nuclear industry has reached full maturity, the "Génie Atomique" diploma is still unique in the French educational system, and affirms its mission: to train engineers who will have an in-depth, global vision of the science and the techniques applied in each phase of the life of nuclear installations from their design and construction to their operation and finally, their dismantling.

The INSTN has committed itself to publishing all the course materials in a series of books that will become valuable tools for students, and to publicise the contents of its courses in French and other European higher education institutions. These books are published by EDP Sciences, an expert in the promotion of scientific knowledge, and are also intended to be useful beyond the academic context as essential references for engineers and technicians in the nuclear industrial sector.

Joseph Safieh
"Génie Atomique" Course Director 2000–2014

Authors

Graduated from *École des Mines de Nancy*, retired from CEA and AREVA, **Bertrand Barré** teaches Nuclear Engineering at *Institut National des Sciences et Techniques Nucléaires*, and Sciences-Po. He was Nuclear Attaché in Washington DC, Director of Engineering at Technicatome, Head of the Nuclear Reactors Directorate at CEA, R&D Vice-president at COGEMA, Scientific Advisor to AREVA, and Member of many Scientific Committees in France and abroad.

Graduated from the *École Centrale de Paris*, France, **Pascal Anzieu** made his career at CEA on nuclear reactors design and safety. He led the Superphenix research program from 1994 to 1998 and then conducted research programs on future nuclear systems: sodium, gas, molten salt reactors, accelerator-driven systems, etc. He currently teaches at the National Institute for Nuclear Science and Technology and several engineering schools and universities.

Doctor of Orsay University 1982, currently coordinator of a CEA PWR expert group, **Richard Lenain** teaches Reactor Physics and Nuclear Engineering at INSTN, *École Polytechnique* and *École Centrale de Paris*. He was formerly head of Applied Mathematics and Reactor Studies Section in CEA/Saclay.

Graduated from *École Centrale de Paris* with a postgraduate degree in Theoretical Physics (atomic and nuclear) at Orsay University, currently scientific advisor to the Nuclear Energy Director in CEA, **Jean-Baptiste Thomas** teaches Reactor Physics and Nuclear Engineering as a professor at INSTN (in charge of the Nuclear Reactor Systems course created by Bertrand Barré). He was formerly Director for ADS studies in CEA and Director for Simulation and Experimental Facilities in the Nuclear Energy Directorate.

Contents

Foreword — XVII

Chapter 1. Introduction

- 1.1. General introduction — 1
- 1.2. The ebullient beginnings — 2
  - 1.2.1. Prehistory [1–10] — 4
  - 1.2.2. Uranium enrichment, the deus ex machina — 4
- 1.3. Bases for comparison [12, 13] — 5
  - 1.3.1. Fertile and fissile isotopes — 5
  - 1.3.2. Moderators — 6
  - 1.3.3. Coolants — 6
- 1.4. The driving forces of selection — 7
- 1.5. Today (and tomorrow) — 8
  - 1.5.1. Gas-cooled reactors — 9
  - 1.5.2. Graphite-moderated and boiling water-cooled reactors RBMK — 9
  - 1.5.3. Heavy water reactors CANDU — 10
  - 1.5.4. Light water reactors PWR, BWR and VVER — 10
  - 1.5.5. High temperature reactors — 10
  - 1.5.6. Fast breeders [14] — 11
  - 1.5.7. Molten salt reactors [1] — 12
- 1.6. Biotope, domination and selection — 12
- 1.7. From spontaneous selection to a formalized process [14, 15] — 13
  - 1.7.1. GIF, the Generation IV International Forum — 13
  - 1.7.2. INPRO, International Project on Innovative Nuclear Reactors & Fuel Cycles — 14
- 1.8. Fusion — 15
- 1.9. Conclusion — 15

Chapter 2. CO_2 gas cooled reactors

- 2.1. Introduction — 17
- 2.2. General architecture — 18
- 2.3. General features of graphite-moderated reactors — 20
  - 2.3.1. Fuel: natural uranium and magnesium clad (UNGG & Magnox) — 20
  - 2.3.2. Graphite moderator — 21
  - 2.3.3. General physical properties of graphite moderated reactors — 23
- 2.4. UNGG — 25
  - 2.4.1 The French UNGG program — 25
  - 2.4.2 St Laurent A example — 28
- 2.5. Magnox — 31
- 2.6. Advanced gas cooled reactor AGR — 35

Chapter 3. RBMK (Reactor Bolchoi Mochtnosti Kanali)

- 3.1. General 43
- 3.2. General description 44
- 3.3. Core physics 53
- 3.4. Chernobyl accident 56
- 3.5. Changes made to improve RBMK core behavior 58

Chapter 4. Heavy water moderated nuclear reactors

- 4.1. Introduction 61
- 4.2. General 63
  - 4.2.1. Heavy-water 63
  - 4.2.2. Natural uranium 64
  - 4.2.3. Pressure tubes 66
- 4.3. Description of a CANDU 6 68
  - 4.3.1. Reactor 68
  - 4.3.2. Primary system 72
  - 4.3.3. Moderator system 74
  - 4.3.4. Fuel 74
  - 4.3.5. Reactivity control systems 75
  - 4.3.6. Safety systems 76
  - 4.3.7. Fuel cycle 79
  - 4.3.8. The vacuum building 79
  - 4.3.9. Difficulties and incidents in the Canadian programme 81
  - 4.3.10. Economy 83
- 4.4. Fuel cycle possibilities 83
  - 4.4.1. CANFLEX fuel 83
  - 4.4.2. Slightly enriched uranium 84
  - 4.4.3. Recycling of the LWR fuel 84
  - 4.4.4. Perspectives 84

Chapter 5. Nuclear marine propulsion

- 5.1. Introduction 93
- 5.2. Main properties required for propulsion 93
- 5.3. History and development 95
- 5.4. Naval reactor development 96
- 5.5. Civilian fleet 98

Chapter 6. Experimental reactors

- 6.1. Different types of experimental or research reactors 101
- 6.2. Materials irradiation reactors (MTR, TRIGA...) 102
  - 6.2.1. OSIRIS, in Saclay 102
  - 6.2.2. TRIGA 104
- 6.3. MTR Fuel, RERTR Programme 105
- 6.4. Neutron source reactors 105
- 6.5. Spallation sources 106

| 6.6. | Materials irradiation facilities in Europe, the JHR project | 108 |
| 6.7. | Myrrha, Pallas | 109 |

Chapter 7. Advanced "Generation III" reactors

7.1.	Introduction: Genesis of "Generation III"	113
7.2.	Evolutionary or Revolutionary?	114
7.3.	EPR, the Evolutionary Power Reactor [1–6]	114
	7.3.1. Genesis of the EPR	114
	7.3.2. EPR General Characteristics	116
	7.3.3. Primary and secondary circuits	116
	7.3.4. Systems architecture	118
	7.3.5. Mitigation of severe accidents	118
	7.3.6. Future economics of the EPR	119
	7.3.7. EPR status in 2014	121
7.4.	The Korean APR 1400	121
	7.4.1. S 80+ basic options	122
	7.4.2. General characteristics	122
	7.4.3. Primary circuit	123
	7.4.4. The APR 1400	123
7.5.	The AP 600 and AP 1000 by Toshiba-Westinghouse [12–14]	124
	7.5.1. General characteristics	125
	7.5.2. Core and primary circuit	126
	7.5.3. Emergency systems	127
	7.5.4. From the AP 600 to the AP 1000	129
7.6.	Other generation III PWRs	130
	7.6.1. The ATMEA	130
	7.6.2. The APWR	131
	7.6.3. The AES 92	131
7.7.	Japanese and American ABWRs [17–22]	132
	7.7.1. General characteristics	133
	7.7.2. Architecture simplification	133
	7.7.3. Simplification of the primary circuit	135
	7.7.4. Additional improvements	136
7.8.	General Electric Simplified BWRs [24–29]	136
	7.8.1. General characteristics	138
	7.8.2. The SBWR (600–670 MWe)	138
	7.8.3. The ESBWR (1300–1550 MWe)	138
7.9.	The KERENA [30, 31]	140
7.10.	SMRs [32, 33]	142
	7.10.1. SMRs' potential advantages and drawbacks	144
	7.10.2. Short description of four SMRs	144
	7.10.3. Prospects for SMRs?	149

Chapter 8. High Temperature Reactor

| 8.1. | Obsolete or futuristic | 151 |
| 8.2. | HTR fuel [1–3] | 151 |

8.3. HTR demos: Dragon, AVR, Peach bottom 153
   8.3.1. Dragon 153
   8.3.2. The AVR 154
   8.3.3. Peach bottom 155
8.4. The "Astronuclear" Saga [6, 7] 156
8.5. Fort St Vrain and THTR Prototypes, the Thorium Cycle 158
   8.5.1. Fort St Vrain 158
   8.5.2. The Schmehausen (or Uentrop) THTR 160
   8.5.3. The thorium cycle [8–10] 160
8.6. False start in the USA 161
   8.6.1. General atomic's 1160 and 770 project 161
   8.6.2. The French HTR programme (first period) 163
   8.6.3. An assessment of HTR programmes, as seen from 1980 . . . 163
8.7. Why a renewed interest for HTRs? 165
   8.7.1. A changing environment 165
   8.7.2. The GT-MHR, Gas turbine modular high temperature
          reactor [11–14] 166
   8.7.3. ESKOM PBMR pebble bed modular reactor [15] 167
   8.7.4. The VHTR and ANTARES 168
   8.7.5. The Chinese HTR-PM 169

Chapter 9. Molten Salt Reactors

9.1. Liquid fuel reactors [1–6] 171
9.2. MSRE, Molten Salt Reactor Experiment 171
9.3. The Breeder MSR Projects 172
9.4. Generation IV MSRs 172
9.5. AHTR 174

Chapter 10. Liquid metal cooled fast neutron reactors

10.1. Introduction 177
   10.1.1. Breeding 177
   10.1.2. Waste incineration 179
   10.1.3. Situation of the industry 180
10.2. Description of Superphenix 180
   10.2.1. Principles 180
   10.2.2. General design 182
   10.2.3. Core and fuel 184
   10.2.4. Handling the assemblies 186
   10.2.5. Reactor block 188
   10.2.6. Sodium circuits 188
   10.2.7. Steam generators 189
   10.2.8. Decay Heat Removal systems 189
   10.2.9. Main Superphenix characteristics 191
10.3. Fast reactor fuel 192
   10.3.1. Special characteristics 192
   10.3.2. Operating criteria 192

	10.3.3.	Stresses in service 192
	10.3.4.	Fuel material 193
	10.3.5.	Clad materials and effects of irradiation 194
	10.3.6.	Characteristics of fuel elements and behaviour problems 195
	10.3.7.	Fuel behavior 195
	10.3.8.	Reprocessing 197
10.4.	Fast reactor safety 197	
	10.4.1.	Containment 197
	10.4.2.	Reactivity control 200
	10.4.3.	Decay Heat Removal 201
	10.4.4.	Considering accidents involving fuel melting 201
10.5.	Sodium technology 203	
	10.5.1.	Sodium 203
	10.5.2.	The choice of sodium 203
	10.5.3.	Sodium chemistry and purification 204
	10.5.4.	Compatibility of sodium with materials 205
	10.5.5.	Circuits and instrumentation 205
	10.5.6.	Interventions, inspection, repair 206
	10.5.7.	Safety 207
	10.5.8.	Overall assessment of the use of sodium 208
10.6.	Alternatives to sodium 208	
	10.6.1.	Liquid metals 208
	10.6.2.	Corrosion by heavy liquid metals 209
	10.6.3.	Lead-bismuth reactor feedback experience 210
	10.6.4.	Lead-cooled reactors 210
	10.6.5.	Conclusion 214
10.7.	Development prospects 214	
	10.7.1.	Current context 214
	10.7.2.	Economy of sodium-cooled FRs 215
	10.7.3.	FR plutonium burner and radioactive waste transmuter 215
10.8.	Conclusion 216	

Chapter 11. The gas-cooled fast reactor

11.1.	Introduction 219	
11.2.	History 219	
11.3.	The GFR, a Generation-IV system 220	
11.4.	GFR design options 224	
	11.4.1.	Fuel element 224
	11.4.2.	Core design and performance 225
	11.4.3.	Primary system 225
	11.4.4.	Power conversion system 228
	11.4.5.	Towards a demonstration reactor 228

Chapter 12. BWR: specific features, trends

12.1.	History, principles and architecture 231
12.2.	Neutronics, absorbers, fuel 235

	12.2.1.	BWR vs. PWR: moderation ratio 235

        12.2.1.   BWR vs. PWR: moderation ratio 235
        12.2.2.   Core structures and fuel assemblies, Reactor Pressure
                    Vessel (RPV) 236
        12.2.3.   Distribution of enrichment and of poisons 238
12.3.   Thermal-hydraulics and its tight coupling with neutronics 240
        12.3.1.   Recirculation ratio 240
        12.3.2.   Coupling between neutronics and thermal-hydraulics 240
        12.3.3.   Thermal-hydraulic instability 241
        12.3.4.   Stability loops; conceptual scheme of a sequence of feedback
                    effects 244
12.4.   Operation 244
        12.4.1.   Principles 244
        12.4.2.   Operating envelope 245
        12.4.3.   Operation, fuel and plutonium 245
12.5.   Chemistry of water and materials 247
        12.5.1.   Radiolysis 247
        12.5.2.   Cladding 247
        12.5.3.   Intergranular stress corrosion 248
        12.5.4.   Activation and gamma-emitting deposits, radiation protection
                    in the turbine hall 248
12.6.   Safety 248
        12.6.1.   Containment barriers 248
        12.6.2.   Containment pressure reduction 249
        12.6.3.   Safety injection, core meltdown and long-term containment . . 250
12.7.   Trends 256
        12.7.1.   Safety, in the aftermath of Fukushima 256
        12.7.2.   Fuel cycle improvements 259

Chapter 13. The place and the potential of Light Water Reactors in the transition from Gen-III to Gen-IV

13.1.   Introduction 261
13.2.   The stable and plentiful ground of physics and a changing world 262
13.2.   The Gen-IV vs Gen-III specification gap: the specifications
        for sustainable nuclear power 264
        13.3.1.   Introduction 264
        13.3.2.   The basic specifications: formulation and discussion 264
13.4.   The physical basis of sustainable nuclear power: high nuclear efficiency
        and the conditions required to achieve it 269
13.5.   Fast spectrum: the main constraints and specific issues 272
        13.5.1.   The design constraints related to the fast neutron spectrum . . . 272
        13.5.2.   From the past to the future 274
13.6.   "Smart" plutonium multi-recycling in LWR: The natural uranium saving
        context issue 276
13.7.   Energy scenarios and nuclear power worldwide: a prospective
        framework for the century 278
13.8.   Affordable natural uranium resources 280

- 13.8.1. Rising natural uranium prices as ore of decreasing uranium concentration has to be used 280
- 13.8.2. The strategic risk of preclusion of access to natural uranium is latent and may take form for a number of reasons 283
- 13.8.3. Shortages and price fluctuations in the short and long term uranium market 283
- 13.9. Light Water Reactors, the current situation: Strengths, Weaknesses, Opportunities, Threats 284
  - 13.9.1. Current situation 284
  - 13.9.2. LWR strengths: robust options, wealth of experience 289
  - 13.9.3. Weaknesses 290
  - 13.9.1. Opportunities 290
  - 13.9.2. Threats 291
- 13.10. LWR: further improvements in fuel cycle efficiency by spectral hardening 292
  - 13.10.1. LWR: an overview of the present fuel cycle performances, of the trends and of some possible improvements 292
  - 13.10.2. The last decades: fluctuations in the objectives, shooting on a mobile target 296
  - 13.10.3. The state of the art regarding the limits and the trends for the burn-up and for the recycling of plutonium 298
  - 13.10.4. What could be the next step? 299
- 13.11. A stepwise transition, a synergistic cohabitation: defining a flexible scheme for a sustainable nuclear fleet growth rate, worldwide, and transferring fissile material to the future through continuous valorization 303
  - 13.11.1. Introduction 303
  - 13.11.2. How to manage, from the uranium extraction rate viewpoint and from the nuclear plant type viewpoint, a strong nuclear energy growth after 2025/2030? 304
  - 13.11.3. Competing options around 2040–2050 for the utilities and for the countries launching a large fleet of nuclear reactors 307
  - 13.11.4. Best available technologies for "thrifty" Gen-3$^+$NSSS 310
  - 13.11.5. Thorium and related strategies (basically, it is a ^{233}U issue) ... 312
  - 13.11.6. An "exotic" enabler from "Nuclear Energy Synergetics": fusion-fission hybrid as fissile plutonium (and ^{233}U) factories 313
  - 13.11.7. FBR fleet breeding doubling time: estimates and sensitivity analysis 314
  - 13.11.8. Conclusion 315

Chapter 14. Nuclear fusion

- 14.1. Introduction 323
- 14.2. Principles and basic data 324
  - 14.2.1. General 324
  - 14.2.2. More on physical principles and basic data 325

		14.2.3.	Plasma 328
		14.2.4.	The ignition criterion 329
	14.3.	Fusion by magnetic confinement 331
		14.3.1.	Principles. 331
		14.3.2.	Confinement and the Tokamak principle 333
		14.3.3.	Heating of magnetized plasma 336
		14.3.4.	Findings: principles and noteworthy facts 338
	14.4.	Fusion by inertial confinement 343
		14.4.1.	Introduction: orders of magnitude 343
		14.4.2.	Target ignition by hot point 344
		14.4.3.	Instabilities 345
		14.4.4.	Findings 346
	14.5.	Reactor and associated technology 348
		14.5.1.	Reactor principle 348
		14.5.2.	Tritium production 348
		14.5.3.	Materials 352
	14.6.	The reactor: magnetic fusion 353
		14.6.1.	Energy efficiency 353
		14.6.2.	Superconducting electromagnets 355
		14.6.3.	Divertor 355
	14.7.	The reactor: inertial fusion 356
		14.7.1.	The positive energy balance criterion 356
		14.7.2.	Energy source 356
		14.7.3.	Reaction chamber 357
		14.7.4.	Targets 357
		14.7.5.	In summary 358
	14.8.	Nuclear safety 358
		14.8.1.	Normal operation: containment of toxic substances 358
		14.8.2.	Accident situations: a few remarks 358
	14.9.	Waste 358
	14.10.	Costs 359
		14.10.1.	Composition of costs and orders of magnitude 359
		14.10.2.	Ecological impact and external costs 360
	14.11.	Historical trends, current challenges; R&D ways and needs 361
		14.11.1.	Historical trends and current challenges 361
		14.11.2.	R&D trends and needs 363
	14.12.	Conclusion 365

Chapter 15. Futuristic systems: ADS, Space Nuclear propulsion and power generation, ADNIS

	15.1.	Accelerator Driven Systems (ADS) 367
		15.1.1.	Introduction 367
		15.1.2.	The physics of ADS. Basic principles and first design consequences 368
		15.1.3.	Technology and design: main specific components, challenges, and key points for feasibility 374

	15.1.4.	Preliminary techno-economic assessment 381
	15.1.5.	Defining a role for the ADS in the nuclear fleet: elements for a rationale 382
	15.1.6.	The R&D programs 382
	15.1.7.	The future in the world, in Europe, in France 383
15.2.	Nuclear space power and propulsion 383
15.3.	Advanced neutron irradiation sources (NIS) 389

Chapter 16. A few questions fostering further thought on some key issues

16.1.	The designer's carrousel 393
16.2.	Entering a new era or circling around a carrousel? 393
16.3.	Main questions to be addressed (combining innovation, design, marketing and acceptance issues) 394
16.4.	Some answers coming from past and recent history 395
16.5.	Design as a conceptual approach: design wheel and "helix" 397
16.6.	Beyond the incremental improvement of LWRs (safety, flexibility, fuel cycle (plutonium), lifetime, availability, uprating), what are the main achievements of recent (in the last three decades) design and operational qualification for power reactors? 397
16.7.	Other examples 399
16.8.	The coolant issue: updating some questions 400
16.9.	As for the coolant choice, there is no single merit index 401
16.10.	Main topics involved in the coolant issue 402
16.11.	Multi-criteria assessment: the representation and computation issue; a tentative representation diagram 404
16.12.	Making a positive contribution to the qualification of Gen-IV "enablers" . 405
16.13.	Knowledge bases and tools 405
16.14.	"War" is (or should be) over 406
16.15.	Optimisation of a multi-strata nuclear fleet achieving "smart recycling" is the new frontier 407
16.16.	Qualification (including substantial operation feedback) of all efficient enablers, with an updated design fulfilling the post-Fukushima requirements, must be started ASAP 407

Foreword

This book incorporates the core knowledge of the lectures given in the framework of the "Atomic Engineering" specialization, as well as of the Nuclear Energy / Nuclear Reactor Physics and Engineering (M2), under the heading: **Nuclear reactor systems: *a technical, historical and dynamic perspective*.**

Prerequisite knowledge is about neutronics, core and system thermal-hydraulics, fuel and fuel cycle, as well as about PWRs, which are considered as the reference system. This is the logic of the integration of the lectures in the curriculum, where a series of book [1] is dedicated to PWRs.

The conferences (and thus the present book) combine four approaches.

(i) **The descriptive one**. The benefit is a better understanding of how reactors are designed by combining "genetic chunks" taken from a common "library", just like living beings are. Compatibility constraints play a major role in the designer's choice. This gives a first insight into the design issue. It also gives an overview of the main strengths and weaknesses of the different reactor systems, tightly related to the design choices.

A series of examples show how recent choices are connected to events like those at Three Mile Island, Chernobyl and Fukushima, and how some problems can be fixed by incremental adaptation, while some generic solutions are proposed for future designs, with an increasing emphasis on the thorough implementation of defense in depth principles.

Finally, one can get a fresh look on biodiversity, in the worldwide current fleet as well as in Generation-4 systems. The latter resemble the earlier ones but fulfil an extended set of specifications, thanks to a few but significant innovations in technology and design. There are well known pedagogical limits of the plain descriptive approach: it can lead to get lost in the maze of the details. To get the whole picture, the book provides also the following approaches, tightly intertwined and complementing each other.

(ii) **The axiomatic approach**. Starting with a focus on a specific set of long term criteria concerning the fuel cycle sustainability, a conceptual solution is established, and then a family of reactor systems is selected for development and qualification. That is the way fast breeders were selected from the early days of nuclear engineering and are still dominating the Generation IV competition. When combined with the "market pull, technology push" paradigm which led to the supremacy of LWRs, this approach gives a binocular view, thus a perception of the depth of the landscape, even if the set of axioms has not to be trusted in blindly, as the past history has shown.

(iii) **A historical approach**, from the 1940's to nowadays, with an extrapolation to the near future. This approach, extensively developed in the Introduction, sheds light on the Heraclitean/Darwinian process which is at work on the "market", as well as on innovations or efficient industrialization considered as game changers. This has been the case with the emergence of a competitive enrichment capability paving the way to the triumph of the LWR. The LWR dominance being firmly established, what is the next step? This is the purpose, for instance, of the Generation IV International Forum, launched in 2001, as well as of the INPRO programme at the IAEA. The driver is a set of specifications requested for the nuclear energy sustainability. A worldwide cooperative effort is needed to achieve a jump from the "business as usual" stimulation by the market to the higher level rules fixing the roadmap for the future. Actually, combining competition and cooperation is a big challenge.

(iv) **A dynamic approach**. In the early 2000s, the prevailing image combined a "nuclear renaissance", a quick implementation of rules - enabling the worldwide energy consumption increase to keep compatible with a strong limitation of the greenhouse gases (GHG) concentration - and a strong growth of the world economy. Undertaking a deep mutation of the energy paradigm thanks to a high and sustainable investment would clear the way for low carbon power production systems, including a high share of competitive base-load nuclear. The last decade main events and trends damped some hopes and dissipated some illusions.

In a few years, the financial then economic crisis has slowed down the global growth and made the capital less available for long term, moderate Return On Equity (ROE) ratio projects. The construction costs of nuclear plants soared in most occidental countries, when compared with those of fossil-fuel plants.

The Fukushima accident put temporarily a cap on nuclear growth. Moreover, Europe seems to overreact when compared to other regions, and the "new energetic paradigm" is often (up to now) oriented against nuclear despite the strong pro-nuclear commitment of many European countries and the fact that nuclear is the most "scalable" low carbon power source, competitive in base-load operation, and flexible enough to accommodate a large share of intermittent renewables in the power fleet. The risk assessment and perception are the core topics of a fierce debate.

This series of events has fostered a dynamic approach involving, beyond the "market" viewpoint on the one hand and a long term, cooperative approach on the other hand, a deeper knowledge of actors and forces at work. This approach relies on strategic prospective studies, on "humanities", as well as on the design process from a conceptual viewpoint. The design key point, at the moment, is the coolant issue. These topics are not addressed in depth in the present book. They provide a framework for future investigation and modeling.

As a consequence of this fourfold approach of the nuclear reactor system issue, **the plan is as follows.**

Chapter 1 – Introduction

The evolution of nuclear reactors since the 1942 Fermi experiment can be described along the lines of natural history, with an initial flourish of uninhibited creativity followed by a severe selection process leading to a handful of surviving species, with light water reactors occupying most of the biotope today. The criteria which drove this selection have evolved with time and might not all be relevant in the future. The recent interesting development of the Generation IV International Forum and INPRO comes from a desire to rationalize and formalize the selection process for the future nuclear systems. Will these attempts be successful? Will "natural" selection still prevail? In any case, in the context of a growing demand for energy for the developing world and of the need to reduce greenhouse gas emission, the nuclear reactor "species" is here to stay.

Chapter 2 – CO_2 Gas-cooled Reactors: MAGNOX, UNGG, AGR

This chapter dedicated to first generation of European reactors: MAGNOX, UNGG and AGR, is aimed at focusing on major architecture characteristics imposed by knowledge and industrial capabilities reached in the 1950's; solutions were found based on natural uranium fuel, graphite moderator, gas coolant, concrete caisson, large core and continuous fuel reloading and permanent in-core instrumentation. A very brief overview of some reactor architectures (dramatically different from LWR ones) envisaged at the early stage of nuclear power development can be found through examples presented. Some important common core physic parameters are described according to their general design options such as: heterogeneous lattice design, reactivity coefficient, power density... AGRs present an effective solution for graphite gas reactors compatible with the mainstream fuel concept share with other reactors types.

Chapter 3 – RBMK

RBMKs combine different design options that can be found in CANDU (water channel type reactor), in BWRs (boiling water coolant and energy production systems) and in AGRs: (moderated with graphite and continuous fuel reloading). The chapter presents the general reactor design and emphasizes on some design options that involved core physics parameters peculiar to RBMK. Attention is given to RBMK design modifications carried out during the decade after Chernobyl disaster that is shortly reported. In Russia RMBK remain workhorse reactors for electricity production even if no more reactors will be ever built in the future.

Chapter 4 – Heavy Water Moderated Nuclear Reactors

Heavy water is the best neutron moderator. It allows operating a nuclear reactor with natural uranium and eliminates the need for fuel enrichment industry, thus simplifying the fuel cycle to the simplest technology, and reducing proliferation risk.

Nevertheless, the volume of heavy water needed is 10 times the one with light water. This generates several design constraints that the design of a CANDU™ reflects. This technology is the main representative of heavy water moderated reactor systems. It is based on reactors of large volume and moderate power, separated horizontal fuel channels with pressure and vessel tubes, a big amount of static moderator in a calendar. Each fuel channel is filled with several rods assembly made of Zircaloy cladded oxide pellets. Fuel is loaded during operation.

The size of the core and the costly heavy water induce an important construction cost which reduces the advantage gained by the good loading factor. This reactor system is the third to produce nuclear electricity worldwide.

Chapter 5 – Nuclear Marine propulsion

Nuclear marine propulsion was the first motivation for the development of a nuclear reactor industry, this chapter illustrates how large was the field of a priori solutions. This period permits to put in place theoretical core physics basis that oriented design solutions, experimentation and mock-up approach was systematically followed. This period brought a decisive input in reactor nuclear development.

Chapter 6 – Experimental reactors

If the mainstream of nuclear reactors is devoted to electric power generation and besides those which are specialized in naval propulsion a number of experimental reactors are in operation in the world with a rating from zero to a few tens of megawatts They have many diverse uses from operators training to isotopes production or basic research and some are dedicated to material and fuel irradiation in support of the power reactor R&D Though most were initially fuelled with highly enriched uranium almost all are now using fuel enriched below 20% in ^{235}U

Chapter 7 – Advanced Generation III Light Water Reactors

Many lessons were learned from the Three Mile Island accident and used to improve the safety of nuclear plants around the world The specific lesson learned from Chernobyl was: no matter how unlikely to happen a severe accident is possible but must not result in a massive radioactive contamination of the environment This new requirement was the basis of the design of advanced LWRs dubbed "Generation III" plants Designed in the 90s built during the following decade these Gen III plants are now starting to operate

Chapter 8 – High Temperature Reactors HTRs

Invented in the 1950s around a very specific fuel design the coated particle HTRs have a high degree of passive safety and can co-generate power and process heat with a very good efficiency Two varieties of HTRs have been developed: the prismatic and the pebble bed A number of demonstrators of both varieties have operated successfully but it remains to be seen whether they can reach commercialization

Chapter 9 – Molten Salt Reactors MSRs

MSRs are designed to use a liquid both as fuel and as primary coolant which allows for a wide variety of fuel cycles A small demonstrator was operated in the 1960s Most recent designs are based on the thorium cycle either in thermal or fast neutron spectrum MSRs are probably the most futuristic of the 6 concepts selected by the Generation IV Forum

Chapter 10 – Liquid Metal-cooled Fast Neutron Reactors LMFRs

Fast neutrons bring the theoretical capability to breed plutonium. They provide more neutrons per fission than in thermal spectrum, one of them being used for a capture on a fertile isotope. So they allow transmuting non-fissile uranium (^{238}U) in plutonium (^{239}Pu) during the core irradiation, thus multiplying the available resource by a factor of 100.

In the 1970s, Sodium-cooled fast reactors have been developed all around the word with the fear to lack of uranium. But their high construction and electricity cost did not lead to an industrial deployment, even if some countries are still building prototypes. Lead-cooled fast reactors are also studied based on military experience from Soviet submarines.

To make it possible, a fast neutron nuclear reactor must use no moderator. Liquid sodium has been used for all relevant demonstrator and industrial applications. It's a cheap, efficient, non-corrosive coolant. Unfortunately, liquid sodium burns in air and violently reacts with water, so sodium technology necessitates dedicated equipment and components that make it expensive.

Most of the reactors are pool-type ones where the non-pressurized primary circuit is contained in the reactor vessel. Stainless steel cladded oxide fuel has been used for all significant demonstration. To benefit from breeding, it's a high content plutonium (around 30%) and natural or depleted uranium fuel that necessitates a closed fuel cycle.

Chapter 11 – Gas-cooled Fast Reactors GFRs

To overcome the chemical reactivity of liquid sodium or the corrosion of liquid lead, a fast neutron reactor cooled by helium has been chosen as a potential Generation IV system. It gathers fast neutron sustainability and high temperature inert coolant efficiency. However, the primary circuit must be pressurized and the core remains sensitive to temperature rises. A complex layout to ensure core cooling whatever happen and a high temperature resistant fuel, cladded for instance with a composite ceramic, make the development of the gas-cooled fast reactor very dependent of the success of future technical research and development.

Chapter 12 – BWR: specific features and trends

The BWR was first considered as a gifted descendant of the naval propulsion PWR and as a talented competitor in the commercial power production arena: without secondary system and, after a series of daring innovations, without recirculation pumps external to the vessel and with the "drywell-wettwell" containment design, it is the most possible compact and integrated – moderate pressure – Nuclear Steam Supply System. On the other hand, a few but questioning drawbacks balanced these advantages: potentially disruptive reactivity induced accidents; no permanent materialized second barrier and the related threat on operation resulting from a potential contamination of the whole system, including the turbine; a complex and costly fuel.

Further Innovations/improvements led to a success story. In 2011, the Fukushima accident seemed to put an end at the BWR attractiveness, even if no specific weakness – but a lack of updating the defense in depth implementation – was highlighted. One can even notice that the design from the sixties was equipped with passive safety systems for decay heat removal. The current Generation III BWR design, showing smart safety options, looks competitive and attractive. Moreover, benefiting from an adaptive fuel assembly potential thanks to the core physics decoupling provided by the casing, BWRs possess a structural competition edge related to the Gen-III$^+$ challenge about multi-recycling capability.

Chapter 13 – The role of LWR in the transition from Gen III to Gen IV

The Generation IV set of specifications holds for the nuclear fleet. As long as a single reactor system does not possess all the required capabilities, the worldwide fleet will be composed of several functional "strata", made of diverse reactor systems: a majority of competitive "power reactors" for "flexible base-load" power production (acting as a back-up for intermittent renewables), with a few tens of % of power being generated by " enablers " achieving a specific function. For a long time, power reactors will be LWRs.

The LWR limits/challenges are related to the fuel cycle: multi-recycling efficiently plutonium and reducing their own natural uranium consumption when compared to the "lazy" once-through cycle. LWR reduced moderation ratio cores are able to comply with a low plutonium isotopic composition degradation criterion, established by comparison with plutonium "on-the-shelf" degradation. The best way would be to "only change the core portion" and to get "adaptable" LWR capable of dealing with stepwise evolving fuel assemblies. BWRs get an advantage in this perspective.

Meanwhile, the LWR fleet will produce the huge amount of plutonium needed for launching a large fleet of fast breeders in U-Pu fuel cycle, as well as ^{233}U (involved in synergistic strategies using diverse neutron spectra in diverse reactor systems).

Chapter 14 – Nuclear Fusion

The research in the field of controlled nuclear fusion has not to be compared with the R&D on the fission systems. The time scales are too different. Anyway, both depend on the success of the reactor design (and reliable operation), as well as of the fuel cycle, with a strong breeding constraint for the tritium (for the D-T fusion process). The next achievement in magnetic confinement fusion research will be the ITER experiment ([2]). Improvements are necessary in many specific fields (plasma heating and control, instrumentation, on-line software, off-line simulation, tritium management) as well as on the system integration issue. Moreover, some breakthrough is requested in magnetic confinement, as well as in the field of materials, to clear the way for the next step: the DEMO. In the inertial confinement area, the results of the NIF and of the LMJ facilities will give a deep insight into some key scientific and technological points. No "commercial" extrapolation can be considered at the present time. A strong synergy with the R&D on – cyclopean – accelerators could emerge.

Finally, with DT fusion, 14 MeV neutrons have an enormous value in the "Nuclear Energy Synergetics" [3] perspective: about 14 MeV, the neutron reproduction factor η is close to 4.5 for all the heavy nuclei, including ^{238}U and ^{232}Th…

Chapter 15 – Futuristic systems will not belong to the current century fleet of reactors. The related designs, R&D and projects are the way to deep innovation, mainly because most of the solutions to the current issues will come from design, not from radical innovation (basically due to its long "time-to-nuclear-market"). Futuristic systems are thus the nuclear designer's adventure playground. And possibly more…

They come from age (see [3], for instance, with its attractive title : "Nuclear Energy Synergetics") : fusion, fission and accelerated particles, and the various types of related hybrids address fuel cycle issues: producing more fissile nuclei from fertile nuclei by tapping the huge amount of energy - easily - produced by fission to deliver complementary neutrons.

Space nuclear applications, as well as the ADNIS (Accelerator Driven Neutron Intense Sources), incorporate challenging innovations. These innovations rarely lead to industrial transposition for power generation, even if HTRs share original fuel design features with the nuclear rocket project NERVA. New materials benefit from Futuristic Systems projects, but not exactly with specifications well suited to industrial power production.

Sometimes, unexpected cross-fertilization can occur, just like in the case of VASIMR, a nuclear space propulsion project benefitting from the – magnetic confinement fusion - Tandem experiment.

Chapter 16 – A few questions fostering further thought on some key issues

There is no clear cut, black-and-white conclusion to this presentation of the nuclear reactor system history, assessment and dynamic-prospective approach. The afterword focuses on three main challenges that we have to meet in the future and that require investing in the knowledge and expertise management, as well as in some methodology and tool development.

- The design issue and the related expertise and creativity, as well as the required improvements in terms of methodology, with a focus on the coolant issue, the most important at the moment where we are facing the Generation-4 system selection and development
  challenge.

- The strategic prospective issue, requiring the simulation of the game, seeking to gauge the balance of power between actors and to study their convergences and divergences when faced with a new trend, context, event. The risk assessment vs. perception issue is a core challenge.

- The appropriate way to manage the expertise and the creativity, with adequate methods, tools, data and knowledge bases, as well as the opportunities, through projects, to train these skills and tools, in a collective improvement strategy, locally and internationally.

These challenges give raise to further work, debate and teaching.

References

[1] "Génie Atomique" collection, EDP Sciences: H. Grard, *Physique, fonctionnement et sûreté des REP, le réacteur en production;* B. Tarride, *Physique, fonctionnement et sûreté des REP – Maîtrise des situations accidentelles du système réacteur;* P. Coppolani, N. Hassenboehler, J. Joseph, J.-F. Petretot, J.-P. Py, J.-P. Zampa, *La chaudière des réacteurs sous pression;* N. Kerkar, P. Paulin, *Exploitation des cœurs REP*.

[2] R. Aymar, P. Barrabaschi, Y. Shimomura, *The ITER Design*, Plasma Phys. Control. Fusion **44**, 519–565 (2002).

[3] A.A. Harms, M. Heindler, *Nuclear Energy Synergetics*, Plenum Press, 1982.

1

Introduction

B. Barré

1.1. General introduction

A nuclear reactor is a device where a sustained fission chain reaction, during which heavy nuclei are split by neutrons and heat results from this fission, takes place. The "deliverables" of such devices can be a neutron flux (for scientific and medical applications, radioisotope production, production of fissile or fusible materials, neutron-radiography, testing of materials under neutron radiation, and so on), heat (for urban heating or process heat), mechanical power (e.g. for naval propulsion) or electrical power (nuclear power plants or NPPs). A last class of reactors is intended for simulation, from critical mock-ups to facilities dedicated to safety experiments. Most of the present book will be devoted to NPPs.

If one excludes a few very specific reactors designed for space applications, nuclear power plants, or units, present two main subsets of components. In France, we call them the "conventional island", where steam powers a quasi-conventional turbo-alternator, and the "nuclear island", which supplies this steam. This does not differ much from the subsets called BOP (balance of plant) and NSSS (Nuclear Steam Supply System) in the USA. The nuclear island is centered around the "core" of the reactor, where fissions occur in the "fuel assemblies". The core reactivity and the power level of the reactor are controlled by devices which introduce neutron poisons in the core. Those devices are activated by a Command & Control System. The heat generated in the core is evacuated by a "coolant". This coolant may directly produce steam, as in BWRs and RBMKs, or it is contained in a closed *primary* circuit and transfers its calories to a *secondary* coolant circuit through a *steam generator*. In some reactor types, such as the FBRs, another intermediate circuit exists between the primary coolant and the steam generator. The nuclear island is usually contained within a bulky *containment building* which acts as a tertiary barrier against the possible dispersion of fission products in the environment (the first two barriers being the fuel cladding and the primary circuit [8]).

The steam, once it has lost its momentum in the turbines of the conventional island, is condensed in a *condenser* and the liquid water is pumped back to the reactor or to the steam generator. In the condenser, another coolant circuit evacuates the waste heat. This circuit is either open, or closed on a final heat exchanger connected with the ultimate heat sink.

The history of nuclear reactors covers a very short fraction of mankind's saga: it starts on December 2, 1942, under the tribunes of Staggs Field, Chicago, where the first fission chain reaction was sustained in the CP1 atomic pile. The first, quite modest, production of electricity dates back to 1951, and one had to wait the middle of the 1950s to witness the first "commercial" NPPs in the Soviet Union and Great Britain, as well as the first US nuclear submarine.

In 2010, some 440 NPPs generated about 2600 billion kWh, slightly less than 14% of the world's electricity production, almost equivalent to the total production of all the world's hydraulic dams. Generating the same amount of power in thermal plants would require burning close to 650 million metric tons of oil, more than the oil production of Saudi Arabia. Even though the 2011 accident in Fukushima impeded the growth of nuclear electricity production, jumping from 0 to 14% of the world power generation in a mere half century is not a trivial achievement. This brings into proper perspective the quasi-stagnation experienced in Europe this last decade.

But this is not the history of nuclear power development which we will describe later – it has been widely covered by others [1–10]. We shall try to address the topic from a slightly different angle, and to sketch the natural history of the phylum "Nuclear Reactor". This will lead to evoke vital ebullience, evolutions and mutations, on a background of rigorous selection.

There cannot be natural history without some kind of taxonomy, without classifying the species according to more or less arbitrary criteria. Nuclear reactors are usually classified according to the main components of their core: fissile/fertile couple, physico-chemical nature of their fuel, nature of their moderator (or lack of it), nature and thermal-hydraulic condition of the cooling fluid. As the taxonomy is very much diversified, we shall call "species" only those classes comprising a significant number of successive specimens.

Even though, almost two billion years ago, there were already seventeen "natural" nuclear reactors operating near what is now Oklo (Gabon), we shall start our natural history only from December 2, 1942, with the CP1 Pile in Chicago, where the first man-made sustained fission chain reaction started.

1.2. The ebullient beginnings

After the end of the Manhattan Project, and in the era of "Atoms for Peace", a period of unbridled creativity, of vital ebullience started – essentially in the United States. In retrospect, it appears hardly believable. All possible reactor types, and again some more, were dreamed, designed, and built, and most of them have actually been operated! All possible combinations of fissile and fertile materials, of moderators and of cooling fluids have been tried and corresponding facilities built, generally at a rather small scale and, let's admit it, with a relative respect for safety and environmental impacts. But all these machines have worked, if only for a very limited time, and without any major accident if one excludes SL1, in January 1961, which might have been a bizarre suicide.

Uranium, plutonium and thorium as metals, oxides, carbides or more exotic compounds, air and other gases, light water, heavy water, graphite and beryllium, rods, pins, spheres, particles, suspensions and fluidized beds, liquid metals and molten salts, everything has been tried at least once between 1950 and 1965 (Table 1.1). More than

Table 1.1. A few of the many possible combinations (names in bold italic refer to "species").

Spectrum	Fissile/fertile	Fuel	Moderator	Coolant	Example
Fast	Pu	liquid Pu/Fe	–	= fuel	LAMPRE
	HEU	oxide	–	liquid Na	BN 600
				liquid Pb	Submarines
				mercury	BR 1
	Pu/^{238}U	U/Pu/Zr	–	liquid Na	EBRE 2
		oxide	–	liquid Na	*LMFBR*
	Pu/Th	carbide	–	liquid Na	FBTR
Epithermal	U natural	metal	graphite	liquid Na	Hallam
	HEU/Th	particles	graphite	He	*HTR*
	MEU	particles	graphite	He	Dragon
	HEU	particles	graphite	hydrogen	Nerva
	HEU	oxide	BeO	air	Airplane
Thermal	HEU	solution	H_2O	= fuel	Los Alamos
	HEU	suspension	H_2O	= fuel	Kema
	HEU*	molten salt	graphite	= fuel	MSRE
	U natural	oxide	D_2O	H_2O	*CANDU*
	LEU	oxide	D_2O	CO_2	EL 4
				H_2O	SGHWR
	U natural	metal	graphite	H_2O	Hanford
				CO_2	*UNGG*
				air	G1
	LEU	oxide	graphite	CO_2	*AGR*
				H_2O	*RBMK*
	LEU	oxide	H_2O	H_2O (B)	*BWR*
				H_2O (P)	*PWR*
	^{233}U/Th	oxide	H_2O	H_2O	LWBR
	HEU	U/Al	organic	organic	Piqua
			D_2O	D_2O	RHF
			H_2O	H_2O	*MTR*
	MEU	silicide	H_2O	H_2O	OSIRIS
	HEU	oxide	H_2O	H_2O (P)	BR 2
	HEU	U/Zr/H	H_2O	H_2O	*TRIGA*

50 different reactors have been built on the single "National Reactor Testing Station" (now INEEL) near Idaho Falls, and then many more in Oak Ridge, Hanford, Los Alamos, not to mention the Kurchatov Institute, IPPE, Harwell, Saclay and others.

1.2.1. Prehistory [1–10]

At the very beginning, though, the range of possible choices was rather narrow. The only fissile material available was natural uranium, with its given 0.7% ^{235}U. Sustaining a chain reaction with natural uranium required to thermalize the fast fission neutrons as quickly as possible to avoid their capture in the resonances of ^{238}U: the reactor could only be heterogeneous and moderated with light elements. Hydrogen being too absorbent, the choice of moderator lay between heavy water, graphite or possibly beryllium. At the time, the world stockpile of heavy water (painfully produced by electrolysis in Norway) was small, while pure graphite was industrially produced in the USA in large quantities for electric furnace electrodes.

Consequently, CP1 could not be anything other than what it was: a regular and roughly spherical array of uranium pebbles (metal and oxide) within a pile of graphite bricks, hence the name "atomic pile" used to designate early nuclear reactors, while electrochemical piles retain the name of the original Volta pile of wafers. For the same reasons, a similar design was adopted in 1946 for the first Soviet reactor, still visible today in the Kurchatov Institute. The first British pile, GLEEP (1947) was also an array of natural uranium/graphite (with some heavy water as well), while Canada and France chose heavy water (D_2O) to moderate their first reactors ZEEP (1945) and ZOÉ (1948).

For their first power reactors, while they were developing the technologies for isotopic enrichment, the USA continued with the "natural uranium – graphite" species, with the air-cooled Oak Ridge 1 pile (1943), followed by the large plutonium-production reactors of Hanford, directly cooled by water from the Columbia River. But to leave no stone unturned, they also started in 1944 a D_2O moderated pile at Argonne.

Unable to enrich uranium and unwilling to depend entirely on the United States for their fuel supply, the United Kingdom and France based the first phase of their nuclear development on "Magnox" and "UNGG" reactors of similar technology, using rods of metallic natural uranium clad in a light alloy, disposed in an array within a graphite pile and cooled by pressurized gas under pressure (air first, then CO_2). On the other hand, Canada – today the second largest producer of uranium in the world – stuck with the natural uranium-D_2O combination, developing the "CANDU" species.

1.2.2. Uranium enrichment, the deus ex machina

As soon as the technology for uranium isotopic enrichment was developed in the framework of the Manhattan Project, access to enriched uranium, from 1% to 93%, freed reactor designers from one of the main constraint they were facing, neutron scarcity, and liberated their unbridled creativity. The drivers of this ultra-fast initial evolution were the search for more compactness, higher thermal efficiency and better use of the fissile materials.

Low enriched uranium (LEU), containing around 3% ^{235}U, notably allowed the use of ordinary water H_2O as moderator, the simplest of them all, despite its drawbacks which were already well known in conventional thermal plants: low boiling temperature under atmospheric pressure and corrosiveness with ordinary steels. In order to reach a decent thermal efficiency, high pressures, typically 7 to 15 MPa, were required.

In order to illustrate this initial creativity, here is a non-exhaustive list of reactors built in the USA before 1955:

- natural uranium-graphite reactors for plutonium production (Hanford, 1944);
- homogeneous solution of enriched uranium salts in ordinary water (Los Alamos, 1944);
- Clementine, unmoderated sphere of metallic plutonium (Los Alamos, 1945);
- EBR 1, liquid metal fast neutrons reactor (Idaho, 1951);
- MTR, pool-type water reactor with plates of highly enriched uranium alloy (Idaho, 1952);
- STR, land-based prototype of the Nautilus, first PWR (Idaho, 1953);
- Borax 1, first BWR (Idaho, 1953).

At the same time, the Soviet Union developed a reactor fuelled by LEU, moderated by graphite and cooled by boiling ordinary water circulating in pressure tubes. This Obninsk prototype was then the first true power reactor to operate. It was also a dual purpose plutonium production reactor, forerunner of the RBMK species.

It was to be followed by the Magnox reactors of Calder Hall, dual-purpose as well, inaugurated in 1956 by Queen Elizabeth II, by Shippingport, the first power PWR in 1957, then by G2 in Marcoule (1959) and, in 1962, both NPD, ancestor of the CANDU species, and the Windscale AGR prototype.

It can be seen that, as early as 1962, a mere twenty years after the Chicago experiment, all the present "species" were in existence, be it in embryonic form. Finally, in 1966 the prototypes of what are still today would-be species started: Dragon, the first HTR, built in the United Kingdom in the framework of an OECD Project, and the MSRE, Molten Salt Reactor Experiment in the United States.

At the end of the 1960s, the most advanced Western state in terms of civilian use of nuclear power was the United Kingdom, the only country to have developed a true species. Wylfa, the last of those early Magnox, is still in operation in 2014. Even more than the difficult development of the AGR species, it is the discovery and exploitation of the North Sea oil fields which have stymied the UK's nuclear development in the 1970s and led to the loss of their leadership position. As for the United States, in the 1960s, their nuclear supremacy was mostly in Defense applications, including a monopoly on naval propulsion and uranium enrichment.

1.3. Bases for comparison [12, 13]

Without interfering with the following chapters, devoted to individual reactor types, the following tables summarise the bases for comparison among basic reactor constituents.

1.3.1. Fertile and fissile isotopes

Table 1.2 compares the characteristics of the main fissile and fertile isotopes.

There is a great similarity between the ^{238}U/^{239}Pu and ^{232}Th/^{233}U couples, even though the half-life of the intermediate ^{233}Pa is far greater than that of ^{239}Np.

Table 1.2. Main fissile and fertile isotopes.

		^{233}U	^{235}U	^{239}P	^{232}T	^{238}U
2200 m/s	fission (barns)	527	579	741	0	0
	capture (barns)	54	100	267	7.6	2.7
	# neutrons/fission	2.5	2.4	2.9	-	-
Fast	fission (barns)	2.8	2.0	1.9	0.01	0.05
	capture (barns)	0.3	0.5	0.6	0.35	0.3
	# neutrons/fission	2.5	2.5	2.9	2.3	2.75
	ßeff (%)	0.28	0.64	0.21	2.3	1.5

Table 1.3. Elements for comparison.

Moderator	moderating ratio	# collisions	free path (m)
D_2O	5 000	19	152
H_2O	62	35	0.5
graphite	218	115	29

1.3.2. Moderators

A moderator has two roles:

- to slow-down fast neutrons with the maximum energy loss per collision (compactness);
- to cause the fewest possible parasitic captures during slowdown (neutron economy).

The choice of a moderator therefore results from a compromise between the smallest possible atomic mass, the highest possible density and the smallest possible absorption cross section, not to mention chemical and thermodynamic properties.

Table 1.3 gives three elements for comparison: the moderating ratio, the number of collisions to thermal equilibrium, and the mean free path before capture. The *moderating ratio* is the ratio of the macroscopic slowing down power to the macroscopic cross section for absorption. The higher the moderating ratio, the more effectively the material performs as a moderator.

1.3.3. Coolants

The ideal coolant should exhibit high heat capacity and low viscosity, be transparent, be insensitive to radiolysis, be chemically inactive[1] and non-toxic, not capture neutrons, stay monophasic even at high temperatures and be available in large quantities at an affordable price. Depending on whether the reactor is "fast" or "thermal", the coolant's moderating

---

[1] At least, it should be non corrosive to the core and circuits materials, and non reactive to air and water.

ratio will be a nuisance or an asset. There is no coolant exhibiting all these qualities together: we must make it with existing fluids!

Referring only to the thermodynamic properties, a relative "merit factor" can be defined by combining in one empirical formula [12] the specific heat, density and viscosity, with sodium acting as a reference value of 1.

Table 1.4. Coolants.

Coolant fluid	(°C)	P (MPa)	merit factor (relative to Na)
water (D_2O or H_2O)	300	15	60
organic liquid	300		7
sodium	300		1
lead	500		0.25
CO_2	300	5	0.003
helium	300	5	0.002

1.4. The driving forces of selection

Of all possible designs and many promising prototypes, we know today that only a handful made it to industrial series, to become what we call here a species. This was not the result of any deliberate process but rather "natural" or spontaneous selection. The reactors presently operating were, indeed, the fittest to survive; were they the fittest to safely and sustainably generate competitive electric power? This is an open issue…

There have been various forces driving this selection, some positive and some negative. Let's start with the negative drivers which led to elimination.

Some branches aborted because of technological difficulties (such as the lack of compatibility between sodium and graphite in the Hallam reactor), because the project was overly ambitious (the "air breathing" reactor for airplane propulsion) or too cumbersome (the reactor of the Q244 submarine, fuelled by natural uranium), or even as the result of a choice made by one key individual (the sodium-cooled Seawolf, rejected by Admiral Rickover in favour of the PWR to equip the whole Navy).

Safety considerations played an increasingly significant part in the elimination process. Nowadays, one would not dare submit the design of LAMPRE, with its liquid plutonium fuel/coolant, to any regulatory authority. The Windscale fire (1957) forever eliminated air cooling for graphite reactors. The Babcock & Wilcox subspecies did not survive the Three Mile Island 2 accident, and nobody would order a new RBMK after Chernobyl. One example illustrates both the safety concern and the energetic methods in use during the pioneer era: to complete a series of reactivity excursion tests in Borax 1, AEC deliberately provoked the destruction of the reactor in a spectacular vapor explosion following a core meltdown (1954). This voluntary accident remains to date the basis for dimensioning the containment of pool type reactors. The precautions taken to carry out severe accident tests in the Phébus-PF reactor make Borax seem like a dream.

We have seen that mastering enrichment rendered useless the "fine tuning" of the neutron balance required by natural uranium fuelled reactors. Similarly, the discovery that, with enough exploration efforts, uranium was relatively abundant, even though in less rich ores than the initial deposits of Bohemia and Katanga, reduced the competitive advantage offered by uranium-thrifty species. Fast Neutron Breeders have evidently suffered from this evolution which consistently makes them "necessary... but later". So have the Molten Salt reactors.

In some occurrences, selection may have been a matter of bad luck, bad timing. Such is notably the case for the High Temperature Reactors HTRs, which made their commercial debut in the USA at the beginning of the 1970s, just before the rush of cancellations of nuclear plant orders triggered by the first oil crisis. On a side note, it is a very intriguing fact that the same initiating event, the 1974 oil shock, launched the French programme while it nearly killed the American programme...

This rather long list should not suggest that selection was influenced only by negative factors leading to the elimination of one competitor or the other: positive factors have also played a key part. To illustrate this point, it is obvious that the emergence of the PWR species was considerably helped by the fact that these reactors had been previously chosen to power the nuclear Navy because of their compactness, a quality not so essential for land-based power plants. In the United States, they could therefore benefit from a strong industrial development which allowed Westinghouse to outrun General Electric, its giant competitor. Similarly, Magnox, UNGG and RBMK directly evolved from reactors designed and developed to produce plutonium for weapons.

1.5. Today (and tomorrow)

The results of this selection process are summarised in Table 1.5 [11].

Table 1.5. Year 2011.

Species	In operation		Construction	
	GW	#	GW	#
Magnox, AGR	9	18	-	-
PWR + VVER	250	271	53	55
BWR	81	88	4	5
RBMK	10	15	1	1
CANDU	23	47	2	3
FBR	0.6	1	1	2
Total	374	440	63	65

1.5.1. Gas-cooled reactors

The gas-cooled graphite-moderated species, once flourishing, was successively abandoned by France, Italy, Spain and Japan. The only survivors are the British Magnox and AGR, roughly 4% of the installed capacity today. It should be noted, though, that the most recent reactor operating in the UK is a PWR.

Magnox and the French UNGG were fuelled by natural uranium but, because graphite is a mediocre moderator, the neutron balance was very tight and they could not afford neutron capture in cladding. Therefore, the fuel was metallic uranium clad in light alloy, whose melting temperature is unfortunately low. These designs also required as few captures as possible in the fission products, a constraint which made on-line refueling compulsory.

The factors which favored their early selection were the relatively unsophisticated technology required to build them – the technology of post-World War II France – and the combination of natural uranium + on-line refuelling allowing the production of almost pure ^{239}Pu for military applications without significantly interfering with their role as power reactors.

Conversely, the factors driving their elimination were the low power density in their core which led to facilities of very large size when the unit power rating started to increase above a few hundred MWe, their sensitivity to ^{135}Xe poisoning which limited their flexibility… and their capability to produce weapon-grade plutonium, once proliferation became undesirable.

The AGR were designed by the UK to overcome some of the weaknesses of the Magnox, notably the low melting point of the cladding: the switch to slightly enriched uranium allowed for an oxide fuel with stainless steel cladding, which in turn allowed for a reduction in core volume by a factor 2, an increase in the exit gas temperature and excellent thermal efficiency. Their success was limited by a rather troublesome start-up, due in part to bad industrial organization and mostly to corrosion of steel by the higher temperature CO_2. They operate very well today, but too late to save the species: in the meantime, LWR have won the race.

The case for gas cooling is not closed yet, as we shall see when turning to future generation reactors (HTR, GFR), but their first period is over.

1.5.2. Graphite-moderated and boiling water-cooled reactors RBMK

The "Big water channel reactor" RBMK was one of the two nuclear workhorses of the former Soviet Union, limited to Russia, Ukraine and Lithuania. Fuel bundles of low enriched uranium oxide are cooled by boiling water circulating in pressure tubes vertically inserted in a massive graphite pile.

The initial success factors of this species were identical to the former examples: accessible technology and dual-purpose (electricity and weapon-grade plutonium). This latter quality restricted their use to the Soviet Union itself and only VVER were exported to the other countries of the former "Eastern Block".

The reason for their progressive elimination has one name: Chernobyl (even though later improvements have significantly reduced their instability domain).

1.5.3. Heavy water reactors CANDU

The CANDU (for CANadian Deuterium Uranium) is today the only species maintaining an active niche in a Light water dominated world. From the point of view of pure neutron physics and excluding cost considerations, heavy water is the best possible moderator. Using D_2O as both moderator and coolant, CANDU can be fuelled with natural uranium oxide and reach significant burnups. The fuel bundles, refuelled on-line, are cooled by pressurized heavy water circulating in horizontal pressure tubes. Heavy water must be periodically re-enriched and de-tritiated.

Canada exported the CANDU in many countries (India, Pakistan, South Korea, Argentina, Romania and China) based on the following factors: accessible technology (pressure tubes are easier to manufacture than big pressure vessels), no dependence on a very limited number of uranium enrichment suppliers, and smaller sizes as they offer 300 and 600 MWe plants where most LWRs are rated at 1000 MWe and above.

On the negative side, the specific investment is high and heavy water is very expensive, even though fuel cycle costs are low. On-line refueling allows for high availability but, together with natural uranium, it also makes it technically easy to produce weapon-grade plutonium.

It should be noted that the "3rd Generation" CANDU being developed by AECL, called ACR-750, uses ordinary water as coolant and low-enriched uranium fuel.

1.5.4. Light water reactors PWR, BWR and VVER

In terms of installed power, PWRs alone account for more than half of the world nuclear capacity; together, the three LWR "cousins" represent over 85% of the market. LWRs really dominate the reactors biotope. In reality, this domination is even more absolute, as most of the reactors used for naval propulsion are small PWRs (their thermal rating on the order of 100 MWth instead of 3000).

As ordinary water is the best to slow neutrons down, the cores are compact. On the other hand, water is absorbent and LWRs must use a fuel with a significant concentration in fissile isotopes, typically 4% ^{235}U or 8% plutonium, refueling being done off-line. Even with high pressure in the primary circuit, these reactors exhibit mediocre thermal efficiency (~35%) and a rather low conversion factor, around 0.6. But they are sturdy and more flexible than initially expected. Relatively large units are, in most countries, quite competitive to generate baseload power [12–13], even without any credit for carbon emission.

In addition to the intrinsic qualities mentioned above, the key factors for this immense success were the technology transfer from submarines to power plants, the sheer power of the American industry in the Western world in the 1960s and 1970s, and the driving force generated by a then huge nuclear power programme in the USA. One of the reasons behind the French switch from UNGG to the "American" species was: "because everybody else chose them" – even though we are expected to cultivate the French exception. "Vive la différence" as the British say.

1.5.5. High temperature reactors

The HTR concept was developed around a very innovative fuel of British origin, the "coated particle". This finely divided fuel, whose proportion can easily be adjusted within a bulk graphitized moderator, allows for very refractory cores, cooled by high-temperature helium. This, in turn, allows for high thermal efficiency, comparable to the best fossil-fuelled plants. Mixing different particles allows adjustment to almost any fuel cycle.

The HTR equivalent of a "fuel assembly" exists in two varieties: prismatic block with coolant channels or spherical pebble.

Their large thermal inertia and significant margins with regards to their operating conditions (the space propulsion program NERVA saw several HTR reactors heat their hydrogen coolant above 2500 °C) make HTRs very safe reactors. On the other hand, their low power density results in high specific investment costs.

After promising demos, HTRs had a false start in the beginning of the 1970s in the USA and during the 1980s in Germany. They are presently reconsidered as small modular reactors with direct helium cycle cooling. Their small size allows for "passive" decay-heat cooling after shutdown. The GT-MHR project and the South African PBMR, Pebble Bed Modular Reactor, are examples of such reactors. The GT-MHR is optimized for the consumption of "weapon grade" plutonium.

Further on, the VHTR belongs to "Generation IV" (see Chapters?).

1.5.6. *Fast breeders* [14]

Fast breeders have fascinated reactor designers since the beginning of the 1950s because successive recycling allows them to use almost all the potential energy of the uranium ore, making them a hundred times more efficient than LWRs. Even better, they can feed on the depleted uranium leftover from the enrichment process.

This property is explained in Figure 1.1.

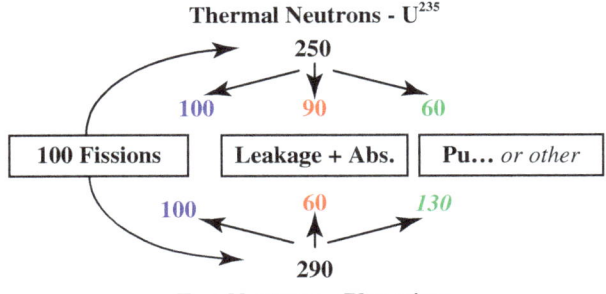

Figure 1.1. Diagram of neutron balance.

Liquid water is too good a moderator: the choice of coolant for FBRs can only be between liquid metals and gases. The only significant return of experience comes from sodium cooled FBRs. Used between 300 and 500 °C, sodium is an excellent coolant and, if pure, offers good compatibility with stainless steels. But it burns in air and strongly reacts with water.

In order to prevent any reaction between the activated sodium of the primary circuit and the water going to the turbines, present FBRs have an intermediate circuit of inactive sodium. This additional circuit and the use of "noble" materials in contact with sodium increases the investment cost as compared to LWRs. FBRs will therefore develop only when uranium economy becomes a key factor.

Industrial size FBR plants have been put into operation in France (Superphénix, 1200 MWe) and the Federation of Russia (BN-600), and significant demonstration plants have been built in France, Great Britain, Japan, Kazakhstan and the United States… Smaller experimental facilities are or will soon be in operation in India and China.

Managing the high level radioactive waste may lead to eventually choose to partition and transmute some long-lived nuclides, notably minor actinides (Am, Np): in this case, FBRs will be well suited to the task thanks to their high neutron flux and the possibility to locally adjust the neutron spectrum to a specific capture resonance.

Because they save natural resources, FBRs have the lion's share in the Generation IV concepts.

1.5.7. Molten salt reactors [1]

For the sake of completeness, we should mention molten salt reactors MSRs, which can be either in thermal spectrum (molten fluorides and graphite moderator) or fast spectrum (molten chlorides).

A very small MSR facility was operated in Oak Ridge from 1965 to 1969, demonstrating the specific qualities of this concept: fuel cycle flexibility, on-line reprocessing and recycling (which eliminates transports) and small breeding potential.

1.6. Biotope, domination and selection

The LWR species will continue to dominate in the near future. With one exception, all "Generation III" reactors are LWRs: the ABWR, EPR and AP1000 plants under construction and projects like ESBWR, AES 92, ATMEA, KERENA, APWR, etc. The only exception, the ACR 750 project, is no longer a pure CANDU, as seen above…

In nature, the success of one species makes it very difficult for competing species to penetrate the same ecological niche. Without the mass extinction of dinosaurs, would mammals have thrived? Would one of their branches have evolved into Man? The same applies in the reactors phylum: construction facilities are made to build LWRs, fuel cycle facilities are optimized for LWRs, nuclear R&D is devoted in majority to LWR improvements, and most Safety Authorities around the world "think" LWR. An anecdote of uncertain truth tells that one NRC staff member once asked where the emergency water injection system was in the gas-cooled Fort St Vrain, and the story is believable. I know for a fact that one of the heaviest constraints imposed on the sodium-cooled Superphénix, the hypothesis of a guillotine double break of the largest secondary piping, came only from some kind of PWR contamination.

One can reasonably ask if there is room for any species other than PWR and BWR. Both technologies are close enough to co-exist without problem, and both are different enough to meet the desire for minimum diversification which may explain today's CANDU niche. Shouldn't we simply go on improving these two species?

Let's put it another way. No new candidate species will find a niche unless it exhibits a significant advantage in at least one of the domains where LWRs have an undisputed weakness. But the very notions of advantage and significant can vary with time! As an example, let us consider the automobile.

In the beginning of the motor car technology, there was also a flourish of creativity, and the current domination of the internal combustion engine over its electrical and steam-driven counterparts did not occur overnight. The first automobile to break the 100 km/h barrier was an electric car, C. Jenatzy's "Jamais Contente" in 1899, a performance that we could liken to the first electricity production by the Fast Neutron reactor EBR-1 in 1951. In 1906, the world speed record belonged to a steam car which reached 196 km/h! But a key factor established the success of the internal combustion engine: the unmatched autonomy offered by the gas tank. To push the comparison to its limits, one can note that internal combustion engines exist in two varieties, petrol and diesel, as close and as different as PWRs and BWRs. Today, the electric car survives in a niche (apart from its railroad avatar, the TGV). However, as soon as 1900, it was obvious that the electric car was very superior to its competitors in terms of atmospheric pollution and noise. At the time, those undisputed advantages were not significant, but the winds are changing nowadays. The environment which enabled the domination of the LWR species may also be evolving.

1.7. From spontaneous selection to a formalized process [14, 15]

It is precisely this kind of thinking which explains two very interesting initiatives launched just at the turn of the century. At any given moment, and considering objects whose lifetime (from design to dismantling) spans the century, the weight of the existing species hinders all attempts to make significant breakthroughs. Even the so-called "revolutionary" advanced LWR designs, which exhibit rather modest changes in the basic LWR technology, meet some diffidence from many regulators, because they are – factually – losing part of the return of experience from their forerunners. Obviously, any breakthrough, any introduction of a brand new reactor species must be prepared decades in advance. With this kind of delay, one can hope to design not only "reactors", but nuclear systems better suited to the selection criteria of the 2040s or 2050s, which may not be identical to the criteria having led to the present selection.

Philosophically, this is a real change. Instead of designing reactors optimized for today's market, one tries (as is the case with Generation IV) to anticipate what the market will be four decades from now, to imagine the best possible concepts to answer these faraway demands and to launch today all the required actions, in R&D and beyond, to make this desired future possible. In effect, one tries to replace the past "natural" spontaneous selection by a formalized process.

1.7.1. GIF, the Generation IV International Forum

At the initiative of the USDOE, since 1999, twelve countries and the European Union have worked together to select a few model concepts for future nuclear systems, and to define and conduct the required R&D to make them ready for possible commercialization after 2040. Criteria for formal selection included Sustainability (fissile resources utilization, waste minimization, proliferation resistance and physical protection), Safety & Reliability (radio-protection, reactivity control, heat removal, mitigation features) and Economics. A first phase was open to the unbridled creativity which led to revisit all former reactor designs and invent a few more. The result was a list of roughly 120 concepts. Then, within the Generation IV International Forum, each concept was passed through the sieve, and six of them survived the ordeal. The following six model concepts were ultimately selected:

- Supercritical Water Cooled Reactor System, SCWR;
- Very High Temperature Reactor System, VHTR;
- Sodium Cooled Fast Reactor System, SFR;
- Lead Alloy Cooled Fast Reactor System, LFR;
- Gas Cooled Fast Reactor System, GFR;
- Molten Salt Reactor System, MSR.

Without going into details, there are a few comments worth making.

All these systems do not follow the same development timeframe. One could reinvent Superphénix – or, rather its successor design EFR – in little time. Prototypes could be developed rather quickly for the VHTR and SCWR. The GFR is farther away, and the MSR even farther (some GIF participants do not even include it).

The weight of the "sustainability" criteria appears in the fact that at least 4 and possibly 5 of these systems are based on closed fuel cycles for better use of the fissile natural resources, 3 of them being fast breeders.

The VHTR and (later) the GFR aim not only at generating electric power but also at co-generating power and hydrogen with the goal of making transportation fuels from fission power. This is certainly necessary if we want nuclear power to significantly contribute to the reduction of greenhouse gas emissions on a world scale.

Finally the future may not be limited to these 6 species. Some intriguing concepts, such as the AHTR, a molten salt-cooled HTR, are under development. Some teams keep working on the Accelerator Driven subcritical Systems, whose niche could be the transmutation of minor actinides, if deemed useful at some point. And these are only a few examples.

1.7.2. *INPRO, International Project on Innovative Nuclear Reactors & Fuel Cycles*

In 2000, the IAEA initiated the INPRO Project gathering fifteen (now over twenty) member states to define "User Requirements" for innovative nuclear energy systems in the area of Economics, Sustainability and Environment, Safety, Waste Management, Proliferation Resistance and some cross-cutting issues. The time horizon for this exercise is 2050. An assessment methodology was also developed for such systems. Phase 1A of INPRO was completed in June 2003, and the methodology is being tested on 6 "national cases". Started entirely as an extra-budgetary initiative, INPRO is now partly funded by the IAEA's regular budget.

Based on similar analyses and motivations, the works of GIF and INPRO are, however, not identical: GIF partners are mostly suppliers, and their work will steer the R&D, whereas INPRO expresses mostly the requirements of potential future users. Each group is quite aware of the other's results, and they have cooperated on some issues (like proliferation resistance).

1.8. Fusion

Regarding fusion, it is too early to see if the "natural history" approach is relevant. ITER, the big international facility currently being built, might not yet be the true equivalent of the December 1942 CP1 experiment which was our starting point… One can nevertheless remark that, from a flourish of early configurations (Yin-Yang, stellarator, Z-pinch…), the Tokamak species dominates today fusion's prehistory. I am referring here only to Magnetic Confinement Fusion, because Inertial Confinement Fusion does not seem to offer promising prospects for power generation.

1.9. Conclusion

From 0 TWh 57 years ago to some 2400 today, exceeding the primary energy equivalent of the oil production of Saudi Arabia or Russia, nuclear power has come a long way. This is where the "natural selection" has led us but this is not enough.

Over the next 50 years, the most dramatic problem which we will have to address is probably the energy/climate dilemma. One third of the 7 billion people with whom we share this planet Earth have not enough energy to lead a life considered decent under any modern standard. Within fifty years, mankind will probably exceed 9 billion people. Even under the most optimistic assumptions concerning increases in energy efficiency and "thrifty" development, the demand for primary energy will then at least have doubled since the 10 billion tons of oil equivalent we consumed in the year 2000.

On the other hand, it is scientifically established that, since the industrial revolution, we have managed to significantly affect the composition of our atmosphere, where the concentration in greenhouse effect gases (GHG) is today twice higher than the highest level the Earth has ever experienced since the dawn of Man. There is still some residual controversy about just how much of the global warming which occurred throughout the 20th century is due to this proven increase in GHG concentrations, but the best computer models reviewed by the IPCC (Intergovernmental Panel on Climate Change) teams throughout the world agree: at least half of the warming was due to the increase in greenhouse effect. And what those models tell us about our 21st century is far from reassuring. Any precautionary principle would require that we cut our GHG emission by half.

To complete the picture, 82% of the primary energy we use today comes from fossil fuels, Oil, Coal and Gas, whose combustion is the main source of GHG releases. This value of 82% takes into account the 10% coming from (non-commercial) traditional biomass which remains the only source of energy for some 1.6 billion people today.

To face this tremendous challenge, there is no "magic bullet", not even the nuclear bullet. We will have to take all the measures we can take, and then some more…

1. We must address the demand side and not only the supply side of the equation. We must limit our energy consumption, increase our energy efficiency, and conserve energy. This applies first and foremost to our presently industrialized economies in the OECD, but emerging giants like China, India and Brazil (to name the largest) must also find thrifty development paths, because of the sheer weight of their demography.

2. We need to "sequester" CO_2 whenever and wherever technically feasible. The development of sequestration is still in its infancy, and the long-term safety assessment of disposal sites will not be much easier than that of radwastes, but I do not believe that we can do without it.

3. In our energy mix, we must increase the share of non GHG-emitting sources, namely nuclear power, hydropower and other renewable energies.

It is my firm belief that to tackle such a problem, all of these measures will be necessary. It would be dramatically unrealistic to hope to succeed without a significant increase in the contribution of nuclear power: for electricity generation, of course, but also for other uses such as water desalination, process heat, hydrogen production…

Let us hope that the formalized selection process which the nuclear community is presently attempting will prove efficient, and allow nuclear power to meet our hopes.

References

[1] A.M. Weinberg, *The First Nuclear Era / The life and Times of a Technological Fixer.* AIP Press, 1994.
[2] J.W. Simpson, *Nuclear Power from Underseas to Outer Space.* ANS, 1995.
[3] *Controlled Nuclear Chain Reaction: The First Fifty Years.* ANS, 1992.
[4] B. Goldschmidt, *L'Aventure Atomique.* Fayard, 1962.
[5] B. Goldschmidt, *Les Rivalités Atomiques 1939-1966.* Fayard, 1967.
[6] B. Goldschmidt, *Le Complexe Atomique.* Fayard, 1980.
[7] B. Goldschmidt, *Pionniers de l'Atome.* Stock, 1987.
[8] B. Barré, P.R. Bauquis, *Nuclear Power: Understanding the Future.* Hirlé éditions, 2007.
[9] P. Reuss, *L'épopée de l'énergie nucléaire.* EDP Sciences, 2007.
[10] *Repères chronologiques de l'histoire du nucléaire 1896-1995.* CEA, 1998
[11] PRIS Database from IAEA website, June 2011.
[12] *Projected costs of generating electricity.* OECD, 2010.
[13] R. Tarjanne, A. Kivistö, *Comparison of electricity generation costs.* Lappeenranta University of Technology, 2008.
[14] A Technology Roadmap for Generation IV Nuclear Energy Systems. GIF 002-00.
[15] *INPRO, IAEA 1362.* TecDoc, 2003

2 CO_2 gas cooled reactors

R. Lenain

2.1. Introduction

Just after the end of the Second World War, it was decided in France and the UK to launch a nuclear electric program. The main objectives were rapidly decided upon according to the technical and industrial capacities of these countries:

- fuel: metallic natural uranium;
- moderator: graphite;
- coolant: gas, air and pressurized carbon dioxide (CO_2).

France had sufficient uranium resource for this program, but no enrichment technology, no heavy water, and no industrial feedback for manufacturing large steel components such as pressure vessels. However valuable graphite experience was gained from electro chemical and mechanical industries. Then the UNGG reactor type "Uranium Naturel Graphite Gaz" was developed according to French government plans (Table 2.1).

At the same time, the UK launched its own program following the same rationale: as a first stage, a Magnox type reactor (metallic uranium, graphite moderated, CO_2 cooled) and following this the AGR (Advanced Gas Reactor). For this second reactor type, the British solution joined the fuel cycle main stream: oxide and enriched uranium. Calder Hall was the world's first commercial nuclear power unit. The connection to grid was made on the 27 August 1956 and the station closed on 31 March 2003. Calder Hall Reactor 1 is to become a museum site.

The development of UNGG and Magnox was stopped in the 70's and the UK pursued its AGR program until the end of the 80's. In the 60's, France and the UK exported some reactors (Vandellos in Spain for UNGG and Tokai Mura in Japan, Latina in Italy for Magnox).

At the opposite end of the spectrum, the USA that had developed uranium enrichment technology in the framework of the Manhattan project, having considerable experience in steel industry; PWR and BWR reactor designs reaped the benefits of the naval propulsion reactor program.

By the end of 1998, the graphite-moderated reactor type was the third nuclear reactor type produced in terms of nuclear electricity (including RBMK). Figure 2.1 presents the

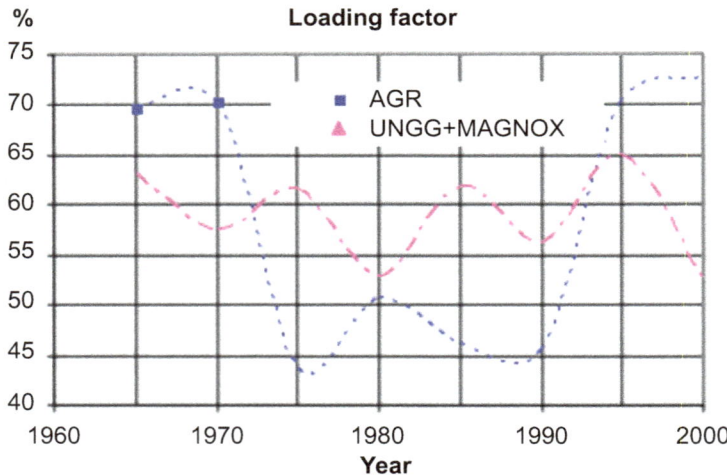

Figure 2.1. CO_2 cooled reactor capacity factor.

capacity factor (energy effectively produced during the year divided by energy produced if the reactor worked at full power during all the year). This figure illustrates the AGR starting phase where some major technical challenges had to be addressed.

In 1994, all French UNGG units were shut down. In the UK, the last two Magnox NPP are Oldbury and Wylfa; Wylfa will be shut down in 2015. EDF operates the 14 remaining AGR reactors in operation in the UK.

2.2. General architecture

The reactor core is made of graphite stacks where holes receive fuel elements and cooling channels or control rods and detectors. Vertical and horizontal arrangements have been built and horizontal and vertical coolant flows have been used.

During the period of time where these reactors were built, various general architectures were used, dependent mainly on the unit power (MWe). For the first units (40 MWe), a steel caisson (leak-tight housing) could be manufactured but for the most powerful unit (above 500 MWe), a concrete housing has been considered, both in France and in the UK.

The main components are the caisson (housing), the blowing machines, heat exchanger, purification system, auxiliary cooling system, fuel handling system, control system (reactivity control rods and detectors), control and monitoring system and fuel non-integrity detection system (Figure 2.2).

There are two types of general architecture: non-integrated systems (in which blowers and heat exchangers are outside of the caisson) and integrated systems with blowers and heat exchanger inside the caisson (Figures 2.3 and 2.4).

Figure 2.5 shows the graphite reactor structure during construction.

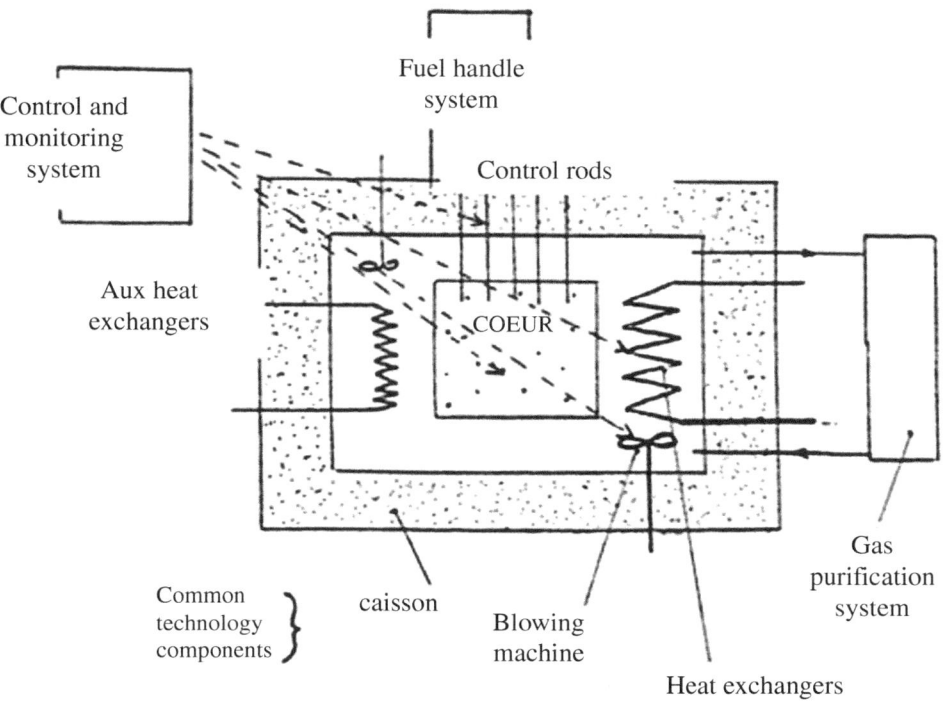

Figure 2.2. General architecture, main gas cooled reactor components.

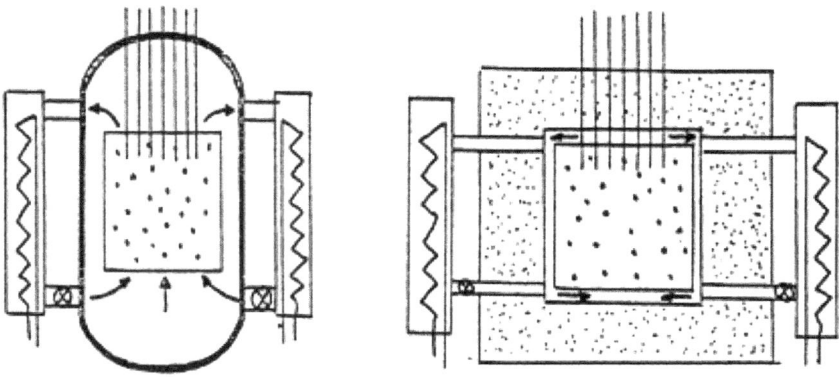

Figure 2.3. No-integrated architecture: small power unit steel caisson and, for large power unit, concrete caisson.

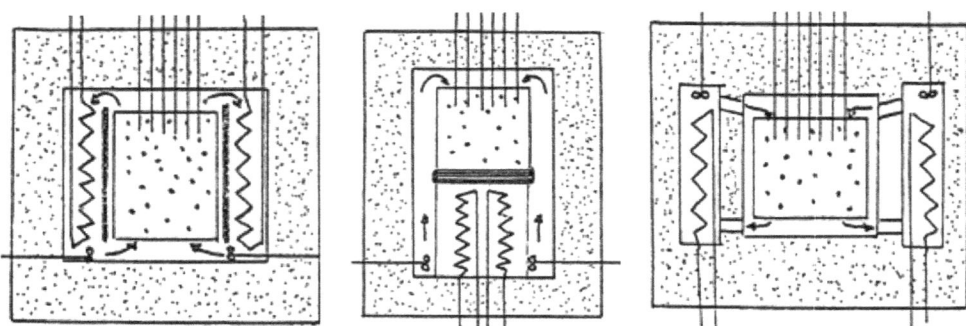

Figure 2.4. Integrated architecture: annular, pile and alveolar types.

Figure 2.5. Graphite reactor during construction phase at Chinon A3 in 1965 (source EDF).

2.3. General features of graphite-moderated reactors

2.3.1. Fuel: natural uranium and magnesium clad (UNGG & Magnox)

Criticality can be reached with natural uranium if the major part of neutrons generated by fission in the fuel is thermalised. This leads to a "heterogeneous" reactor type design where neutrons leave the fuel just after the fission and are slowed down in the moderator far from the uranium 238 whose resonances can be avoided. It is also necessary to exclude sterile neutron absorption and consequently metallic uranium with low alloy grade is considered. In France and the UK, UFeAl was developed (0.02–0.05% Fe, 0.05–0.09% Al). This alloy presents better mechanical resistance than uranium and quite good stability under neutron flux. UMo alloys (0.1–1.1% Mo) have also been developed with this solution being satisfactory at high temperature. The fuel is in bar or annular shapes. For the annular shape, graphite or coolant can be located in the center part of the fuel element.

The clad is made of magnesium-based alloy and striated in order to increase heat exchange with the coolant. The clad also defines the coolant channel and permits dimensional change during the irradiation time.

2.3.2. Graphite moderator

Associated with natural uranium, two moderators can be used: heavy water and graphite (Beryllium is also considered for specific purposes). The efficiency of various moderators is discussed in chapter 4. In graphite-moderated reactors, the reactor structure is made up of graphite blocks (moderator) where holes receive fuel elements, control rods and channels for the coolant.

In order to reduce activation under neutron and gamma fluxes, the graphite used is of a specific nuclear quality in which impurities are extracted; a dedicated manufacturing process that limits air content is applied. Non-irradiated nuclear graphite has an initial density in the range 1.6–1.8 g/cm^3. This can be compared with a theoretical density for natural graphite of 2.265 g/cm^3, the difference being due to internal porosity in the manufactured blocks.

Wigner effect (from Wikipedia)

The Wigner effect is the displacement of atoms in a solid generated by neutron radiation. To create the Wigner effect, neutrons that collide with the atoms in a crystal structure must have enough energy to displace them from the lattice. This amount (threshold displacement energy) is approximately 25 eV. For example, a 1 MeV neutron striking graphite will create 900 displacements, however not all displacements will create defects. The atoms that do not find a vacancy come to rest in non-ideal locations and they have an energy associated with them. When large amounts of interstitials have accumulated, they pose a risk of suddenly releasing all of their energy, creating a temperature peak. This energy build-up is referred to as Wigner energy. It can be released by heating the material, a process known as annealing. In graphite this occurs at 250 °C. [An accident during this controlled annealing was the cause of the 1957 Windscale fire].

Dimensional change

During reactor operation, the graphite component dimensions change and, in some cases, this can lead to considerable whole core deformations. Nuclear graphite components are polycrystalline in nature and their physical irradiation property changes are dominated by irradiation-induced changes to the graphite crystallites. The effect of irradiation on the crystallites is to expand in one direction and shrink, to a lesser extent, in the other direction. The consequence of this crystal dimensional change on the polycrystalline graphite component is critically dependent on the manufacturing process and the irradiation temperature. For nuclear design, the determination of the dimensional changes due to the neutrons bombardment with neutrons is very important and is taken into account in the design by controlling mechanical gaps between graphite blocks maintained together by a graphite key system (Figure 2.6 illustrates two typical French designs).

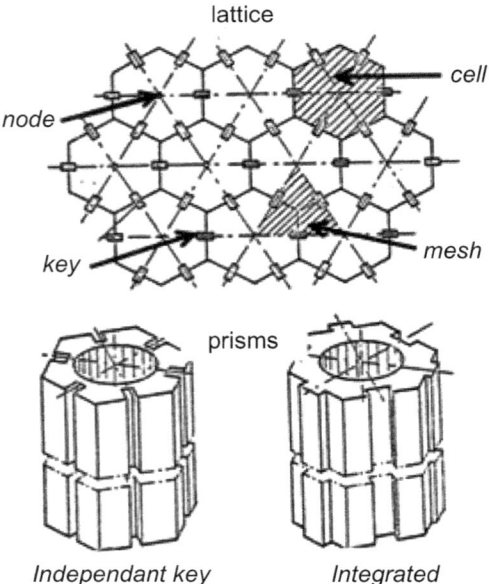

Figure 2.6. Graphite block system (source CEA).

Activation

Graphite contains impurities that have to be reduced in order to limit the neutron absorption and graphite activation. The absorbing characteristics of impurities are defined in equivalent boron (some ppm). The main causes of graphite activation come from Li, Cl, N, C and Co.
Li under irradiation produces Tritium (period 12.3 years) ^{6}Li(n, α)^{3}H
The natural chlorine compounds are ^{35}Cl (76%) and ^{37}Cl (24%); the ^{36}Cl is radioactive (period of 3×10^5 years) and is produced through three different mechanisms:

- neutron capture ^{35}Cl (n, γ) $\rightarrow$ ^{36}Cl (main production route); activation ^{39}K (n, α) $\rightarrow$ ^{36}Cl, radioactivity β of ^{35}S;

- ^{14}C is an important part of the graphite activation and comes from ^{13}C (n, γ);

- ^{14}C (88%); ^{14}N (n,p) ^{14}C (9%); ^{17}O(n, α)^{14}C (3%); reference [1].

Figure 2.7 presents an example of activity found in graphite.

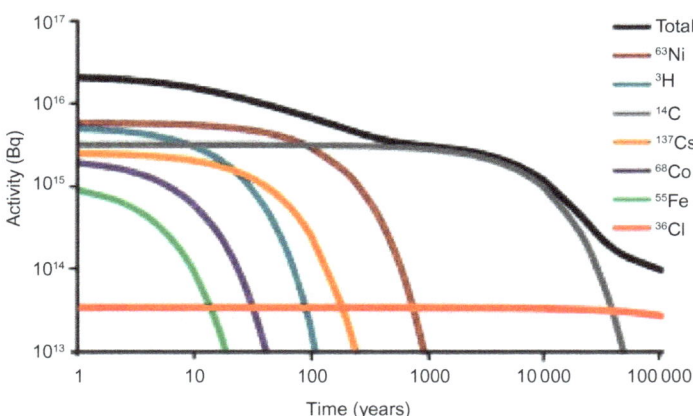

Figure 2.7. Activity(Bq) versus time (years) (source ANDRA).

Coolant

Taking into account the absolute necessity to reduce the neutron absorption in the reactor, the coolant has to be transparent to neutrons. The first reactors were cooled by air (Windscale and G1 at Marcoule) which required low graphite temperature, and CO_2 was rapidly chosen permiting operation above the Wigner temperature and excluding important Wigner energy storage. The main properties of the coolant have to be: no activation, no significant neutron absorption, stability under gamma and neutron fluxes, compatibility with reactor structures (graphite and steel), optical transparency, easy purification and capability to be used at high temperatures. The gas as coolant strongly impacts the reactor characteristics and when associated with graphite as moderator, this leads to a low power density (kW/m^3), and associated with metallic fuel, the specific power is also low (kW/kgU). The accidental loss of coolant determines the limits of specific power and power density because, during this accident heat cannot sufficiently be removed from the reactor and should be transferred to graphite without reaching the fuel temperature limit; this leads to a low specific power. Gas circulation requires efficient blowers and significant associated power; specific clad designs were defined in order to increase clad-coolant heat transfer.

2.3.3. General physical properties of graphite moderated reactors

Neutron spectrum

Figure 2.8 presents neutron spectra for PWR900, RBMK and UNGG reactors. Some general trends can be seen: for graphite-moderated reactors (UNGG and RBMK), a larger quantity of thermal neutrons than for PWR. A regular shape at high energy can be observed in the

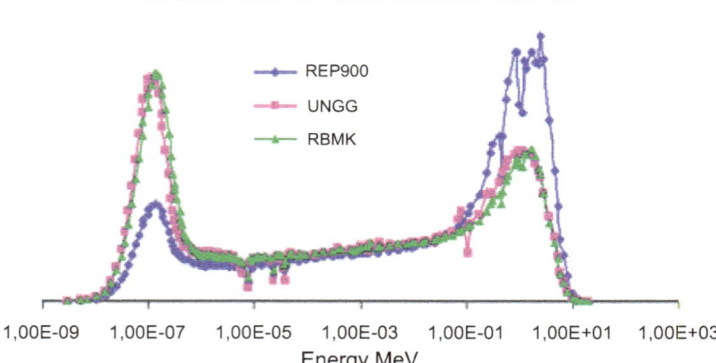

Figure 2.8. PWR900, UNGG and RBMK neutron flux shapes, normalized values.

range a of 0.1 up to 4 MeV, this regular behavior is mainly due to inelastic diffusion on UO_2. For all reactors, the U8 resonance effects are easily found in the range of eV.

Reactivity changes

At low ^{235}U content, fuel depletion leads to a maximum in reactivity due to two main effects: ^{235}U depletion and Pu buildup. The maximum depends on graphite temperature that plays a role on ^{238}U absorption and then on Pu production speed (Figure 2.9). The life time of the fuel in terms of reactivity reserve is also affected by the graphite temperature. As a result, there is an important coupling between fuel technology (burn up, temperature capacities), reactor architecture (graphite cooling) and reactivity control.

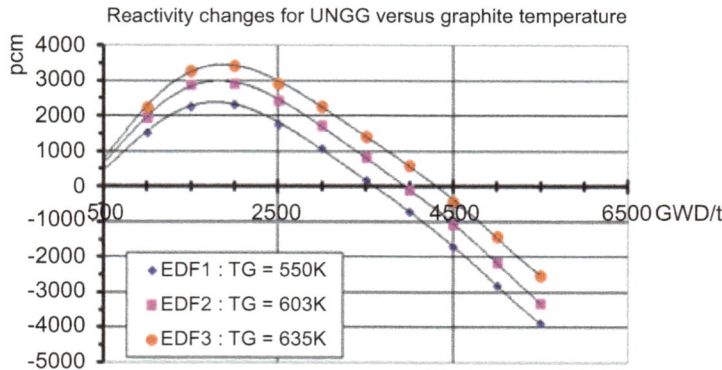

Figure 2.9. Reactivity changes given for UNGG lattice at different graphite temperature: 550 K, 603 °K, 635 K.

The reactivity control system has to compensate for some reactivity changes mainly due to:

- graphite temperature effect that depends on fuel burn up, at about 1500 pcm,
- increase of reactivity due to fuel depletion, at about 3500 pcm,
- xenon effect, at about 2500 pcm and fission products accumulation.

Graphite reactivity coefficient

The moderator coefficient (pcm/°C) that results from the reactivity change due to graphite temperature modification is negative with fresh fuel and becomes positive after 200 MWd/t (Figure 2.10).

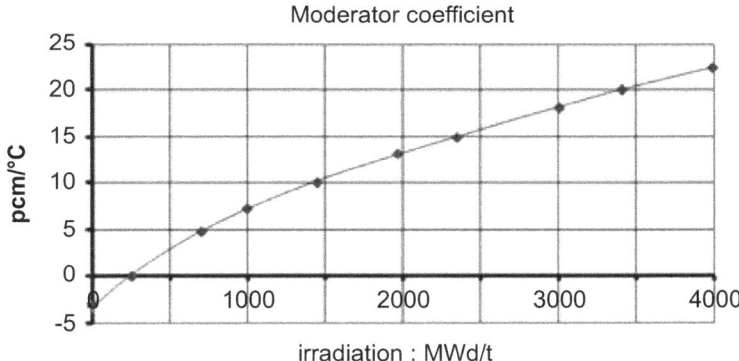

Figure 2.10. Moderator coefficient (pcm/°C) versus irradiation (MWd/t).

The graphite temperature modification is a slow process governed by heat transfer from the fuel (heat transfer coefficient and flow regime) and graphite inertia (specific heat and mass). The fast power coefficient that results mainly from the Doppler Effect and, for a small part, from graphite effect is always negative.

2.4. UNGG

2.4.1. The French UNGG program

The French nuclear program was initiated in the 1950's. Table 2.1 shows a complete view and an open approach of nuclear electricity production. The first UNGG reactors were developed for Pu production needs and then for commercial electricity. In parallel, research programs were launched on heavy water and sodium cooled reactors. PWR research activities were first dedicated to naval propulsion associated with experience gained through buying commercial Westinghouse PWR with Belgium (Chooz A).

Table 2.1. French nuclear planning.

Plan number	Reactor	Power (MWe)	UNGG (MWe)	others	total
II	G2	40	150	0	150
	G3	40			
	Chinon 1	70			
III (1957–1961)	Chinon 2	210	690	210	900
	Chinon 3	480			
	1/2 Chooz*	140			
	EL4 (Mont d'Arrée)**	70			
IV (1962–1965)	Saint Laurent 1	480	1020	0	1020
	Bugey 1	540			
V (1966–1970)	Saint Laurent 2	515	635	668	1303
	1/4 Vandellos	120			
	1/2 Tihange*	435			
	Phénix***	233			
total			2495	878	3373

*PWR; **DO$_2$ moderated; ***Na cooled.

The construction time was very short and various designs were successively built without large operation feedbacks from previous ones. Tables 2.2, 2.3 and Figure 2.11 give the calendar and the main characteristics of the UNGG EDF fleet.

Table 2.2. French UGG fleet characteristics.

Unit	Thermal power MW	Electric power MW	Net electric power MW	Thermal efficiency %
G2	260	45	40	15,4
G3	260	45	40	15,4
Chinon 1	300	80	70	23,3
Chinon 2	850	230	200	23,5
Chinon 3	1560	500	480	30,8
S Laurent 1	1650	500	480	29,1
S Laurent 2	1700	530	515	30,3
Bugey 1	1920	560	540	28,1
Vandellos	1670	500	480	28,7

Table 2.3. EDF UNGG fleet program.

Unit	Construction start	Criticality	Grid connection	Closed	Caisson	Architecture	Fuel type
G2	1955/09	1958/07	1959/04	xx/80	concrete	Non integrated	bar
G3	1956/03	1959/06	1960/04	xx/84	concrete	Non integrated	bar
Chinon 1	1957/03	1962/09	1963/06	xx/73	steel	Non integrated	tube
Chinon 2	1958/01	1964/08	1965/02	xx/85	steel	Non integrated	tube
Chinon 3	1961	1966/03	1966/08	xx/90	concrete	Non integrated	tube
S Laurent 1	1963/08	1963/08	1969/03	xx/71	concrete	integrated	tube
S Laurent 2	1965/01	1969/01	1971/08	xx/92	concrete	integrated	tube and internal graphite
Bugey 1	65	1972/03	1972/04	xx/94	concrete	integrated	annular
Vandellos	1967/11	1972/02	1972/05	xx/90	concrete	integrated	tube and internal graphite

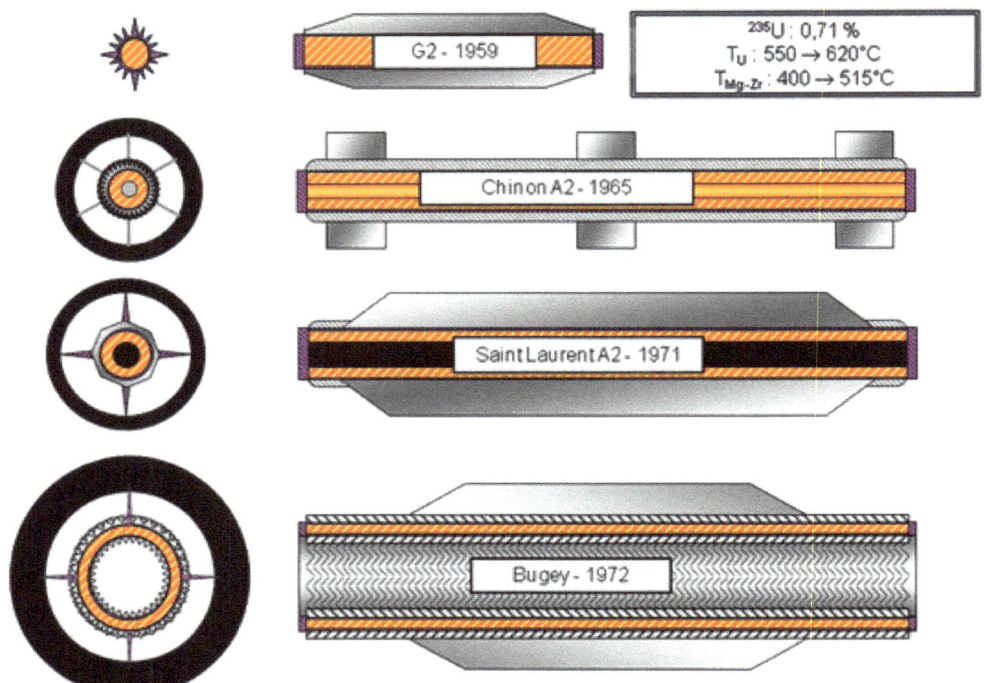

Figure 2.11. Fuel evolution in French fleet (source CEA H Glandais).

2.4.2. St Laurent A example

Caisson

The caisson is a building (48 m height and 33 m diameter) containing the primary circuit with an internal pressure generated by hot gas. The gas pressure at the blowing machine outlet is 29 bars. The gas circulation is driven to firstly cool internal structures and integrated components with cold gas before going through the core. Heat exchangers and blowing machines are placed inside the internal reactor cavity located under the core.

Core

The graphite core structure is approximately cylindrical (10.2 m height and 15 m diameter), the total mass is 2680 tons. The lattice is triangular (pith 225 mm); elementary blocks are prismatic. The radial reflector is made of 828 columns, 3601 for active core including 3256 with a central hole distributed as 181 for control rods and 3075 available for fuel elements. The description of UNGG is given for the Saint Laurent 2 design basis in Tables 2.4 and 2.5.

Fuel elements are replaced each day during operation with 2 or 3 fuel channels (Figure 2.12). The loading machine assumes different functions: replacement of irradiated fuel

Table 2.4. St Laurent reactor characteristics (1).

Power	reactor thermal power	1670	MW
	electric power	495	MW
	net electric power	480	MW
	thermal yield	29.6	%
	net thermal yield	28.7	
Reactor	fuel element: graphite in centre	0.72%	natural uranium
	fuel mass	430	t
	number of fuel elements	41 985	
	element length	570	mm
	inside graphite ring	43/23	mm
	outside graphite ring	137/112	mm
	clad material	Mg-Al alloy	
	average neutron flux in fuel at core centre	3×10^{13}	n/cm2/s
Moderator	Type	graphite	
	reactor graphite mass	2440	t
	active high	9	m
	active diameter	13,7	m
	reflector wide	1	m
	Lattice	hexagonal	
	lattice pitch	225	mm
	channel diameter	140	mm
	available channels	3072	
	loaded channels	2883	
Core support plate	Type	steel	
	Mass	250	t
Support area	type: girder, steel	rectangular mesh	
	Diameter	17.30	m
	High	3.5	m
	mass support area	425	t
	steel shield mass	220	t
	graphite biologic shield mass	630	t

and evacuation outside the reactor hall, introduction of fresh fuel, waste management and control channel services. Figure 2.13 presents the upper part of the reactor hall with the refueling machine.

Table 2.5. St Laurent thermal characteristics (2).

alternator	inlet pressure	33.5	bar
	inlet temperature	387	°C
	group specific conception	11 260	kJ/kWh
	group flow rate	1 977	t/h
CO_2 loop	flow rate	8 664	kg/s
	average reactor pressure	28.5	bar
	output reactor pressure	28.02	bar
	output heat exchanger pressure	27.8	bar
	blowing output pressure	29.06	bar
	total pressure drop	1.257	bar
	input reactor temperature	225	°C
	output reactor average temperature	400	°C
blowing groups	blowing machine yield	79.5	%
	gas flow rate per blowing machine	2186	kg/s
	blowing power	44 584	kW
exchanger	total vapour flow rate	2 175	t/h
	output exchanger vapour temperature	390	°C
reactor	theoretical flux shape factor*	0.876	
	actual shape factor*	0.859	
	blowing shape factor*	0.859	
	hottest channel flow rate	3.6	kg/s
	max clad temperature	473	°C
	max uranium temperature	609	°C
	max channel power	695	kW

*shape factor = inverse of the maximum normalized power over the core.

Figure 2.12. UNGG fuel element (source EDF).

Figure 2.13. UNGG reactor hall and refueling machine (source EDF).

2.5. Magnox

Magnox reactors are based on the same type of technology as French UNGG. Eleven power plants totaling 26 units were built in the UK with one exported to Japan and another to Italy. Calder Hall was the first commercial nuclear power unit to be operated. After Calder

Table 2.6. UK Magnox fleet.

Station	Initial reactor Capacity	Nb units	Closure	Age at closure
Berkeley	200 MW	2	1989	27
Hunterston A	200 MW	2	1990	26
Trawsfynydd	390 MW	2	1993	28
Hinkley A	470 MW	2	2000	36
Bradwell	246 MW	2	2002	40
Calder Hall	194 MW	4	2003	47
Chapelcross	196 MW	4	2004	45
Dungeness A	450 MW	2	2006	41
Sizewell A	420 MW	2	2006	40
Oldbury	434 MW	2	2012	41
Wylfa	590 MW	2	2012	41

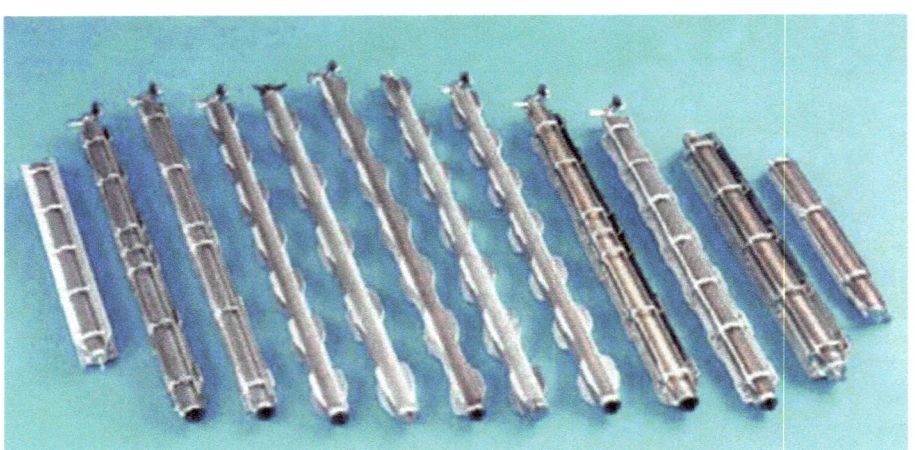

Figure 2.14. Magnox fuel elements (source British energy).

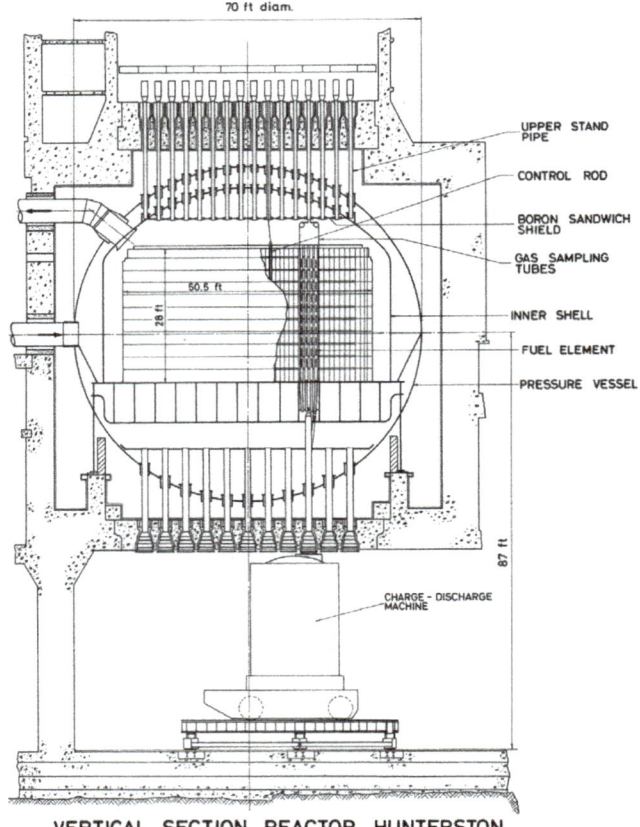

Figure 2.15. Reactor: Hunterston vertical section (source IAEA).

Hall and Chapel Cross plants operated by UKAEA, nuclear stations with higher power were operated by CEGB (Central Electricity Generating Board). On the basis of operational feedback (CO_2 radiolysis and corrosion), the nominal power of some units was reduced.

Table 2.6 presents the UK Magnox fleet.

The name Magnox comes from Magnesium non oxidizing. This alloy, mainly made of magnesium and small amounts of aluminum, has a low neutron capture cross section but limits the maximum temperature of the fuel and reacts with water thereby prohibiting long term irradiated fuel storage under water. The improvement of Magnox fuel was an ongoing activity. Figure 2.14 presents various Magnox fuel elements.

Only the latest reactors were built with a concrete caisson instead of a steel pressure vessel. Figure 2.15 presents the Hunterston reactor (initially 200 MWe) with a very different architecture: spherical steel core vessel and fuel loading machine located under the core. Table 2.7 presents some representative figures of this generation of Magnox.

Table 2.7. Hunterston characteristics, pressure 10 bars.

type MAGNOX 150 MWe	
Thermal yield	28%
Critical mass	31 000 kg nat U
Core uranium loading	251 000 nat U
Bu equilibrium	3600 MWd/t
Bu Max	4500 MWd/t
Specific power	2.13 kW/kg Unat
Power density	0.53 kW/l core
Neutron flux	
	graphite 1.4×10^{13} n/cm²s;
	fuel 1.0×10^{13} n/cm²s
Reactivity effects	
	temperature 1900 pcm;
	Xe Pn 1700 pcm;
	max reactivity 4900 pcm
Temperature coefficients	
equilibrium cycle	graphite 12 pcm/°C
	fuel: −2.1 pcm/°C,
Temperatures	
	max fuel: 582 °C
	max clad: 454 °C

Figure 2.16 shows the general architecture of the Dungeness reactor. Figure 2.17 presents fuel element and graphite block lattice. For the latest Magnox reactor type, an example is given for comparison with AGR in Table 2.10.

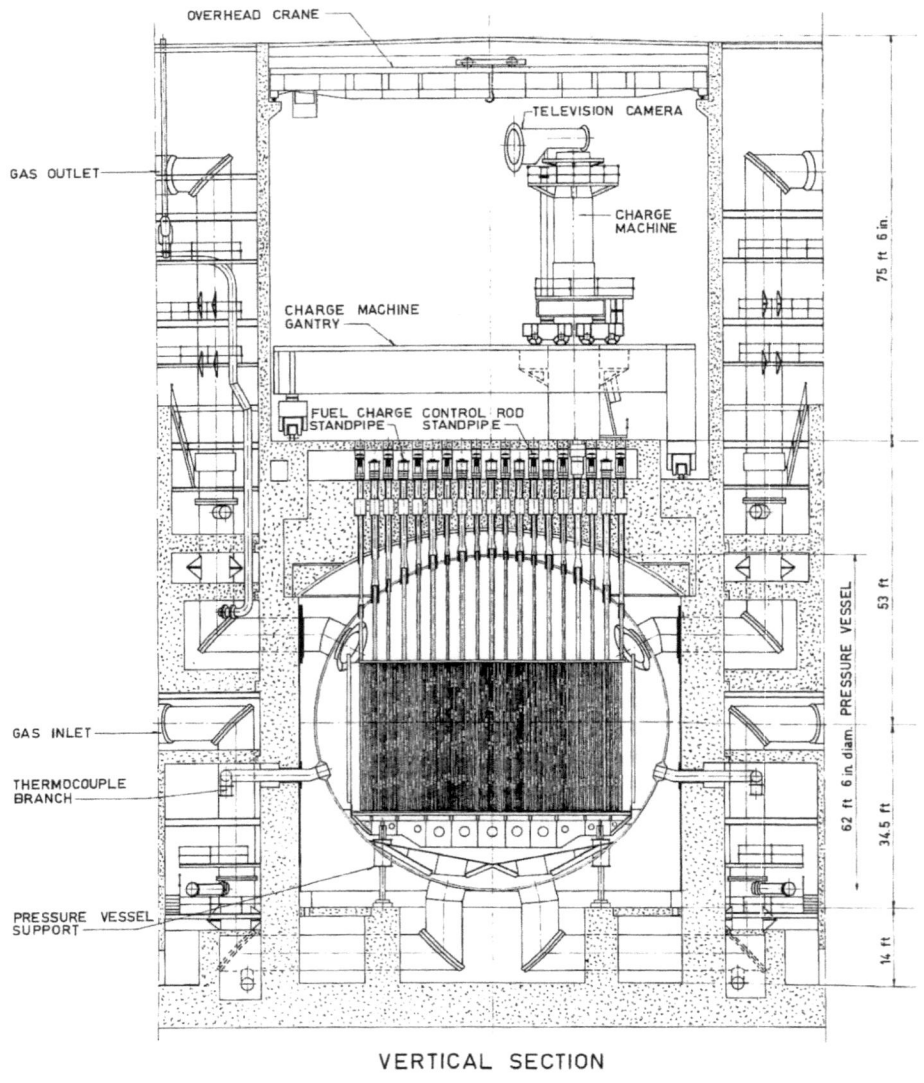

Figure 2.16. MAGNOX Dungeness reactor vertical section (source AIEA).

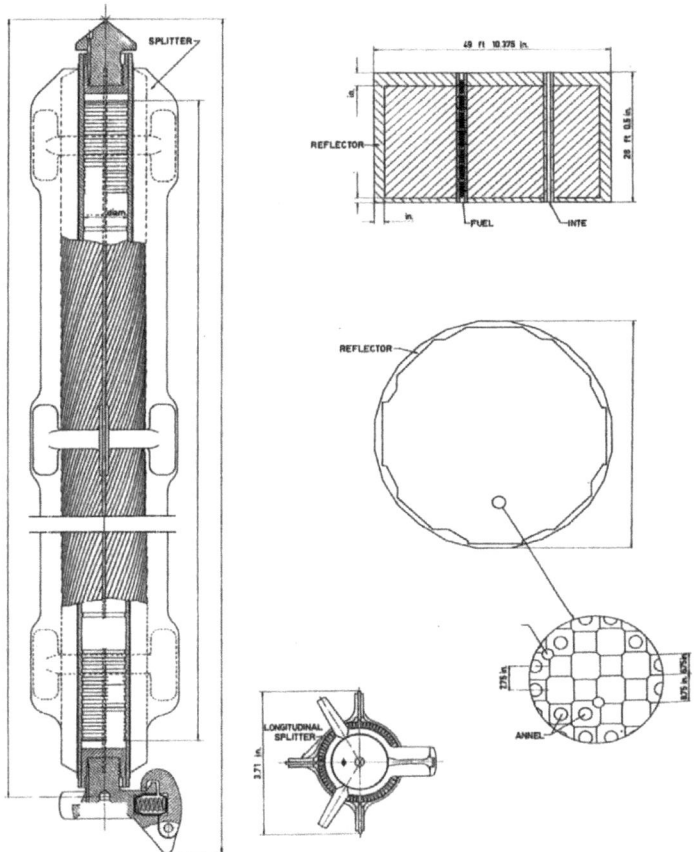

Figure 2.17. MAGNOX Dungeness unit core component, arrangement of graphite bocks, fuel element (source IAEA).

2.6. Advanced gas cooled reactor AGR

In Gas Cooled Reactors (GCR), the moderator is graphite and carbon dioxide is used as the coolant. AGRs use a slightly enriched uranium dioxide associated with stainless steel clad.

This second generation of British power reactors (AGR) results in a political decision to divide not only construction contracts but also design contracts for individual stations between rival companies. Therefore, there is a disparity between the detailed designs of stations, although they do share common general features. All are twin stations, with two AGR reactors, AGR are operated by EDF (Table 2.8). Figure 2.18 gives a map of nuclear sites in the UK.

Table 2.8. AGR fleet; each station received twin units.

AGR Power station	Net MWe	Construction started	Connected to grid	Commercial operation	Accounting closure date
Dungeness B	1110	1965	1983	1985	2018
Hartlepool	1210	1968	1983	1989	2019
Heysham 1	1150	1970	1983	1989	2019
Heysham 2	1250	1980	1988	1989	2023
Hinkley Point B	1220	1967	1976	1976	2016
Hunterston B	1190	1967	1976	1976	2016
Torness	1250	1980	1988	1988	2023

Figure 2.18. Future nuclear sites in UK (source DECC).

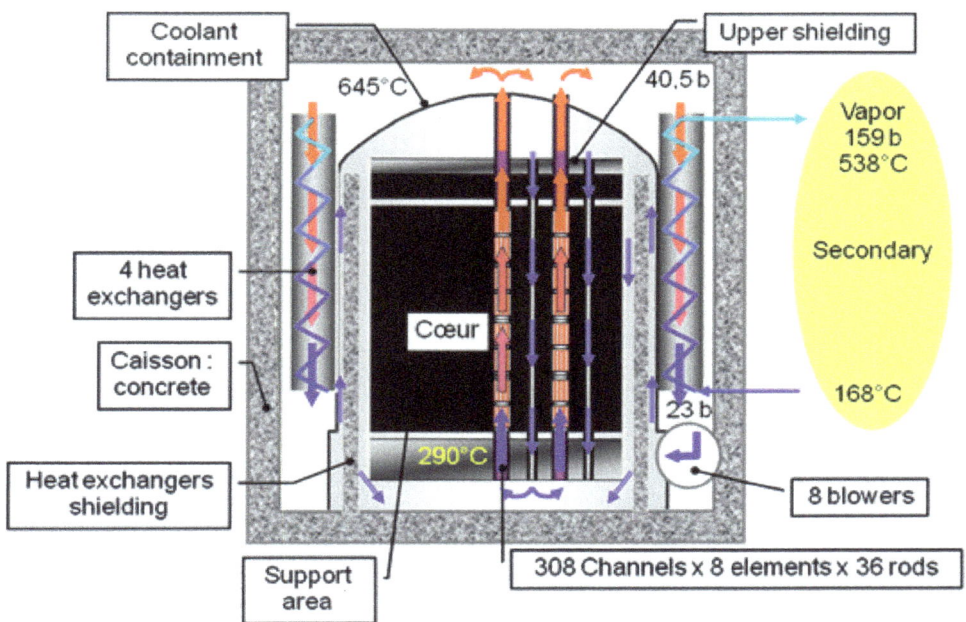

Figure 2.19. AGR scheme source (CEA H Glandais).

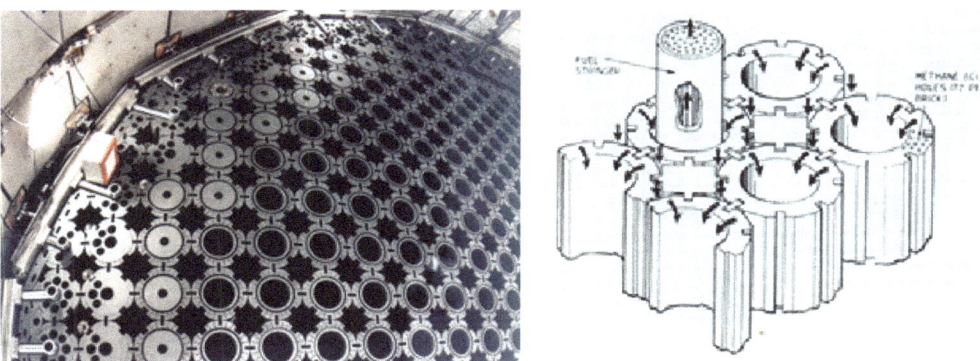

Figure 2.20. AGR core structure: cylindrical holes for fuel element, interstitial ones for control rods and graphite cooling (source Manchester University).

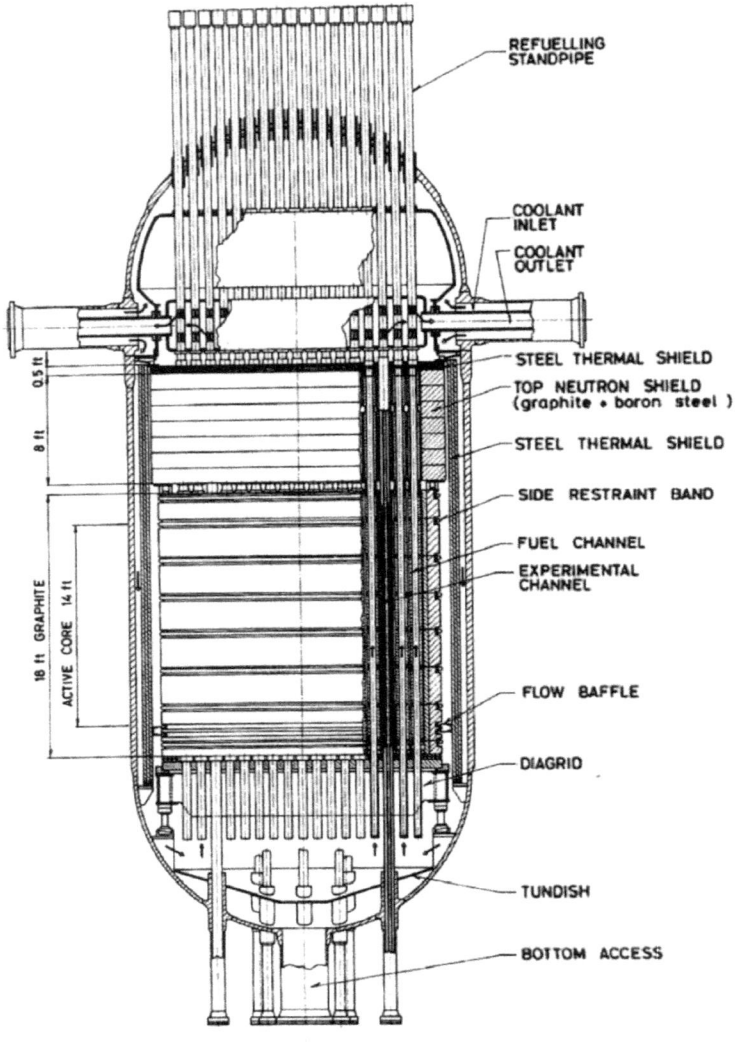

VERTICAL SECTION REACTOR AGR

Figure 2.21. AGR vertical section (source AIEA).

One of the advantages of the AGR design is that the coolant can be heated to higher temperatures which helps in the competition with water cooled reactors. As a result of gas coolant, higher plant efficiency could be attained compared to that of the water cooled design (33-35%). The thermal efficiency of AGRs is about 42%, generating a typical electrical output of up to 660 MW. The reactor and steam generators are housed within a steel

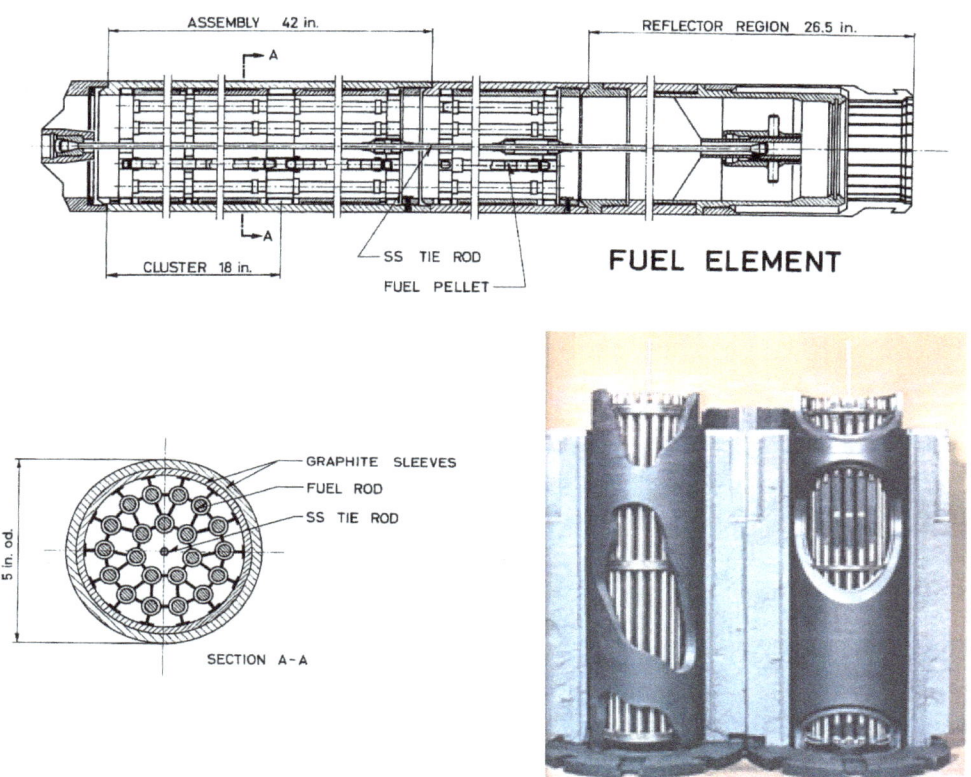

Figure 2.22. AGR fuel schematic, (source IAEA and Manchester University).

reinforced concrete pressure vessel, the walls of which are several meters thick (integrated architecture) (Figure 2.19).

The core structure is made of graphite bricks. AGR cores have two different types of graphite brick (one type in the periphery and another type on the inside): fuel brick where nuclear fuel is located and interstitial brick for control rod. The core has 12 layers of graphite bricks and about 320 fuel channels (Figures 2.20 and 2.21).

Fuel consists of UO_2 fuel pellets encased in 14.5 mm diameter stainless steel tubes roughly 1 m long. Fuel elements consist of clusters of 36 of these tubes, with 8 elements being loaded one above the other in a fuel channel (Figure 2.22). AGRs were also designed to be refueled on-line. Serious difficulties were encountered when operating at full power during fuel loading phases. We can say that AGR maintains gas cooled reactor design and joins the mainstream fuel technology (UO_2 ceramic with LEU). AGR fuel reaches 80'S BWR fuel performances.

Table 2.9. AGR characteristics, pressure 43 bars, LEU: low enriched uranium.

type AGR 660 MWe	
thermal yield	42%
Core uranium loading	114 000 kg LEU
U enrichment	$1.162\% < e < 2.55\%$
Bu equilibrium	18000 MWd/t
graphite mass	1229 t
Bu Max	24000 MWd/t
specific power	13.2 kW/kg Unat
power density	2.73 kW/l core
Thermal neutron flux	8×10^{13} n/cm²s
Reactivity effects	
	temperature 1000 pcm;
	Xe Pn 300 pcm;
	max reactivity 6000 pcm
Temperature coefficients	
equilibrium cycle	graphite 8 pcm/°C
	fuel: –2.1 pcm/°C
Temperatures	
	max fuel: 1700 °C
	max clad: 825 °C
Coolant temperatures	
	inlet 292 °C
	outlet 645 °C

Due to the use of enriched fuel, it is noticeable how much more compact the reactor core of an AGR is (reduction of the moderator mass). It can also be noted that the use of stainless steel cladding allows a significantly higher CO_2 outlet temperature, resulting in a higher High Pressure steam temperature and hence a higher plant thermal efficiency. Table 2.9 gives a comparison between AGR and Magnox for a similar power capacity.

AGR was faced with challenges during the first period of commercial activities. The higher power density induced radiolysis and methane used to manage corrosion needed some modification; cooling of graphite , internal structures and components involved complex circulation schemes. The higher temperature than required for Magnox also required some improvements in metallic characterizations. AGR loading factor was quite low during the period 1976–1990 (Figure 2.1).

Table 2.10. AGR and Magnox main characteristics.

	AGR	MAGNOX	units
net electrical power	625	590	MWe
thermal yield	41.7	31.4	%
graphite mass	1248	3735	t
metal U mass	114	395	t
pressure vessel material	Pre-stressed concrete	concrete	
pressure vessel geometry	cylinder	sphere	
pressure vessel diameter	20.25	29.3	MWe
core diameter	9.5	17.4	m
core height	8.3	9.1	m
nb of channels	308	6156	
nb of fuel element/channel	8	8	
fuel	UO2 oxide	U metallic	
enrichment	LEU	non-enriched	
equilibrium burn-up	18 000	4 000	MWd/t
clad alloy based on	steel	magnesium	
max clad temperature	825	450	°C
CO2 flow rate	3.8	10.2	t/s
outlet coolant temperature	645	414	°C
coolant pressure	41.9	27.6	bar
nb of heat exchanger	12	4	
HP steam temperature	541	396	°C
HP steam pressure	167	48.3	bar

Reference

[1] L. Petit, *French repository concepts for graphite waste.* Nuclear Graphite Technology Course, Carbowaste Eurocourse, octobre 2010.

3 RBMK (Reactor Bolchoi Mochtnosti Kanali)

R. Lenain

3.1. General

The Soviet-built RBMK ("channel-type nuclear power reactor" Реактор большой Мощности Каналый) was designed and developed for both plutonium production and generation of electricity. From 1954 (the year in which a RBMK ancestor, operating in Obninsk, became the world's first nuclear power plant) to 1990, thirty reactors of this type were put into service. Their ratings ranged from 6 MWe to 1500 MWe (at Ignalina, Lithuania, which had the highest output of any unit in service at the time). Of the twenty-three RBMK plants built in the former Soviet Union, 17 were located in Russia, 4 in Ukraine and 2 in Lithuania, giving a total capacity of some 18 GWe. In 2005, RBMKs produced half of Russia's nuclear power, i.e. 15% (or 137 TWhe) of its total electricity (870 TWhe); and, in 2006, their load factor was 62%. RBMK reactors were built according to different Russian safety baselines, corresponding to three "generations". However, because they have no containments, they can all be viewed as "first-generation" (see list of RBMKs in Table 3.1).

Today, only the Russian RBMKs are expected to continue operating. Their Ukrainian counterparts have all been shut down, and negotiations with the European community led to the closure of the Lithuanian units in December 2009. The Chernobyl plant was planned to have received six RBMK units: four of them were operating and two were under construction when, in April 1986, the plant became the scene of the worst ever nuclear disaster, ending in the destruction of reactor 4. Unit 2 was subsequently lost due to a major control room fire; and all reactors on the site have since been definitively removed from service.

The RBMK concept was developed in a climate of intense competition with the United States. Its design eliminated the need to fabricate large components in dedicated workshops and transport them to the site by special convoy: the reactor could be mostly built in situ. The RBMK originated at the same time as Russia's lead-bismuth and water-cooled marine propulsion reactors and the VVER-type PWR.

Before the Chernobyl accident, the West had only scant knowledge of the reactors built in the Soviet Union. In 1972, when Leningrad 1 went critical, a Soviet publication[2] made a few comments about its design. East/West relationships being what they were, western engineers could not form educated opinions on the RBMK, although certain elements of its design options raised questions. After the Chernobyl accident, with political changes taking place in Russia, significant modifications were undertaken. The RBMK was reviewed

Table 3.1. Civilian RBMKs; those now closed down are followed by an asterisk.

Name of plant and country of location	First-generation units	Second-generation units	Third-generation units
Chernobyl, Ukraine	Reactors 1* and 2*	Reactors 3* and 4*	
Kurks, Russia	Reactors 1 and 2	Reactors 3 and 4	
Leningrad, Russia	Reactors 1 and 2	Reactors 3 and 4	
Ignalina, Lithuania		Reactors 1* and 2*	
Smolensk, Russia		Reactors 1 and 2	Reactor 3

by international experts, which facilitated identification of serious safety deficiencies and enabled analysis of the proposed/implemented upgrades.

This chapter is split into three sub-sections: the first provides a brief description of the RBMK, the second is devoted to a discussion of reactor core physics and the third summarizes the Chernobyl accident scenario. For more detailed information on the latter, see references [1–3]. Techniques de l'ingénieur, GRS "l'accident et la sûreté des réacteurs de la filière RBMK," GRS 1996.

3.2. General description

This section is concerned with RBMKs built to generate power for civilian purposes. Table 3.1 below lists these reactors.

Overall design

The RBMK is a fission reactor with slightly enriched uranium fuel and a graphite moderator that acts to slow down neutrons. In this design, light water coolant (in liquid and steam form) removes heat from the reactor core through an array of vertical pressure tubes or "channels". Its channel-based approach, which resembles that of the Canadian CANDU concept, differs from PWR and the BWR technologies, in which cores are enclosed in a reactor vessel. RBMKs can be refueled in service like CANDU, AVR (pebble-bed), GGR, Magnox and AGR models.

RBMK water channels, which also contain nuclear fuel assemblies, run through a stack of graphite moderator blocks that are housed in a leak-proof cavity which also contains a mixture of helium (He) and nitrogen (N2) gas. The complete reactor sits on a structure that is in turn placed in a concrete-lined cavity. The cavity height is 25.5 m, and its sides measure 26.6 m. The RBMK is thus a channel-type, graphite-moderated reactor with a direct-cycle boiling water cooling system. The RBMK has no western-style reactor containment, but features a system for limiting cavity pressure in the event of leakage or pressure tube breaks.

Successive generations of RBMKs are protected by variation on these systems to withstand a maximum of 10 simultaneous tube breaks. Beyond the 10-tube limit, excess pressure on the upper concrete slab can cause it to rupture. This is one of the weaknesses of RBMK design.

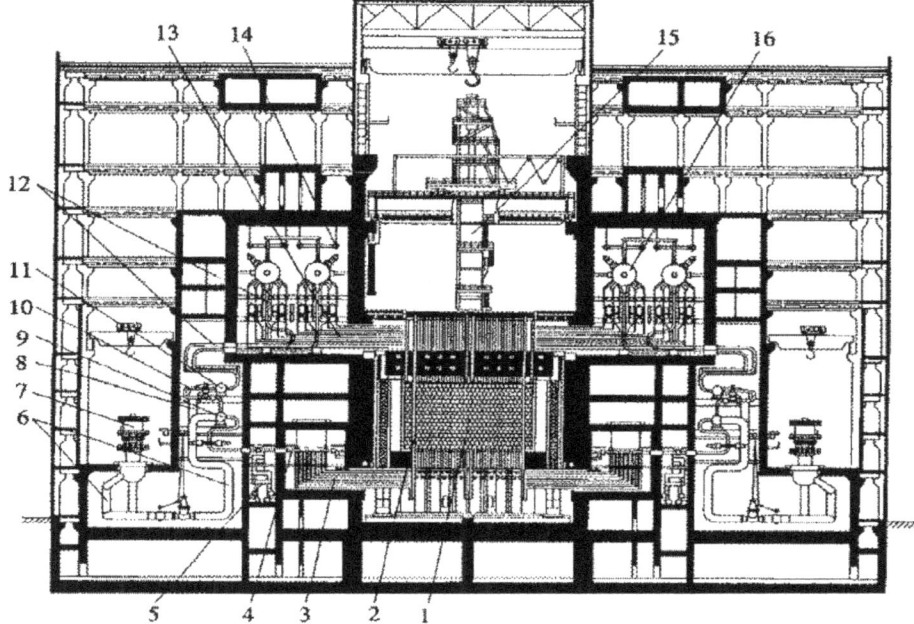

1 - graphite stack; 2 - fuel channel feed pipes; 3 - water pipes; 4 - group distribution header; 5 - emergency core cooling pipes; 6-pressure pipes; 7 - main circulation pump; 8 - suctionpipes; 9 - pressure header; 10 - bypass pipes; 11 - suction header; 12 - downcomers; 13 - steam/water pipes; 14 - steam pipes; 15 - refueling machine; 16 - separator drum.

Figure 3.1. General layout of an RBMK (source Ignalina handbook (LEI)).

Cooling

In the RBMK, heat removal is afforded by an array of 1660 parallel fuel channels. The cooling system architecture includes two half-loops with very few interconnections, each servicing half of the core. A half-reactor loop comprises two steam separators and four coolant circulating pumps, one of which remains on standby. The coolant (in liquid form) is pumped to feedwater distribution headers, which are divided into two groups for distribution of flow to the fuel channels. A flow of 19000 m^3/h is thus distributed by 22 main headers that each supply 40 pressure tubes. Each tube inlet is fitted with a control valve and a flow measuring device that allow crediting of the power supplied by the channel for flow optimization purposes. Sub-cooled (270 °C, 70 MPa) water is injected in liquid form into the core and begins generating steam 1 meter into the tube; the resulting "mixed" flow, whose steam quality is 14.5%, exits the core some 6 meters higher up, and enters the steam separators. The total distance traveled by the water in the channel, starting from the pumps, is some 70 m. Each steam separator consists of a steel drum 30 m long and 2.3 m in diameter. Dry steam leaving the separators is fed directly to the turbines. Having transited through the condenser, turbine outlet flow is circulated as water back to the steam separator, then to the circulating pumps, via 12 lines supplying the pump suction manifolds.

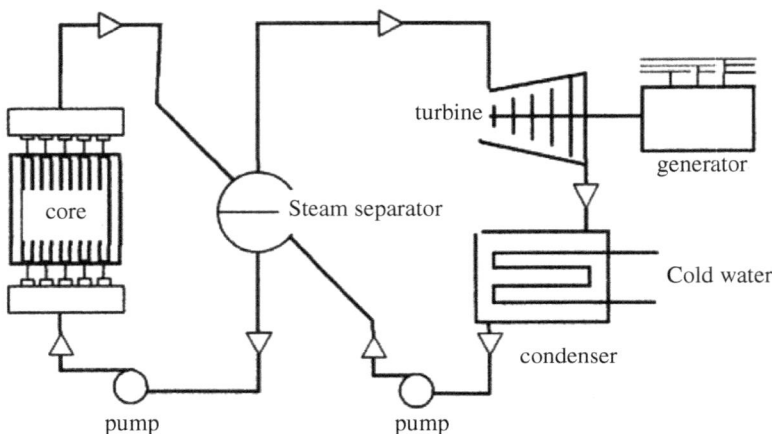

Figure 3.2. Operating principle of a RBMK plant (Technique de l'Ingenieur).

Core

Graphite stack

The RBMK core consists of a stack of graphite blocks each 60 cm high, with a pitch of 25 cm. The complete stack is 8 m high (including 7 m for the fuel) and has a diameter of 13.6 m (11.8 m for the core). Its weight is 1700 tons. Through each graphite block runs a pressure tube containing either fuel rods or control elements (absorber rods or detectors). Water entering the bottom of the core is heated to boiling temperature. The graphite stack, with its 2488 columns, acts as a moderator, slowing neutrons released by fission in the 1660 pressure tubes running through it. More than 2000 core channels are dedicated to the "active" part of the core (reflector, fuel, control rods and instrumentation). The graphite moderator, whose density is 1.65, is cooled by conductive transfer of heat (5% of power) to the pressure tubes. Axial and radial reflectors, which are 0.5 and 1 m thick respectively, are also made of graphite. The stack thus thermalises neutrons and serves as the mechanical structure for the core.

Fuel and control channels are inserted into the graphite blocks through drilled holes (each measuring 11.4 cm) at their centres. Tubes are guided into the blocks by a set of

Table 3.2. Characteristics of RBMK 1000 and 1500 models.

Reactor model	RBMK-1000	RBMK-1500
Thermal output in MW	3200 + 10%	4200 + 10%
Operational reactivity margin (ORM) in equivalent control rods	43–48	53–58
Maximum channel power (MW)	3.0	3.75
Maximum channel linear power density (kW/m)	35.0	42.5
Steam flow rate at full power (t/h)	5440–5600	7400–7650
Water flow rate in the core (m^3/h)	40000–48000	40000–48000
Overpressure in the separator (MPa)	6.9 (+0.1, −0.4)	6.9 (+0.1, −0.4)
Water temperature (°C)	155–165	177–190
Water temperature at core inlet (°C)	265–270	260–266
Maximum graphite temperature (°C)	<730	<750
Control and protection system (CPS) water flow rate (m^3/h)	1030–1220	1250–1350

graphite rings positioned alternately against the blocks or the tubes, to allow enough clearance for temperature and radiation-induced variations in their dimensions. The remaining service life of a stack is determined by the gradual disappearance of these clearances or "gaps" (due to graphite shrinkage and creep-induced increase in pressure tube diameter). Over the past ten years, RBMK graphite stacks have been refurbished by replacement of their guide rings and pressure tubes and by making adjustments to graphite block penetrations (Figure 3.3).

Average graphite temperature is 500 °C, with maximum temperatures in the vicinity of 750 °C. There is therefore no Wigner energy stored. Reactor cavity chemistry control (steam leak detection) is performed by the circulation of an He/N$_2$ mixture that is adjusted to determine the conductivity of gas circulating in the graphite guide rings, thereby also modifying heat transfer from graphite block to the pressure tubes. Likewise, this allows graphite temperature control, notably in power transient phases.

Fuel channels

Pressure tubes that contain the fuel assemblies are 88 mm in diameter and 4 mm thick. They are made of an alloy of zirconium with 2.59% niobium. Pressure tube inlets are connected to austenitic steel pipes that supply them individually with water from the upstream distribution headers; tube outlets are connected to the same such pipes, to convey steam/water outflows to downstream steam separators. Zr/steel transition junctions are obtained by diffusion welding of the relevant sections in workshops located offsite. Since rupture of any pressure tube in the reactor cavity induces overpressure, these junctions are essential to safety. Each fuel channel is fitted at its upstream end with a flow-optimizing control valve

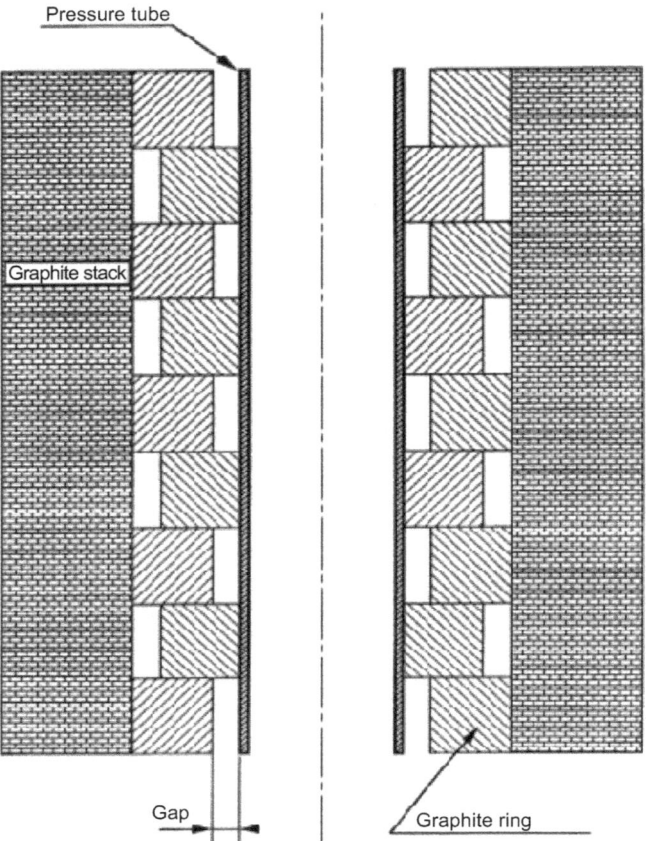

Figure 3.3. Pressure tube positioning in a graphite block (source Technique de l'Ingenieur).

and at its downstream end with a removable plug for in-service fuel handing operations (Figure 3.4).

Fuel assemblies

Each pressure tube contains two 3.5 m fuel assemblies positioned end to end. The RBMK fuel assembly has 18 fuel rods arranged within two concentric rings and is held in place by grids that ensure correct mixing and distribution of the water and steam phases in the channel. A central carrier rod of the same composition as the pressure tube is used to mount and handle the two assemblies; a similar tube provides guidance for measurement and experimental devices at the center of an instrumented assembly. Both assemblies are mounted in the reactor hall; they are separated from each other by a distance of 2 cm. Fuel rods are of the BWR-type with the following characteristics: diameter 13.6 mm, length 3.5 m, cladding material ZrNb (with 1% Nb), clad thickness 0.8 mm. A 17 cm end space is left for release of fission gases. Uranium is pressed into oxide pellets 11.5 mm in diameter with

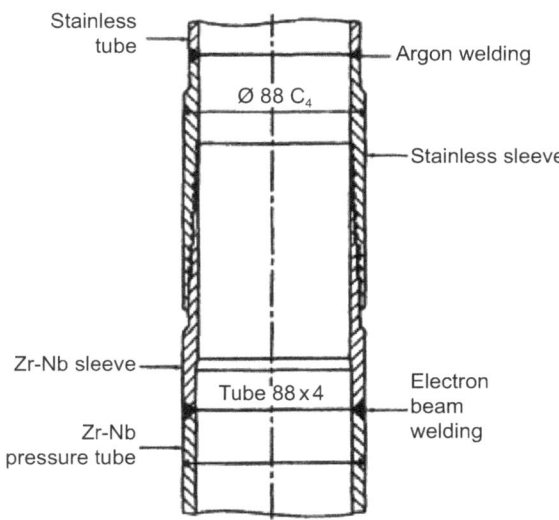

Figure 3.4. Zr/Steel transition junction (source technique de l'ingénieur).

a U235 enrichment that has varied over the years from 1.8 to 2.8%. Some RBMK plants can also use burnable poisons such as erbium (see Table 3.3 and Figure 3.5).

Table 3.3. Characteristics of RBMK cores containing burnable poisons (erbium).

Main characteristics of RBMK cores (fuel and erbium)					
Reactor model	RBMK-1000			RBMK-1500	
Enrichment (%)	2.4	2.6	2.8	2.0	2.4
Erbium (% by weight)	0	0.41	0.6	0	0.41
Number of absorber rods	80	0	0	54	0
Discharge burn-up, MWd/kg U	20.9	25.5	30.0	15.1	20.5
Average void effect, $	+0.6	+0.5	+0.6	+0.7	+0.3

Refueling

Fuel loading and unloading take place while the reactor is in service, with one or two fuel elements replaced daily. Such a refueling scheme minimizes the need for reactivity control and optimizes use of the fuel. The refueling machine is located above the reactor and has its own cooling system.

During fuel handling operations, allowance is made for power distribution management in the affected core region. Fuel assembly loading requires various changes such as adjustment of flow in the channel receiving the new assembly and suitable adaptation of control rod insertion in the zone around it. Control rods are inserted locally to preclude too large

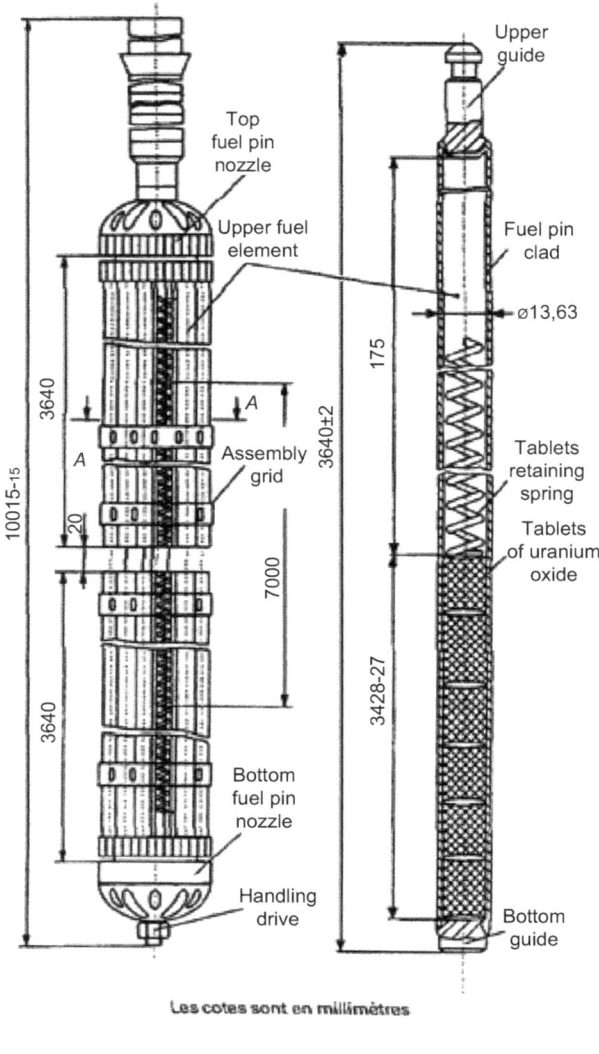

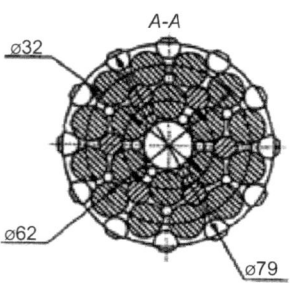

Figure 3.5. RBMK fuel assembly (source technique de l'ingénieur).

a power peak when the new fuel is inserted, and to reduce fuel pin thermal stresses when it penetrates the core. Channel flow is adjusted to allow for heat release under steady state conditions. Such an adjustment is performed manually on the basis of calculated maximum power predictions and is then repeated once or twice over the lifetime of the fuel assembly. Given the physical characteristics of the RBMK core, an automatic power control mode is required, particularly in phases of fuel management that significantly perturb local neutron flux distribution.

Power and reactivity control

An automatic power distribution control system was made necessary by the "thermalised" neutron spectrum associated with the large RBMK core. In such a core, local reactivity upsets can generate spatial instabilities (primarily in radial and azimuthal power distributions). More specifically, the RBMK is very sensitive to xenon poisoning and thermal hydraulic phenomena (neutron absorption by the cooling water).

The solution thus consisted of inducing, via the control rods, a "feedback effect" based on in-core detector signals and an online computer. For power distribution control purposes, the RBMK core is divided into several zones (12, 9 or 7, depending on generation), each with four detectors and one control rod. Power level in each zone is controlled by the in-core detectors, which actuate movement of the control rods via their associated rod control software. Depending on the operating scenario, the system can finely adjust power within a preset range for a given zone, reduce power locally, cut back overall power level, and even initiate a reactor scram.

In the third-generation Smolensk 3 reactor, more than one hundred flux measurements are performed periodically by 117 radial and 12 axial detectors arranged in a square mesh (measuring 1 m per side), then processed by a computer to reconstruct power at any point in the core. Comparison of these measurements with calculated predictions then determines the required control rod movements. The association of in-core detectors and control rods is such as that local correction of power distribution defects is facilitated. Smolensk 3 has a 9-zone control system (control rod and detectors) dedicated to local power control. On detection of a local defect, it is possible to either reduce overall power until the defect disappears, or to initiate a reactor scram. This approach is not specific to the RBMK. (See Smokensk core arrangement (Figure 3.6 and Table 3.3).

Control rods are the only means available for core reactivity control, adapting to changes in overall operating parameters and stopping the reactor. For operational reasons, e.g. to maintain criticality during transients involving changes in xenon poisoning, a minimum number of control rods must be inserted in the core (the amount of equivalent inserted rod is called Operating Reactivity Margin ORM). The upper part of each control rod is an absorber. At the lower part of the control rod, a graphite "follower" is attached to the end of the absorber length which removes the water (absorber) and improves local slowing down of neutrons.

At Smolensk 3, reactivity control is performed by the following:

- 9 automatically-actuated control rods with "graphite followers", for local power control (these rods are inserted from the top of the reactor);

- 146 manually-actuated control rods with displacers (also inserted from the top), 18 of which perform local power control in conjunction with the 9-zone control system;

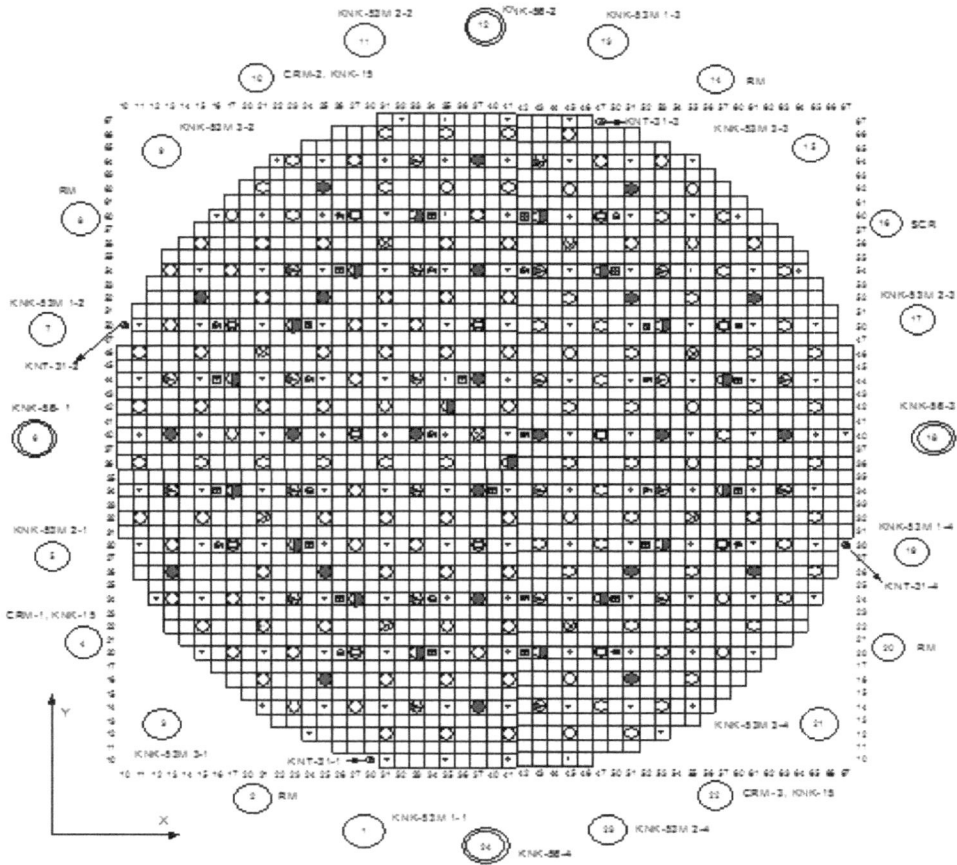

Figure 3.6. Example of RBMK core arrangement (source AIEA).

- 24 fast-acting shutdown rods without followers, added after the Chernobyl accident (inserted from the top to the bottom within 2.4 s);
- 32 short rods designed without followers (inserted from the bottom).

For the control of axial power distribution, Smolensk 3 features 32 manually-actuated, bottom-inserted short rods without graphite followers.

The number of RBMK control rods has increased over the three RBMK generations: Chernobyl 1 (first generation) had 179 rods versus 211 at Smolensk 3 (third generation).

The detectors and control rods have their own cooling system, which is different and separate from the one provided for reactor fuel.

A conventional-type protection system initiates reactor scram in response to signals from ex-core detectors located at the core periphery.

3.3. Core physics

In the following section, the features specific to the RBMK are described, some of which were key to the core behavior observed during the Chernobyl accident and the improvements made thereafter.

Principle of RBMK core design

The basic components of the RBMK core are the graphite blocks whose role is to slow the neutrons released by fission and the fuel assemblies whose rods are cooled with boiling water. This is a heterogeneous reactor concept, which calls for neutrons to be slowed outside the fuel, i.e. in the graphite environment. It saves fissile material by limiting neutron flux in zones with strong uranium 238 resonance absorption.

Void and density effects

If, in a graphite-moderated core whose coolant is ordinary water, enough moderation is provided by graphite, the coolant will act primarily as a neutron absorber. This is one of the sources of RBMK instability: in such a core, loss of water involves a positive contribution to neutron balance. Void effect is a core kinetics factor, and a positive void coefficient reflects the fact that lack of water accelerates the chain reaction changes. During reactor operation, positive void coefficient has a destabilizing effect, since power increase induces a water temperature rise, and thus a fall in density that in turn adds further power. Based on the time constants for fuel-to-water heat transfer, the effect of water loss on the chain reaction must be offset by fast negative feedback effects that exceed, in absolute value, this undesirable increase.

The extent of RBMK fuel assembly void effect is directly dependent on the relative neutron-absorbing capacity of water. In a highly absorbent assembly, void effect is smaller (Table 3.4 depicts the impact of erbium – a burnable poisons – on void effect). Higher uranium enrichments, which lead to greater fuel absorption then reduce the water absorption and the void effect. Plutonium formed during irradiation also increases void effect.

Table 3.4. Void effect on reactivity in a 3 × 3 lattice containing fuel elements (COMB) and control rods with: borated steel absorbers (A); graphite "followers" (G), water (E), and fixed absorbers (AF), pcm = 10^{-5}.

U^{235} enrichment (%)	2.4	2.6
Erbium content (%)	0	0.41
6 COMB + 2 G + AF: $\Delta\rho$ (pcm)	+230	−640
6 COMB + 2 E + AF: $\Delta\rho$ (pcm)	−390	−1170
6 COMB + 2 A + AF: $\Delta\rho$ (pcm)	−3550	−4150
9 COMB: $\Delta\rho$ (pcm)	+420	−800

The core physics behavior of a RBMK core cannot, however, be deduced as simply as for a lattice of fuel assemblies. The leakage term of assemblies and core neutron balance can completely change the analysis given above. Voiding of water leads not only to less neutron absorption in the channel, but also to greater neutron leakage out of it, and since the neutrons released by fission are no longer slowed by water, they are more likely to reach another fuel element. Depending on the environment of a voided assembly, voiding may either positively or negatively impact core reactivity. If this environment contains mainly fissile assemblies, loss of cooling water and its neutron absorption capacity will result in more neutron production. Conversely, if the voided fuel assembly environment contains absorber elements, more neutrons will be absorbed in the reactor. This is an important RBMK core design feature. Table 3.4 shows reactivity changes associated with coolant voidage in a square 3×3 graphite block lattice containing different typical RBMK elements.

The lattice layout of a RBMK core has a 4×4 pattern of graphite blocks (whose sides measure 1 m), penetrated by fuel channels, control rods, fixed absorbers and instrumented channels (Figure 3.6).

In a RBMK core, Doppler effect has a conventional value of some -2 pcm/K. The graphite temperature effect is positive (approximately 7 pcm/K). At low power, the power coefficient, which essentially includes Doppler effect and void effect, is positive.

RBMK void effect, which was highly positive by design, has been reduced through a number of changes, i.e. replacement of 80 fuel elements with borated steel absorber rods, increase in fuel U235 enrichment from 2 to 2.4% (primarily for economic reasons), and increase in ORM (which in itself plays an important role in safety) through permanent insertion of more control rods into the core. While this solution is effective, it nevertheless remains complex, due to the heterogeneities of the modified design, and the necessity to control them. Table 3.5 shows the estimated impact of these changes on void effect.

Table 3.5. Impact of modifications on void coefficient $\beta = 1\$$.

Type of change	Typical result for RBMK cores	Impact on core voidage
Insertion of more control rods	15 additional rods	$-1\,\beta$
Replacement of fuel elements with absorbers	80 replacements	$-2.4\,\beta$
Increase in U^{235} enrichment: 2.% to 2.4%	1660 elements	-1 to $-0.5\,\beta$
Total	All	-4 to $-5\,\beta$

It is not, in fact, advisable to combine reactivity and power distribution control with void coefficient control. Despite removal of 80 fuel channels, there is no loss of RBMK power; this means an increase in average linear power rate that requires better control of power distribution.

Void effect is heavily dependent on core configuration; in all reactors, this coefficient is measured approximately once every 200 days. Such a measurement is a complex process, which relies on measured rod worth and power coefficient values, as well as calculated void fractions and fuel/graphite temperature coefficients.

Instabilities

The RBMK is a large reactor whose volume is about 30 times that of a 900 MW PWR. Reactor size itself is synonymous with intrinsically unstable radial characteristics that are sensitive to perturbed axial power profiles.

The RBMK power coefficient has three main terms: Doppler coefficient (−2 pcm/K) and void coefficient (variable), for the fast component; and graphite temperature coefficient (7 pcm/K), for the slow component. Under certain conditions, the void coefficient is such that a power increase can cause a power excursion.

From a thermal-hydraulic standpoint, the RBMK can be described as a grouping of sets of 40 boiling water channels that are coupled upstream and downstream of the core. These channels operate with an imposed pressure loss, which induces, for any change in power, a corresponding change in flow rate. The water volumes, present (as liquid or steam) in headers supplying the channels and in steam separators receiving their outlet flows, are so large that a perturbation occurring in a channel does not cause pressure variations at its ends. If overpressure exceeds a certain value, steam generation may limit flow to such a degree that cooling is no longer adequate. This can lead to thermal-hydraulic instabilities – a characteristic of boiling water reactors which, when associated with positive void effect, can cause local or generalied neutronic/thermalhydraulic instabilities.

The following are orders of magnitude for reactivity effects associated with the key phenomena described above; they are expressed as delayed neutron fractions:

Control rod:

- Rod worth: −15 **b**

Power and feedback effects between cold and nominal power:

- Graphite temperature: 7 **b**
- Xenon: −4 **b**
- Fuel temperature: −2 **b**
- Void: 5 **b** to 0.5 **b** depending on the configuration

The orders of magnitude for typical RBMK time constants are:

- Prompt neutron lifetime: 0.001 s;
- Fuel temperature: 1 s;
- Coolant flow through the core: 2.5 s;
- Graphite temperature: 30 to 60 min depending on He/N_2 mixture;
- Xenon poisoning: 10 h.

In reactors whose design does not permit homogeneous insertion of absorbers by dilution in the coolant, the reactivity control performed by displacing solid absorbers has first-order impact on power distribution.

Analysis of initial RBMK control rod design

Insertion of control rods with "followers" into the initial RBMK core leads to displacement of both the graphite and the absorber. With a neutron flux distribution mainly located in the bottom of the core, the first effect of rod insertion is to remove water (less absorption) and improve the moderation (add reactivity). Under such flux conditions, at the beginning of the control rod motion, the reactivity can increase dramatically then, once the absorber is sufficiently inserted into the neutron flux zone, the reactivity decreases as expected. This particular behavior, which derives from RBMK control rod design, is partially responsible for the power excursion that occurred in Chernobyl reactor 4. Since the accident, control rod design has been modified so that, when the absorber length is inserted, the graphite "follower" is always at the bottom of the core and is gradually pushed out of it.

Cavity overpressure protection system

The cavity overpressure protection system uses mechanisms intended to foster steam condensation. These include a bubbling pool and spraying of water into a dedicated space to collect steam. This type of device is also used for BWR reactors.

3.4. Chernobyl accident

This section is a summary of the data provided in reference [1]. The analysis below is concerned with design-related aspects, in particular core physics.

Scenario

The Chernobyl accident took place on April 26, 1986, during a routine annual maintenance outage, which involved a test of unit 4's capacity to supply power to main circulating pumps under loss of station power conditions, via residual energy from the turbine, in accordance with Soviet regulations.

The initial test program called for

- power reduction to 20–30% nominal power (NP), and tripping of one turbine generator;

- operation of 8 pumps instead of the usual 6, by waiver, whereby;

- 4 were to continue operating for cooling purposes, using power supplied by the normal grid;

- and 4 were considered to be turbine loads, supplied by the coasting turbine generator.

- It was planned to start the test by tripping the turbine generator still in service, thus initiating a reactor scram.

Accident sequence and analysis

The scheduled shutdown began on April 25, 1986 at 1:00.

- At 3:47, the reactor was at 50% nominal power (NP).

- At 7:10, the ORM fell below allowable limits (xenon poisoning), which should have led to reactor shutdown; later it returned to within the allowable range.

- Until 14:00, the reactor remained at 50% NP, with the grid continuing to demand power; and it was decided to delay the test.

- At about 23:00, power reduction resumed, until 20 to 30% NP was reached; for unknown reasons, it continued to decrease.

- At about 0:28 on April 26th, reactor output was about 500 MWth and there were difficulties in transferring control to the automatic regulating system. Again, for unknown reasons, power fell to 1% NP, which corresponds to that of the pumps. The ORM was below minimum limits; and the reactor should have been shut down.

- By withdrawing the rods still remaining in the core, power could be raised to 7% NP and kept at that level.

- At 0:43, i.e. 40 minutes before the start of testing, a signal that would have led to automatic scram on test initiation was ignored.

- At about 1:23, the actual test began: four reactor coolant pumps coasted to shutdown; and the decrease in water flow in the core caused reactor scram. The emergency button was pressed by the operator and control rods started to enter the core, with the result that power rose, exceeding nominal value by 15%. A few seconds later, a power excursion caused the destruction of the reactor core.

Initial conditions

Design factors

- Core void effect exceeding 2000 pcm.

- Insertion of heterogeneous rods increased reactivity when neutron flux "peaked" in lower core region.

- Very long rod insertion times (14 s for them to move from middle to bottom of core).

- Bottom-inserted short rods excluded from reactor scram system.

- Need for operator to maintain manual scram signal throughout rod insertion process.

- No crediting of ORM by reactor protection system.

Initial conditions arising from operating history

- Reactor exhibiting a positive, i.e. power coefficient at low power was hard to control.
- Reactor containing a large volume of water whose voidage induces massive reactivity insertion (more than five times the delayed neutron fraction) at low power.
- No reactivity control margin due to the amount of water in the core (low power) and xenon poisoning, thus requiring control rod withdrawal, but resulting in a core configuration that again increased the positive void effect.
- Axial power shape with sharp downward peak (due to xenon) inducing an unwanted reactivity-amplifying effect when rods were inserted

All these effects were known to RBMK designers before the Chernobyl accident took place. The main causes of the accident were its unacceptable intrinsic design and the major deficiencies in safety culture observed at all levels (organization, operational feedback, onsite actions). If the modifications described in the next section had been made earlier, there would not have been a Chernobyl disaster.

3.5. Changes made to improve RBMK core behavior

The following measures were taken to ensure that void effect would not exceed the delayed neutron fraction in an RBMK 4 × 4 fuel lattice and core:

- installation of 80 additional absorbers in the core, in place of fuel elements;
- addition of absorbers in the fuel;
- increase in fuel enrichment (for economic reasons), which also contributes toward a reduction in void effect;
- increase in ORM from 30 to 48 rods and installation of an alarm;
- limitation of reduced power operation (prohibited below 700 MWth).

Other improvements were geared to enhancing the control system. They included:

- accelerating the calculation and measurement rates of the online monitoring system;
- improving the system for automatic power distribution control by increasing its core coverage;
- redesigning control rods to exclude the positive effect of rod insertion;
- addition of a fast acting scram system inserting twice the fraction of delayed neutrons in 2.5 s;
- insertion of short control rods in case of scram actuation

References

[1] A.A. Afanasieva et al., *The characteristics of the RBMK core*. Nuclear Technology **103**, July 1993.
[2] IAEA, *Safety assessment of proposed improvements to RBMK nuclear power plants*, Report of the IAEA Extra-budgetary Programme on the Safety of RBMK Nuclear Power, Plants. IAEA-TECDOC-694. Wien, March 1993.
[3] IAEA, *Safety assessment of design solutions and proposed improvements to Smolensk Unit 3, RBMK nuclear power plant*, Report of the IAEA Extrabudgetary Programme on the Safety of RBMK Nuclear Power Plants. IAEA-TECDOC-722. Wien, October 1993.
[4] D. Bastien, Technique de l'Ingénieur, BN3215.
[5] ELECNUC: Nuclear power plants in the word, edition 2012, CEA.

4 Heavy water moderated nuclear reactors

P. Anzieu

4.1. Introduction

Numerous countries, in particular Canada, France, Great Britain and Japan, but also the Czech Republic, Italy, Germany, Slovakia, Sweden, Switzerland and the United States have tried to develop heavy-water reactors and have constructed prototypes. Many forms, often different, have been studied. Of all the varieties, only one developed in Canada has been industrialised. It is therefore the only type of power reactor that, without the benefit of military programmes, can compete with light water reactors. Canada, which possesses abundant resources of uranium, but limited industrial means and no capacity to enrich uranium, was interested early on in heavy-water. The pursuit in this direction has been remarkable. The form of reactor that emerged from the Canadian efforts is usually known as a CANDU[1] – CANadian Deuterium Uranium – reactor (see Figure 4.1). It is a variation of the Pressurised Heavy Water Reactor and is characterised by:

- the use of heavy-water both as a moderator and as a coolant,
- a structure with horizontal pressure tubes,
- natural uranium fuel, refuelled during the reactor operation,
- very efficient uranium usage and particularly low fuel cycle cost.

Canadian prototype reactors have been constructed or are under construction in Argentina, China, India[2], Pakistan, Romania and South Korea (see Table 4.1).

Heavy-water reactors represent, in terms of installed power and accumulated experience, less than 10% of the light water reactor fleet – PWR and BWR – (see Table 4.2),

The ease of manufacture of small fuel elements, their unloading during operation with a relatively low radiation level and the large quantity of military quality plutonium contained in the discharged elements all represent, obviously, a potential for nuclear proliferation.

Of the total of 53 reactors in service or under construction, only two have technology that differs from that of CANDU, in Argentina.

---

[1] CANDU is a registered trade mark
[2] Canada supplied India with two reactors, RAPP 1 and 2 but in 1974, India exploded an atomic bomb which was developed using the plutonium produced in a research reactor provided by Canada. Canada subsequently broke the bond between the two countries. Since then India has proceeded in developing heavy-water reactors, on the basis of clean development, by adapting Canadian technology in the Indian context.

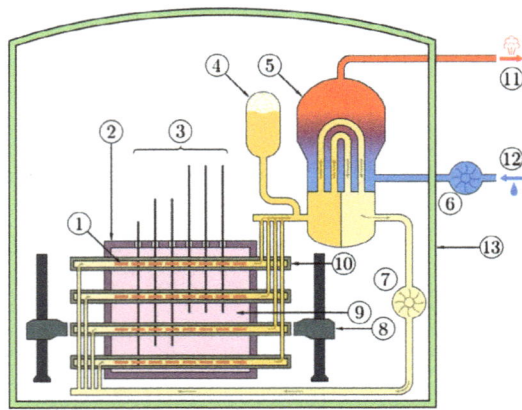

Figure 4.1. Provisional scheme of a CANDU (from Wikipedia Inductiveload). 1 Fuel bundle, 2 Calandria (reactor core), 3 Control rods, 4 *Heavy water* pressure reservoir, 5 Steam generator, 6 Light water pump, 7 Heavy water pump, 8 Fuelling machines, 9 Heavy water moderator, 10 Pressure tube, 11 Steam going to the turbine, 12 Cold water returning from turbine, 13 reinforced concrete containment building.

Table 4.1. Situation in 2012 of heavy-water reactors in the world (number/net MWe).

	CANDU	
COUNTRY	In operation:	In construction:
ARGENTINA	2/935	1/692
CANADA	19/13500	
CHINA	2/1300	
REP. OF KOREA	4/2710	
INDIA	18/4091	4/2520
PAKISTAN	1/125	
ROMANIA	2/1300	
TOTAL	48/23961	5/3212

Table 4.2. Situation of the fleet of water-cooled reactors in 2012.

	Installed Power (MWe net)	Number of unit	Average Unit Power (MWe)
CANDU	23,961	48	499
PWR	252,190	273	924
BWR	78,079	84	929

4.2. General

4.2.1. Heavy-water

Heavy-water is by far the best moderator: its utilisation represents the best compromise between moderation efficiency and neutron capture (see Table 4.3).

Table 4.3. Neutron characteristics of some standard moderators.

		Number of shocks to thermalize	Absorption section (cm^{-1})	Moderation factor (***)
Light water	H^1	19	2×10^{-2}	60
Heavy-water (*)	H^2	35	0.9×10^{-4} (**)	1700
Graphite	C^{12}	115	2.6×10^{-4}	200

(*) 99.7% (or 0.3% Light Water); (**) from which 2/3 are due to residual light water; (***) Combined properties of moderation and absorption.

A certain number of technical characteristics unique to heavy-water reactors are immediately evident in this table:

- in terms of moderation capacity, heavy-water falls midway between light water and graphite. For a given power, the size of the heavy-water reactor is also halfway between the light water reactors (the most compact), and reactors moderated by graphite. For comparison purposes, a 900 MWe reactor requires a 4 m-diameter reactor vessel for a PWR, and 8 m for a CANDU, and a core of a UNGG would have a diameter of about 12 m;

- the quantity of heavy-water used in the moderator circuit is therefore quite large. The most recent CANDU reactors use approximately 0.75 t/MWe of heavy-water distributed in 0.45 t for the moderator (including the radial reflector) and 0.30 t for the coolant;

- considering the absorption of residual light water, very pure heavy-water (at least 99.7%) must be used with natural uranium fuel.

As a coolant, heavy-water has very similar thermal-hydraulic properties to light water (see Table 4.4). The primary system of a CANDU has many features analogous to those of a PWR, notably with regards to components.

Heavy-water is an expensive material (about $600–700 per kilogram, in 2010 economic conditions). This results in high investment costs. Efforts must be made to reduce the amount used by carefully sealing systems and recuperating leaked fluids and gases. These leaks will be then "recycled" in the cooling system if sufficiently pure, or re-concentrated before recycling if the purity is low.

Table 4.4. Physical properties of heavy-water.

	Unit	Light water	Heavy-water
Freezing point	(°C)	0.0	3.8
Boiling point	(°C)	100.0	101.5
Critical temperature	(°C)	374.2	371.1
Critical pressure	(MPa)	22.11	22.14
Density at 25 °C	(g/cm^3)	0.997	1.104

Tritium (H^3) is produced by neutron capture by deuterium. After thirty years of use, the concentration of tritium in the moderator reaches $2-4 \cdot 10^{12}$ Bq/l; it is less in the coolant, but the risks of leakage are greater due to the pressure. This induces operational constraints essentially due to contamination of the containment's atmosphere in the case of a leak. This also results in the existence of a detritiation unit, if not on each site, at least centralised at the level of the reactor fleet.

The production of heavy-water is, in principle, analogous to uranium enrichment: in both cases, there is an isotopic separation. But the industrial problems are not of the same level: for an optimal plant size (about 400 t/year) the investment is moderate and does not represent a very large proportion of reactors to supply electricity. On the other hand, about ten times less energy is needed to produce the heavy water initially required by a CANDU and to compensate the losses during the life of the reactor than for the enriched uranium needed to operate a PWR of the same power.

Appendix 1 shortly describes the production process of heavy-water. All the countries for which heavy water represents an important option in their electro nuclear programme are equipped with heavy-water production facilities.

4.2.2. Natural uranium

The use of heavy-water as a moderator, given its neutronic qualities, makes it possible to use natural uranium directly, with an excellent energetic value (see Table 4.6 below). This essentially explains the interest in this type of reactor.

The use of natural uranium is very structuring for a reactor file and has a strong impact on the general characteristics of the CANDU: the reactivity balance is in fact limited and there is great temptation to give weight to solutions favouring the economy of neutrons to the detriment of technologically simple and robust solutions. But we must be clear that this regards one particular option: the use of natural uranium responds above all to political considerations and does not necessarily correspond to the economic optimum, even for a heavy-water reactor. It is obvious that, in the Canadian reactor type, the characteristics of the reactors are inflected to optimise a neutron balance, and therefore to obtain a satisfactory operation with natural uranium, which in general is to the detriment of investment costs. Natural uranium is considered as a Canadian dogma.

This implies an operation very close to the neutron optimum in terms of moderation ratio ($V_{mod}/V_{fuel} \approx 20$, see Figure 4.2, compared to $V_{mod}/V_{fuel} \approx 2$ at 3 in the light water moderated cores). It demands a significant amount of heavy-water and a large core volume.

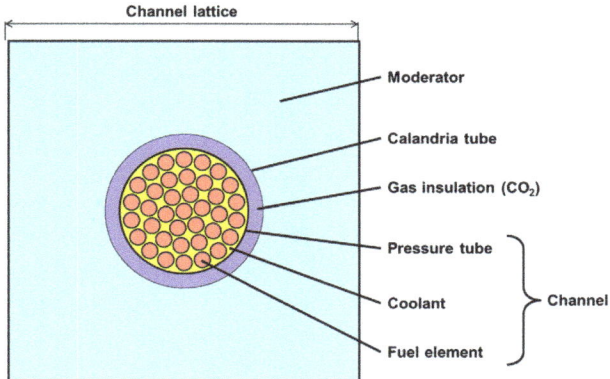

Figure 4.2. Diagram of a CANDU reactor cell.

The concentration of the fuel in the centre of the cell allows the physical separation of the coolant from the moderator. This facilitates power extraction by the coolant, allowing different operating conditions for the moderator and the coolant and making it possible to choose a coolant that is different from the moderator.

To reduce the parasite absorptions due to the structures, materials that are as transparent as possible to neutrons are used under flux (zirconium based alloys), but they are expensive and therefore their use is restricted to zones inside neutron flux. Their thickness drives the pressure level of the primary coolant, and an optimisation must be done.

The absorbers used during the reactor operation compensate for the reactivity variations in the medium and long term; they must also be minimized. As a result, in particular, there is a certain lack of flexibility relative to the transient poisoning with xenon, associated with all transients of power. These transients are important as in all reactors that use natural uranium (about 10% in reactivity at the xenon peak in the heavy-water reactor, compared to 2.8% in a PWR). In this regard, heavy-water reactors have just tens of minutes to restart after a shutdown, and are limited in their flexibility of operation.

The high value of the thermal neutron flux (the consequence of using natural uranium), combined with a large reactor size, makes the reactor unstable in the sense of xenon oscillations: axial instability, in the first radial harmonic and the first azimuthal harmonic, and stability limit of the second radial harmonic. A spatial control of the reactivity of the reactor requires several absorbers and the possibility to use them independently, depending on local power levels that are measured by local sensors.

The small reactivity margin resulting from the use of natural uranium gives a requirement to continually refuel during operation without shutting down the reactor, in order to maintain a good load factor in the facility. Figure 4.3 illustrates the differences in evolution of the fuel reactivity between a heavy-water reactor and a PWR.

The adjustment of the reactivity to the minimum operating requirement available through the continuous renewal of the fuel, yields an excellent neutron economy due

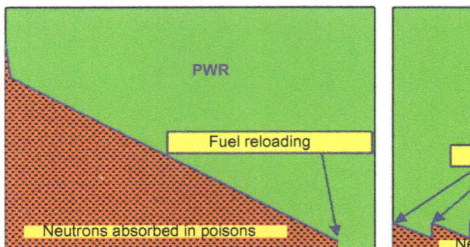 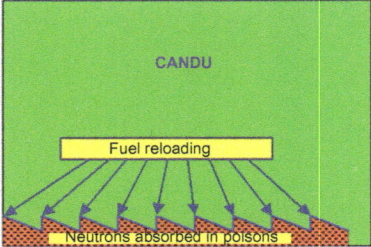

Figure 4.3. Evolution of the reactivity over time.

to the minimisation of their loss, and explains, in part, why natural uranium is well used in heavy-water reactors.

Plutonium production remains high. Half of the energy produced comes from the fission of part of the plutonium produced during the cycle (compared to 33% in a PWR).

The reloading in operation, at full power and without having to stop the reactor, permits a greater availability of the facility and a very high annual load factor. In certain years, it was not unusual to reach 100% availability. This great advantage of heavy-water reactors seems to fail today because of regulatory rules that are imposing a yearly shutdown for inspections, and also because of the longer fuel cycles achieved in light water reactors.

4.2.3. Pressure tubes

The high value of the moderation factor, as discussed above, encourages large dimensions for various components of the reactor (mainly for the main vessel) the construction of which may very rapidly cope with technological difficulties or even impossibilities. Canada recognised early on that it would be difficult to extrapolate such a concept to higher power levels and reoriented its prototype reactor, Nuclear Power Demonstration (NPD), towards a new concept, allowing a grouping of the fuel elements in the centre of the elementary cell resulting from the division of the large pressure vessel, in a large number of thin containments called pressure tubes. The reactor core is therefore made of a group of channels, in parallel, each including a pressure tube, the fuel elements, and a hot pressurised coolant. These channels are immersed in the moderator, which can be maintained cold and at almost atmospheric pressure, and therefore contained in a thin walled vessel called the calandria. In order to permit the renewal of the fuel during operation, the channels have "removable" openings at both ends.

Each channel, with its associated moderator, constitutes an elementary cell of the reactor (Figure 4.2). Typically, the cell pitch is about 28 cm, and the diameter of the pressure tube is about 10 cm.

Germany, unlike Canada, persevered in developing pressure vessel reactors. By extrapolating the demonstration reactor MZFR, they designed and provided a reactor to Argentina, ATUCHA 1 of 357 MWe. The ATUCHA 2 reactor, based on the same concept, but with 750 MWe of power has been under construction for many years. Appendix 2 gives a brief description of this variant of a heavy-water reactor.

A certain number of characteristics arise directly from a design with pressure tubes. The moderator, separated from the hot and pressurised primary system, can be cold and practically at atmospheric pressure. This arrangement has several particularities:

- a cold moderator is, in neutron terms, more interesting as the reactivity is increased;
- because of the absence of pressure, maintaining the seal in the moderator circuit is easier;
- on the other hand, it is more difficult to recover the heat dissipated in the moderator and the reflectors. This energy is dissipated from a large volume and at low temperature (approximately 5% of the power, or 100 MWth for a reactor in the 600 MWe range). The total efficiency of the installation is somewhat reduced.

The channel (pressure tube, fuel and associated coolant) comprises a module of the core, which resembles layers of identical channels, superimposed one above the other. Changing from one power size to another is possible by changing the number of channels. The necessary adaptations are limited and tests on a single prototype channel are very representative.

This modularity, however, results in a great number of "collection pipes" necessary to circulate the primary coolant. These tubes are small in diameter and form a true "forest" that requires good mechanical support (problems of vibrations, seismic resistance) over significant lengths. There are a total of 760 pipes of a few centimetres in diameter for a 600 MWe reactor that must be supported over about 20 meters.

The primary pressure depends on the pressure tube. If this tube, in which the hot primary coolant circulates, is immersed directly in the cold moderator, a significant temperature gradient is created through its thickness. This does not favour mechanical resistance. Constructive measures must therefore be taken to insulate the pressure tube thereby reducing the temperature gradient as much as possible:

- either having a solid insulator on the internal surface of the pressure tube, in order to keep it cold. This solution was used in France for the EL 4 reactor, cooled by carbon dioxide;
- or use of an external concentric tube called "calandria tube" (Figure 4.2) to isolate the pressure tube from the moderator, the space between the two tubes being filled with CO_2. The heat gradient is then deferred to the calandria tube and is not subjected to the pressure.

From a safety point of view, the cold moderator constitutes an important passive cold source.

The possible rupture of a pressure tube would have limited consequences looking like the ones of the rupture of a PWR pressure vessel. The probability that the rupture of a tube involves, by chain reaction, the rupture of the neighbouring tubes is extremely low. "Buffer zones" exist between two neighbouring tubes: a first calandria tube, the moderator, itself protected from excess pressure by a rupture disk, and finally a second calandria vessel.

Continual monitoring each of the pressure tube tightness is possible during operation; it is sufficient to have moisture sensors on the CO_2 that circulates in the space between the pressure tube and calandria tube.

The mechanical and neutronic consequences of a LOCA following the rupture of a pressure tube are limited by the subdivision of the primary system in at least two quasi-independent loops. The modular design of the reactor makes it possible to replace a pressure tube. As illustrated by the Canadians, it is a relatively easy short operation (two days per pressure tube), once the procedures and the tools have been perfected.

The fuel renewal during operation also makes it possible to optimise the power distribution both radially (frequency of renewal of the channel) and axially (renewal of only part of a fuel channel). This last operation is facilitated by the fractioning of the fuel through the design of short elements.

Separation of the moderator from the coolant, as well as the good neutron economy, made it possible to explore many coolant alternatives:

- pressurised heavy-water (the main alternative),
- boiling heavy-water,
- pressurised light water,
- boiling light water,
- organic fluid,
- carbon dioxide.

This proliferation of heavy-water reactor alternatives resulted in dispersing the international efforts, with each country looking to promote its champion. Even Canada has explored three variations. This situation undeniably constituted a heavy handicap with regards to the light water, BWRs and PWRs.

4.3. Description of a CANDU 6

The description that follows concerns the Canadian reactors CANDU 6, heavy-water cooled and moderated (the standard for 600 MWe) whose prototype was Gentilly 2. Table 4.5 gives the main characteristics of CANDU 6 and 9.

4.3.1. Reactor

Vessel and calandria tubes (Figure 4.4)

The calandria or reactor vessel is designed as a cylinder on a horizontal axis, consisting of a ring and two flat bases joined together by 380 calandria Zircaloy 2 tubes. The heavy-water moderator is contained in this vessel. An inert helium gas atmosphere surrounds the free level together with four pipelines provided with rupture disk assemblies to prevent an excess pressure in the moderator. Axial biological protection is obtained by using a mixture of small steel balls and water, located inside the vessel at both ends. This vessel is contained in, and is supported at its ends by a water filled concrete vessel which also ensures radial biological protection.

Table 4.5. Main Characteristics of CANDU reactors.

			CANDU 6	CANDU 9				CANDU 6	CANDU 9
Reactor	Thermal power	(MWth)	2064	2716	**Primary system**	Title Heavy Water	(%)	> 95	
	Gross Electrical Power	(MWe)	715	935		Mass of Heavy Water	(t)	185?	
	Net Electrical Power	(MWe)	665	875		Total flowrate	(Kg/s)	7600	11000
	Net efficiency	(%)	32,2	32,2		core inlet Temperature	(°C)	266,6	267
Reactor Vessel	Material	Stainless Steel (austentic)				Core Outlet Temperature	(°C)	312	312
	External Diameter	(m)	8.09			Core Inlet Pressure	(MPa)	11.04	
	Thickness	(mm)	25.4	32		Core Outlet Pressure	(MPa)	9.9	
	Overall length (Including axial protections)	(m)	7.82	8.2		Average Steam Quality at Core Outlet	(%)	3	?
Calandria Tubes	Material		Zircaloy 2		**Steam Generators**	Number		4	
	Internal Diameter	(mm)	129			Tube Material		Incoloy 800	
	Thickness	(mm)	4.06			Total Steam Flow Rate	(t/h)	1047	1328
Pressure tubes, Channel Extension					**Fuel**	Uranium		Natural	
						Type		Bundle 37 rods	
						Number of bundles per channel		12	
	Number		380	480		Clad material		Zircaloy 4	
	Material		Zr – 2.5% Nb			External clad diameter	(mm)	13.08	
	Internal Diameter	(mm)	103.4			Clad Thickness	(mm)	0.49	
	Thickness	(mm)	4.19			Pellet diameter UO_2	(mm)	12.16	
	Length	(m)	6.3			Bundle length	(mm)	495	
	Pinch	(mm)	286			External bundle diameter	(mm)	102.4	
	Channel extension:					Bundle mass	(kg)	24.1	
	Material	Stainless steel				Total mass UO_2	(t)	95	110
	Total length of channel	(m)	10.82			Average burn-up at discharge	(MWj/t)	7500	8520
Moderator Circuit	Title of Heavy water	(%)	99,75		**Containment Vessel**	Seal skin		epoxy	steel
	Vessel Inlet Temperature	(C°)	49	49		Internal Diameter	(m)	41.45	57
	Vessel Outlet Temperature	(C°)	77	75		External Diameter	(m)	43.59	60
	Heavy Water Mass	(t)	265	?		Total height	(m)	51.21	71.5
						Design pressure	(kPa)	124.1	210

Pressure tubes – Channel extensions

The pressure tube (made of a zirconium-2.5% niobium alloy) is located inside the calandria tube, centred with one or several bracelets and "thermally insulated" by a layer of gas (CO_2) (Figure 4.5). At the end of each channel, there is a biological protection plug (designed to reduce the direct radiation coming from the fuel in order to permit access to the loading area during the reactor shutdown), plus a plug seal which can be removed during the reactor operation, in order to allow access inside the channel during fuel renewal operation.

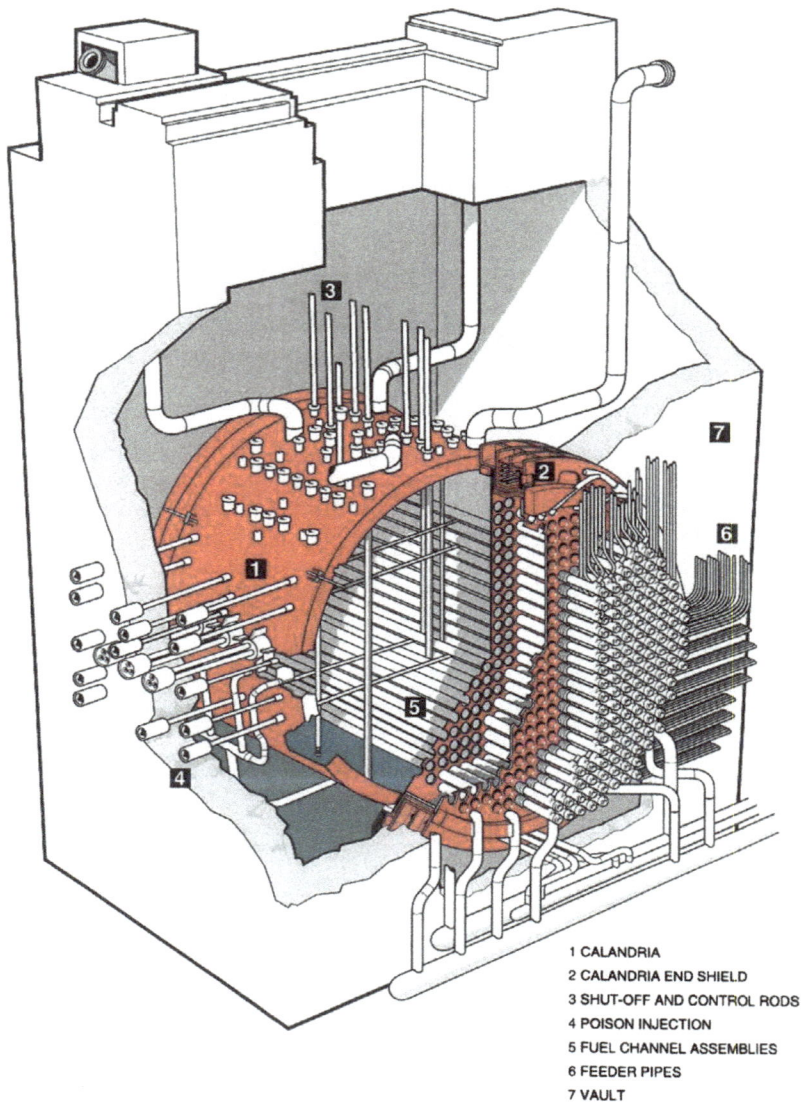

1 CALANDRIA
2 CALANDRIA END SHIELD
3 SHUT-OFF AND CONTROL RODS
4 POISON INJECTION
5 FUEL CHANNEL ASSEMBLIES
6 FEEDER PIPES
7 VAULT

Figure 4.4. Reactor outlook [from W. Snook, 2001, CANTEACH Website].

Pipes that supply the coolant find partially their way between the heads of channels[3], before being gathered in the periphery of the reactor (Figure 4.6), so it is necessary to extend the pressure tube beyond the zone which is under neutron flux. This extension (Figure 4.7) is made out of stainless steel as the use of expensive zirconium alloys is not necessary

---

[3] This obstruction of space between the heads of channels is at the origin of the limitation of the unit power of the CANDU reactors. To increase the power actually leads (except with an increase in the unit power of each channel) to an increase in the number of channels, and therefore the number of connection pipes. In practice, with the current solutions, it seems difficult to exceed 900 MWe.

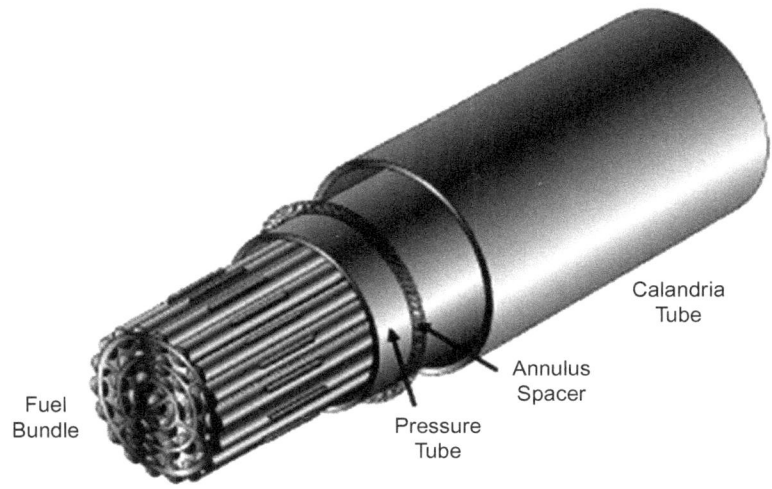

Figure 4.5. Details of a fuel channel (from University of Waterloo).

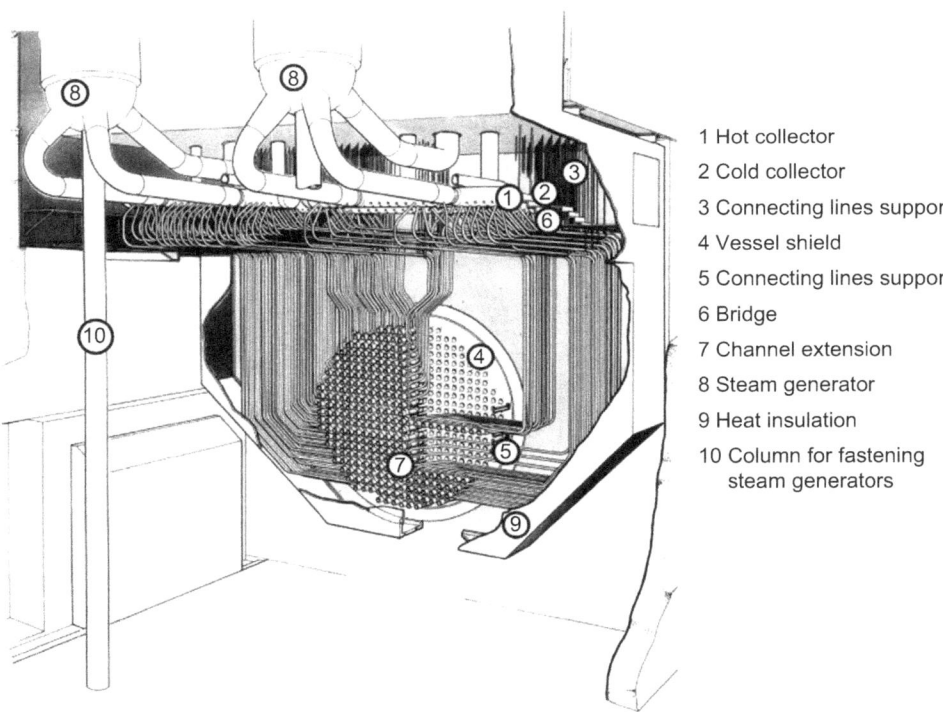

1 Hot collector
2 Cold collector
3 Connecting lines support
4 Vessel shield
5 Connecting lines support
6 Bridge
7 Channel extension
8 Steam generator
9 Heat insulation
10 Column for fastening steam generators

Figure 4.6. Side view of a reactor *[from J. Koclas, 1977, CANTEACH Website]*.

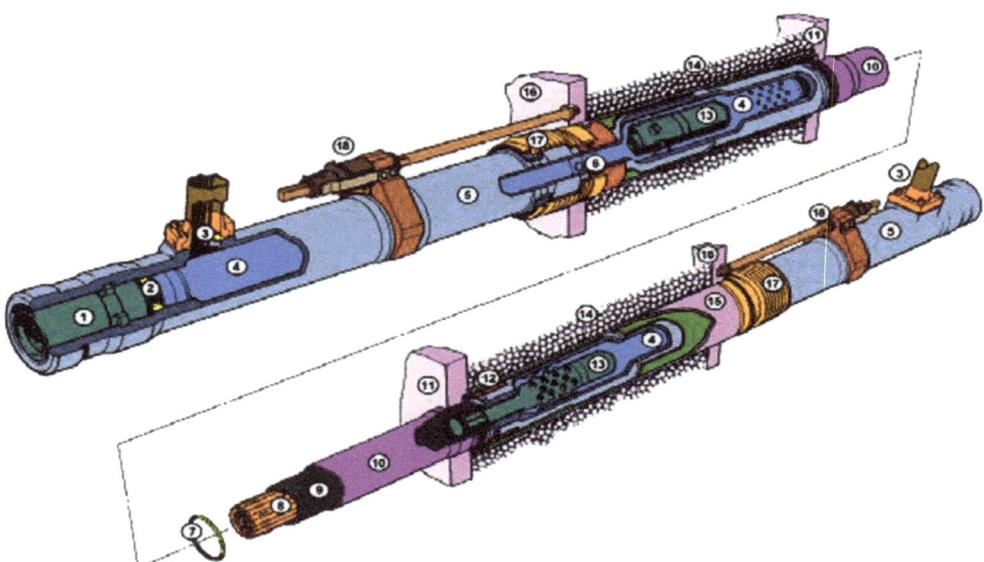

1 Channel closure, 2 Closure seal insert, 3 Feeder coupling, 4 Liner tuber, 5 End fitting body, 6 Outboard bearings, 7 Annulus spacer, 8 Fuel bundle, 9 Pressure tube, 10 Calandria tube, 11 Calandria tubesheet, 12 Inboard bearings, 13 Shield plug, 14 Endshield shielding balls, 15 Endshield lattice tube, 16 Fuelling tubesheet, 17 Channel annulus bellows, 18 Positioning assembly.

Figure 4.7. CANDU 6 fuel channel (from D.A. Meneley and Y.Q. Ruan, 1998, CANTEACH Website [6]).

in zones without neutron flux. Without the possibility of welding these two materials, the junction is made by rolling (or expansion): using a special tool, the metal of the pressure tube is pushed back inside the throat of the extension. This channel design permits, in case of defect, the easy replacement of the pressure tube without having to discharge all the fuel contained in the reactor.

4.3.2. Primary system

The limitation of the thickness of the pressure tubes, so as not to interfere with the neutron flux, confers a limitation of the primary pressure to about 11 MPa, therefore also limiting the operating temperature. The secondary steam pressure thus remains relatively weak, at 4 to 5 MPa depending on the reactors, and consequently the net efficiency of the facility is also weak (30%).

General configuration

Connection pipes coming from various channels are joined together with a hot collector which supplies the steam generator and then the circulating pump, which in turn leads back to a second cold collector. This then distributes water to the various channels of the reactor (Figure 4.1 and 4.8). With the exception of the pressure tubes and steam generators tubes, the entire primary system is made of carbon steel.

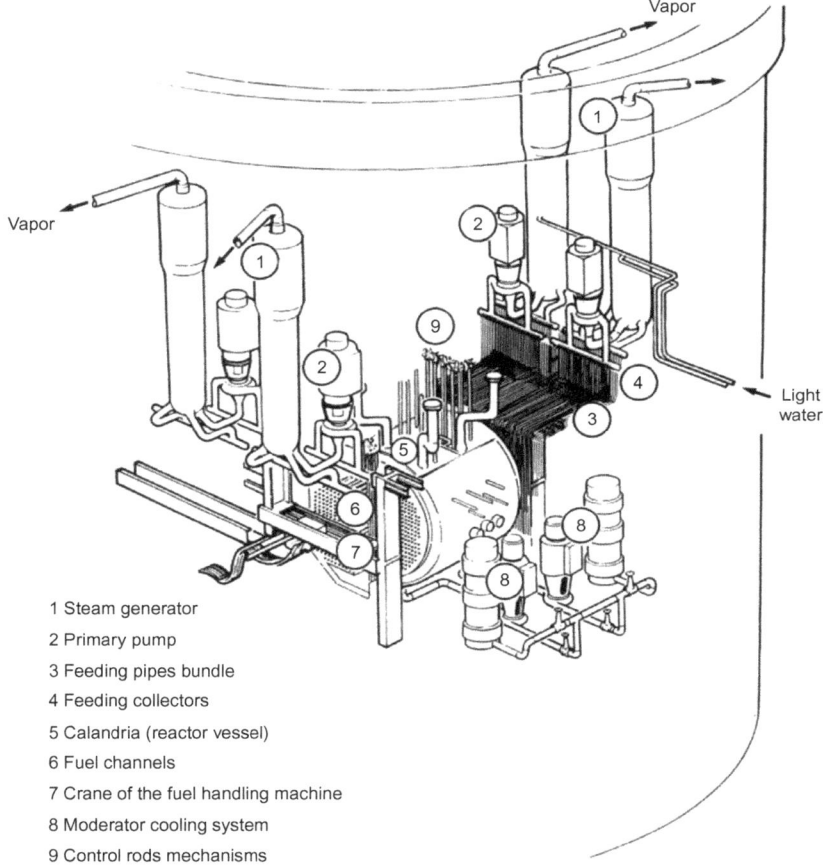

1 Steam generator
2 Primary pump
3 Feeding pipes bundle
4 Feeding collectors
5 Calandria (reactor vessel)
6 Fuel channels
7 Crane of the fuel handling machine
8 Moderator cooling system
9 Control rods mechanisms

Figure 4.8. Configuration of the primary and moderator systems (from R.B. Lyon, E. Price, IAEA Report 407, 2001, CANTEACH website).

The primary loop forms a figure eight and includes four collectors, two steam generators and two pumps. A reactor is comprised of two loops that are practically independent of one another, the only common points being the connections with the pressuriser and the auxiliary systems.

Steam generators

Regardless of the core, the primary system components are, on the whole, similar to those of a PWR. It is especially the case for the pressuriser and the primary pumps. The steam generators are classic devices – U tubes – with some notable differences:

- The existence of a pre-heater integrated into the bottom of the tubes, on the cold side. This arrangement is used to improve the relatively low yield of the power station,

- The use of Incoloy 800 as the material for the exchange tubes,

- Heat exchanges inside the steam generator are improved by the steam coming from the core (about 3 to 4% of the total steam released from the reactor).

4.3.3. Moderator system

Due to part of the energy being produced directly in the moderator (approximately 100 MW of thermal power for a CANDU 6), it is necessary to cool it. There is therefore a moderating system with pumps and exchangers (Figure 4.8). When taking into account the low temperature level of around 60 °C (to avoid water boiling), this energy is not recovered.

4.3.4. Fuel

Description of a fuel element

A fuel element is made up of 37 rods containing natural uranium oxide in the form of pellets cladded in Zircaloy 4 and assembled between two grids with ends welded onto the rod plugs (Figure 4.9). Pads brazed on the clad ensure that there are spaces between rods and ensure the support of the element on the pressure tube. Each channel contains twelve bundles and a CANDU 6 reactor receives 4560 in total.

This fuel element is short in length (50 cm), which facilitates its manufacture and handling. No expansion volume is allowed for fission gases; these remain trapped inside oxide porosities and the scooping out at the ends of the pellets. A thin layer of graphite is deposited on the internal surface of the clad (a fuel concept called CANLUB), with the aim of reducing the pellet-clad interaction.

Thermal performance of the fuel

The average fuel rating is 25.4 W/g, which corresponds to an average linear power of the order of 250 W/cm, the maximum value being 540 W/cm. The maximum temperature of UO_2 is about 1900 to 2100 °C.

Conversely, the power density of the core is much lower in a CANDU ($\approx$11–12 MW/m^3) than that in a PWR ($\approx$100 MW/m^3). This expresses the compactness of light water reactors.

Fuel handling

The handling of the fuel is carried out with two machines, one on each side of the reactor (Figure 4.10).

The fuel renewal operations consist of locking each machine at an end of the desired channel, the distribution machine having been charged with new bundles. Each machine is then put into contact with the primary system in order to balance the pressure and to ensure the circulation of the heavy-water which will cool the channel once it is opened.

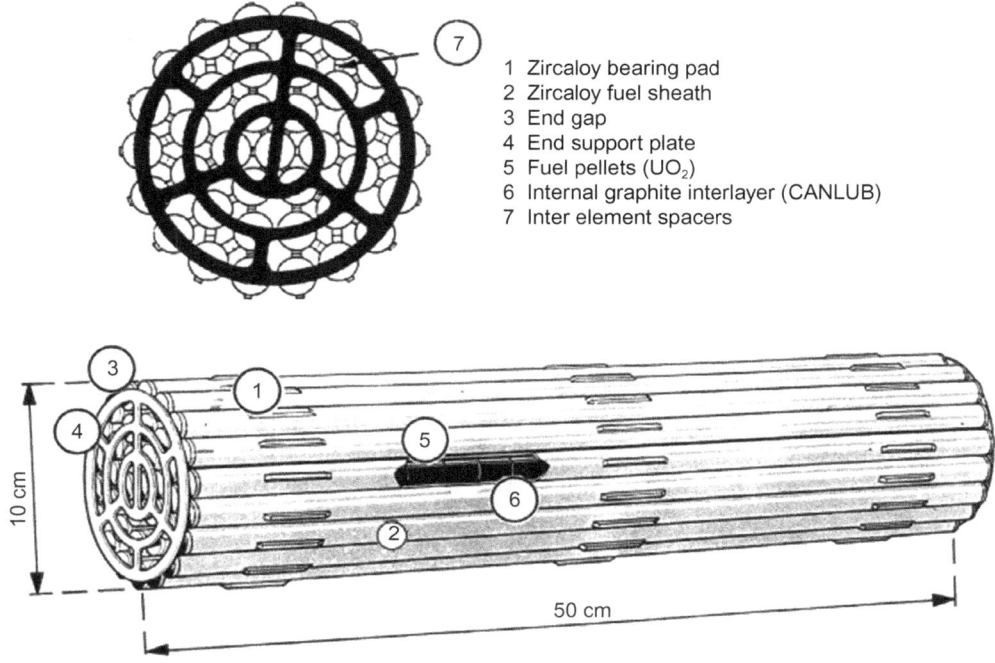

Figure 4.9. Fuel bundle with 37 rods (from R.B. Lyon, E. Price, IAEA Report 407, 2001, CANTEACH website).

This is followed by the extraction of the two seal plugs and protection at each end, with the aid of the rings of the machine. After inserting the new fuel from one side and extracting the spent fuel at the other end, the plugs are again put in place, and finally the machines disconnected. The receiving machine goes to discharge the spent fuel in the storage pool.

The frequency of renewal is about 16 to 18 bundles per day and between 80 to 90 per week. As the reactor sides are not accessible during operation, the renewal procedures are controlled remotely, via specialised computers.

4.3.5. Reactivity control systems

The methods of control of the reactivity are specialised and ensured by different means.

The soluble poison in the moderator (boric anhydride) is used in very low concentrations. In fact, the only consideration is the low reactivity margin (Figure 4.3) needed to cover a temporary unavailability of the handling machine and to recover a limited part of xenon transients.

Some compartments can be filled with light water (thus being susceptible to strongly absorbing neutrons) and placed in the moderator, between the pressure tubes. The variation in the level of water in those compartments changes the reactivity, and provides a way to regulate and pilot the overall reactor. The reactor is divided in 14 zones (Figure 4.11) each of which are associated with a compartment and several detectors.

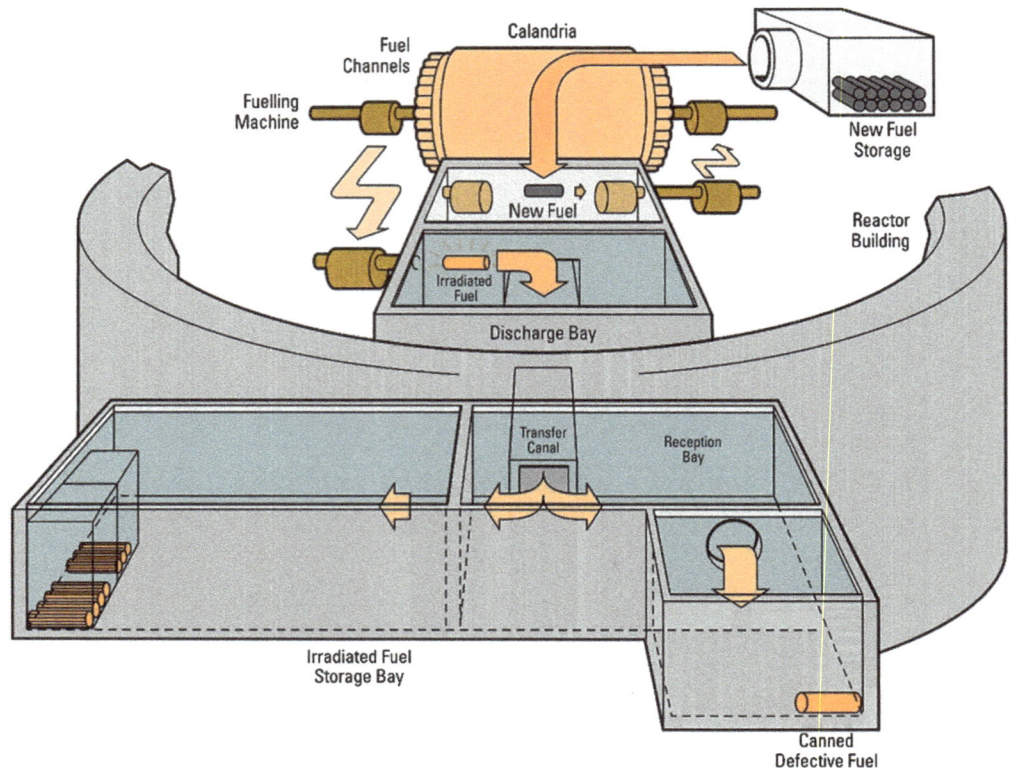

Figure 4.10. Diagram of the fuel handling system (from D.A. Meneley and Y.Q. Ruan, 1998, CANTEACH Website).

Emergency shutdown is carried out by 32 cadmium rods cladded with stainless steel, which penetrate by the top of the vessel (Figure 4.4). A second emergency shutdown device is planned via the fast injection of poison (gadolinium), directly into the moderator vessel.

Stainless steel mechanical compensation rods also take part in stabilising the power distribution. These 21 rods are usually inserted in the core, and extracted when necessary, to release the reactivity, particularly during power transients (compensation of xenon).

4.3.6. Safety systems

Primary loss of coolant accident (LOCA)

The LOCA corresponds, for a CANDU, to the rupture of a cold collector, the largest pipe of the primary system.

When the accident is detected, there is automatic insulation of all the connections between the two primary loops. Emergency water injection comes from a tank located under the dome of the containment vessel (see equipment 8 in Figure 4.12). This light water

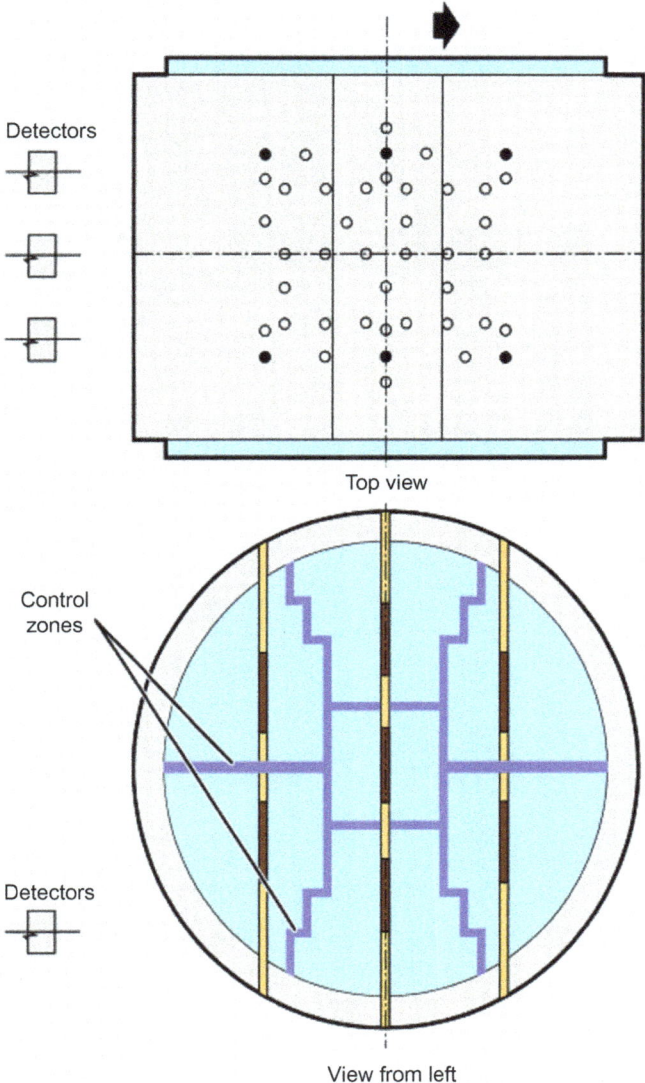

Figure 4.11. Control zones.

is injected into all the collectors in the first phase, with the the second phase consisting of insulating the broken collector. Systems are very similar to those of a PWR:

- high pressure injection starting from accumulators calibrated at 5 MPa,
- medium pressure injection from a tank contained in the containment,
- injection at low pressure by recirculation of the water recovered by the well at the bottom of containment.

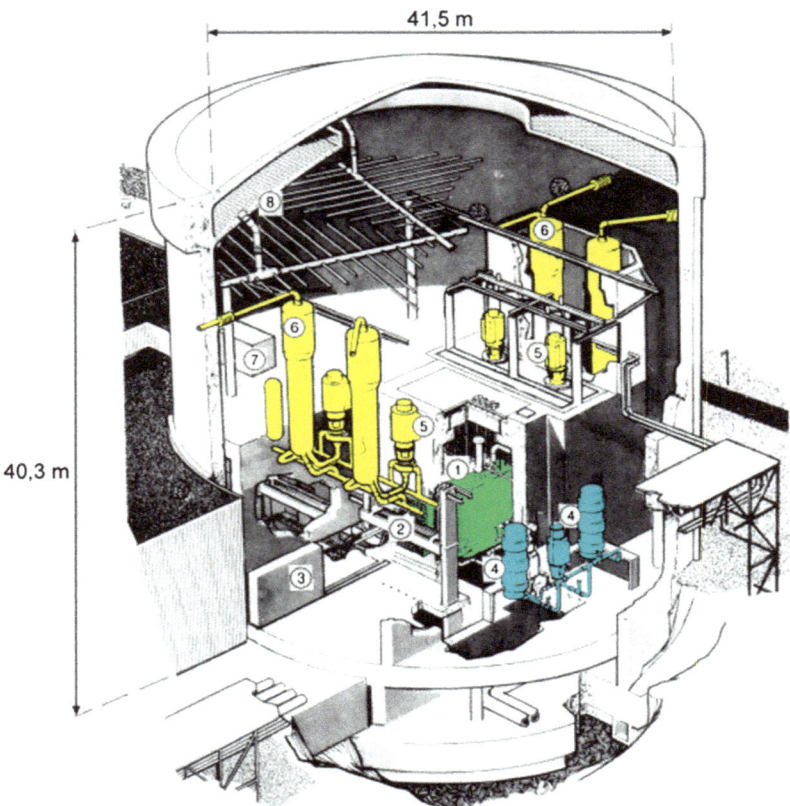

1 Calandria, 2 Fuel loading machine, 3 Reinforced door of fuel machine,
4 Moderator cooling system, 5 Primary pump, 6 Steam generator,
7 Primary pressuriser, 8 Spray water tank.

Figure 4.12. CANDU 6 Building cutaway (from J. T. Brewer, 1983, Nuclear training centre, CANTEACH Website).

Containment vessel

The containment vessel of the CANDU for a site with single plant or several independent plants is rather traditional: it involves a pre-stressed reinforced concrete containment vessel, designed to resist the pressure during an accident (Figure 4.12). The Canadian multi-unit sites (Pickering, Bruce, Darlington) implemented the system known as a vacuum building, which will be described later on.

Safety characteristics of heavy-water reactor

Part of the second barrier (the pressure tubes) is directly under neutron flux. This necessitates a specific design and choice of materials.

The pressure tubes have a low thickness (≈4 mm). There is therefore a leak detection system to detect a possible defect before it evolves quickly enough to cause the tubes to rupture. This is an important asset with regard to safety.

The primary coolant volume is limited, resulting in a weak release of energy in the event of a LOCA, even more so because it is divided into two independent loops. Finally, the high volume of moderator in the vessel constitutes an important cold source (heat sink) and because of continuous fuel reloading, the potential reactivity of the core is low.

The voiding reactivity coefficient of the core is slightly positive. This leads to the subdivision of the primary system in two loops and the existence of a second emergency shutdown device.

The burn-up is weak, resulting in a limited quantity of fission gas, which delays the expansion of the clad in the event of a LOCA.

4.3.7. Fuel cycle

In the reactor core, practically half of the fissions come from uranium 238, either directly through fast fission, or via the plutonium. In the absence of parasitic captures, the average burn-up in a CANDU fuelled by natural uranium is in fact 7500 MWd/t.

The average spent fuel composition is as follows:

$$^{235}U:0.20\%;\ ^{239}Pu:0.26\%;\ ^{240}Pu:0.10\%;\ ^{241}Pu:0.023\%;\ ^{242}Pu:0.007\%$$

One can note the very effective use of the initial fissile uranium (similar to the rejection rate of uranium enrichment plants by gas diffusion), the high percentage of plutonium (despite the consumption of a significant part of produced plutonium during the irradiation), and the high quality of this plutonium.

Table 4.6 summarises the orders of magnitude of the fuel characteristics with regard to the uranium consumption and plutonium production. The data is also compared with a PWR, which makes it clear that CANDU produces more fissile plutonium per MWe and per year, and makes better use of natural uranium, i.e. it uses less to produce a given quantity of energy.

The irradiated fuel is not reprocessed but stored on site, in large-sized pools, with a capacity planned to be sufficient for at least ten years of operation of the power plant. These fuels will be stored later in a dry location above ground.

4.3.8. The vacuum building

The particularity of the Pickering site (Figure 4.13) resides, in fact, in the existence of a "vacuum building" which is an integral part of the confining system on this multi-unit site: in the event of pressurisation of a reactor containment, the containment is immediately put in communication through a concrete channel and valves, with the vacuum building in which a vacuum of approximately 1/10 of the atmospheric pressure is permanently maintained. The vapour there is then aspired and condensed by aspersion, which limits the peak of pressure in the unit, and permits the use of reinforced concrete without having

Table 4.6. Use of fuels in light and heavy water reactors.

		CANDU		BHWR		PWR		
Cycle type		without recycling	Pu recycling	without recycling	U recycling	U recycling	U + Pu recycling	without recycling
Cycle duration / Fractioning	Month – 1/n	12-continuous				12-1/3		18 – 1/3
Initial enrichment	%	Natural U (0.71)	Natural U (0.71) + Pu (0.31)	Enriched U (1.5)	Enriched U (3.25)	Enriched U (3.25)	Enriched U (3.25) + Pu (3.5)	Enriched U (4.0)
Burn-up	MWd/t	7500	16000	22000	33750	33750	33750	45000
Discharge U 235	%	0.20	0.11		0.86	0.86	0.75	0.8
Fissile Pu discharged	g/kg initial uranium							
	g/kg initial uranium	2,84	3,5		6.1	6.1	8.9	7.4
	g/MWe and per year	375	460		160	160	232	130
Need in natural uranium	g/MWh	18.8	9.2	11.5	24.3	21	17	21.4

Hypothesis : Annual load Factor = 80%; Rejection rate of enrichment plant = 0.25%

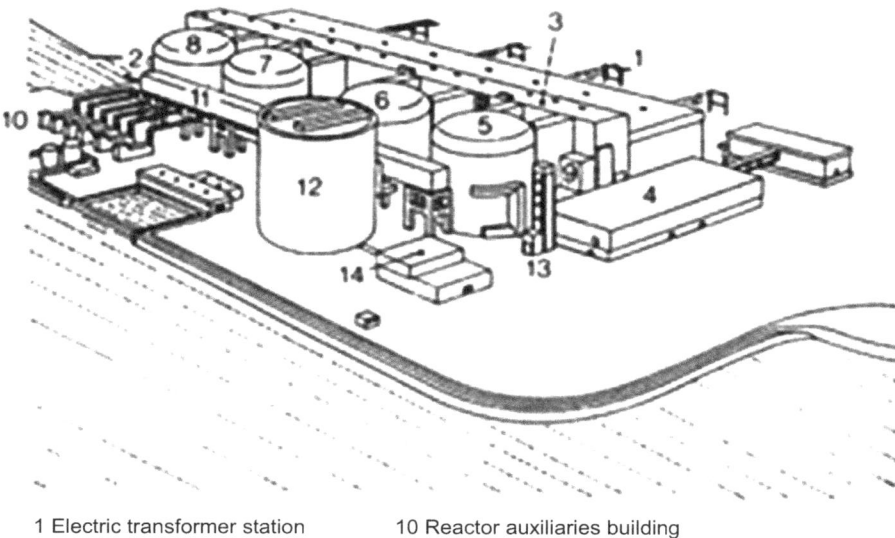

1 Electric transformer station
2 Water discharge canal
3 Power conversion facility
4 Maintenance building
5 to 8 CANDU reactors
9 Turbine auxiliaries
10 Reactor auxiliaries building
11 Confinement building pipe
12 Vacuum building (common to the four reactors)
13 Heavy water re-concentration tower
14 Water treatment building

Figure 4.13. Four stage plant at Pickering A (from Atomic Energy Canada Limited).

to be pre-stressed for the containment of the reactor. The main functions of this device are presented in Figure 4.14.

4.3.9. Difficulties and incidents in the Canadian programme

Until the middle of the 1980's, the Canadian reactors were the world leaders regarding availability (Figure 4.15), which was to be attributed primarily to the renewal of fuel in operation. Since this time, on the one hand the performance of light water reactors has not ceased to improve and on the other hand the performance of the CANDU has deteriorated.

The first difficulties appeared from the very start of the 1970's; the mode of management of fuel resulted in the renewal of two or four bundles with each operation, and not eight. That resulted in the imposition of an important overpower to some bundles, resulting in clad rupture by pellet-clad interaction (the concept CANLUB did not exist at that time).

In 1974, leaks appeared in Pickering 3 in the gap between the pressure tube and the calandria tube. The report showed that a bad positioning of the centering tool had caused important local deformations of the pressure tube and this created excessive constraints in the metal. The pressure tubes in question had to be replaced.

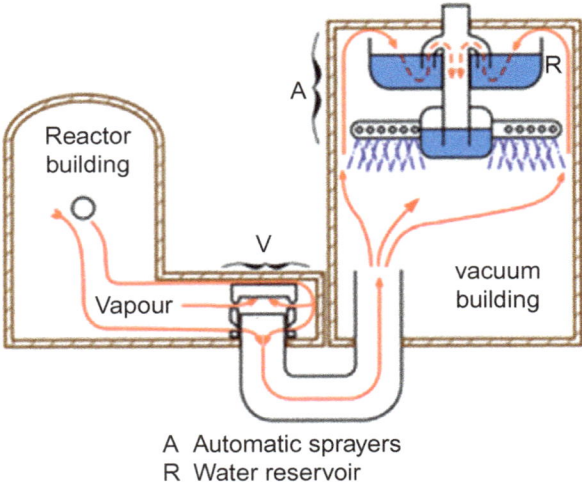

A Automatic sprayers
R Water reservoir
V Automatic pressure valve

Figure 4.14. Principle of the Vacuum Building (from Atomic Energy Canada Limited).

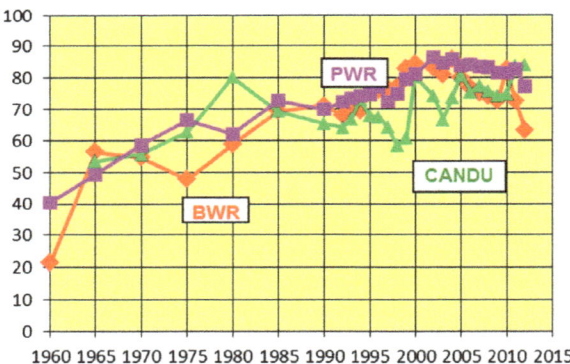

Figure 4.15. Annual load factor (%) (from Elecnuc 2013 CEA).

The phenomenon of lengthening and creep of the pressure tubes gradually led to the shutdown of each reactor for retubing. It is now understood that the phenomenon is generic: the pressure tubes are equipment that must be seen as "consumable". The replacement of a pressure tube has become a routine operation. It requires shutting down the reactor and takes two days per channel. So changing all the tubes in a given reactor takes place over a minimum of two years.

It seems nowadays that most of the difficulties are consigned to the past and CANDU availability is again competing with the one of the light water reactors, as indicated in Figure 4.15.

4.3.10. Economy

Table 4.7 provides a certain number of qualitative but heteroclite elements coming from very diverse sources, and indicates date and origin. Despite all these reserves, this table nevertheless makes it possible to draw some conclusions:

- the cost of the fuel is low if compared to the value of PWR. This is due to the use of non-enriched uranium, the efficient use of this uranium and the low manufacturing cost;
- heavy-water shares a significant part (15 to 20%), of the total investments;
- the structure of the costs can vary significantly year by year, according to the load factor. A high load factor, which for a long time applied only to heavy-water reactors, is likely to compensate for a large part of the capital costs.

Table 4.7. Breakdown of the production cost (in %).

	French Origin. Beginning of the 1970's		Costs at Pickering		Bruce 1986	Canadian Study 1998
	PWR	CANDU	Year 1975	Year 1977		
Investment	57.0	73.0	77.2 (*)	59.2	58.2	63.0
Operation and maintenance	17.0	13.0	10.3	22.7	18.6	28.7
Fuel	26.0	14.0	9.2	12.4	21.7	8.3
Renewal of heavy water			3.3	5.7	1.4	
Total	100.0	100.0	100.0	100.0	100.0	100.0
Factor or annual load = (%)	(70.0?)	(70.0?)	(61.9)	(90.7)	(75.0)	(75.0)

(*) from which 63% outside heavy-water and 14% from the initial investment in heavy-water.

4.4. Fuel cycle possibilities

4.4.1. CANFLEX fuel

In order to reduce the power of the peripheral rods of a bundle, which are hottest and constitute in fact a limitation to operation, one can, by decreasing their diameter, increase the number of rods. This is CANFLEX (CANdu FLEXible fuelling). The maximum linear power is reduced by 15 to 20%. This fuel was the object of a qualification programme in the reactor of Point Lepreau.

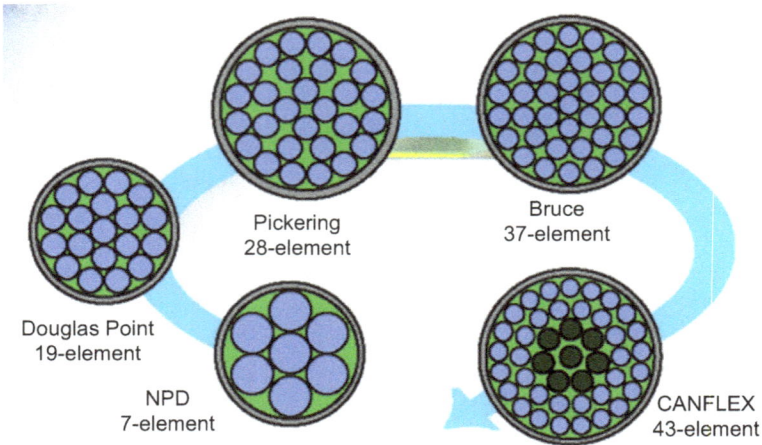

Figure 4.16. CANDU's fuel evolution (from S. Yu, 2002, CANTEACH website).

4.4.2. Slightly enriched uranium

The use of slightly enriched uranium (0.9 to 1.2%) would allow a reduction in the cycle costs in the range of 20%, due to the reduction in the need for natural uranium. Additionally it was seen that this advantage could be combined with a profit on the investments if the cooling heavy-water were replaced by light water. It is an option being studied for future CANDU reactors.

4.4.3. Recycling of the LWR fuel

A way of implementing an enriched fuel is to use fuel discharged from light water reactors (with 0.9% of U^{235} and 0.6% of fissile Pu). Canada has developed this recycling in cooperation with the Republic of Korea. It is known as the DUPIC programme (Direct Use of spent PWR fuel In CANDU). The studies show that recycling allows an energy extraction from the fuel of about 50% additional energy than that already produced in a PWR. The consumption of military quality Pu or the use of thorium is also possible in a heavy-water reactor (see Figure 4.17).

4.4.4. Perspectives

The expressed objective of the current studies of new generation of CANDU such as the Advanced Candu Reactor (ACR 1000, Figure 4.18) is to reduce the capital costs by 40%. The main characteristics of this project are shown in Table 4.8 below.

This project suggests that Canada would renounce the policy of natural uranium and all heavy-water, which would result in:

- The use of light water as a coolant,
- A tightening of the tube pitch, resulting in an under-moderation and thus an almost two-fold reduction in the heavy-water inventory per MW,

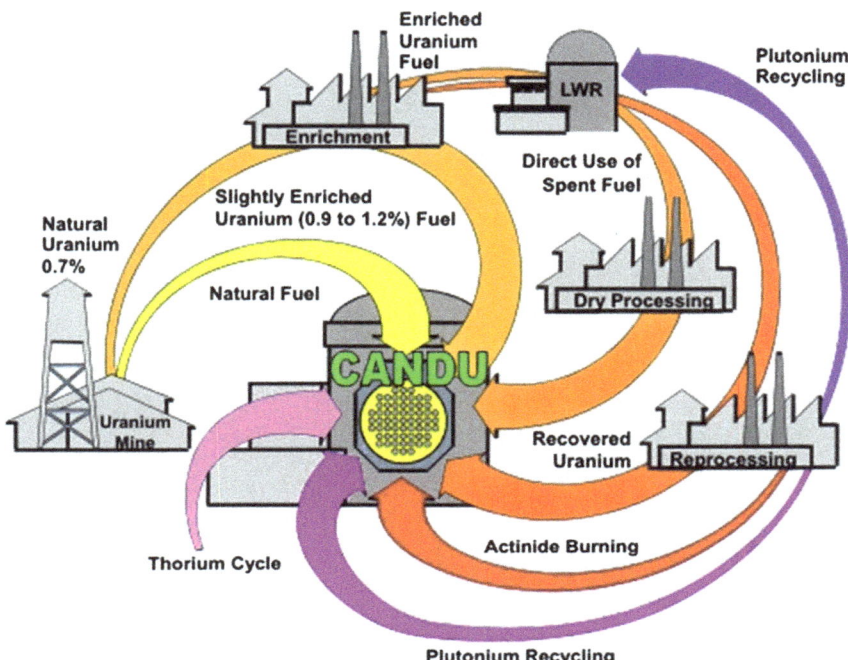

Figure 4.17. Strategies for recycling light water fuel in a heavy-water reactor (from Atomic Energy of Canada Limited).

- An increase in the coolant pressure (thanks to the increase of the thickness of the pressure tube) and therefore an increase in its temperature, from which a good improvement of the vapour characteristics and global efficiency is derived,
- A negative core voiding effect that allows the use of only one primary loop.

Reduction of the investment in heavy-water was (see Table 4.9) obtained by the following three manners: replacement of heavy-water coolant, reduction of the tubes pitch, and increase of the number of channels in the same vessel.

The under-moderation of the core and the use of enriched uranium lead to a negative void coefficient which no longer imposes separation of the primary system in two independent loops. A single primary loop is sufficient and this reduces the investment. The lifetime of the plant reaches 60 years with provision for mid-life replacement of the fuel channels.

Concerning safety, severe accidents are taken into account. The meltdown of a fuel channel is studied together with the absence of propagation to the neighbouring channels. The core melting probability is shown to be around 3×10^{-7}/year/reactor. The containment vessel is reinforced to withstand a crash plane.

It should also be mentioned that, in parallel to ACR-1000, Canada is also developing the ECR, Enhanced Candu Reactor, still fuelled with natural uranium. This reactor is the one currently proposed on the nuclear market.

Figure 4.18. Overview of the ACR-1000 (from AECL in ACR-1000 technical description, 2010, CANTEACH Website).

Table 4.8. Characteristics of project ACR-1000.

Net power	1085 MWe
Nature of coolant	Pressurised Light Water
Primary coolant pressure	12.6 MPa
Steam pressure in the turbine	6 MPa
Fuel element	CANFLEX bundle 43 rods
Nature of fuel	Enriched UO_2 at 1.6–2%
Burn-up rate	20 000 MWj/t
Channel number	520
Average power per channel	6.8 MW
Primary system	1 loop in eight with 4 Steam Generators and 4 pumps
Fuel renewal	2 bundles per channel

Table 4.9. Compared Masses of heavy-water (in Metric Ton) for CANDU 6 & ACR-1000.

	CANDU 6	ACR 1000
Moderator	265	250
Coolant	192	0
TOTAL	**457**	**250**

References

[1] The Nuclear Reactor International journal (GB), presents quite regularly description of nuclear reactors.
[2] The Techniques de l'Ingénieur (F), "Engineering Techniques" includes an article about heavy water reactors which was included in three editions:

- B. Micaux, R. Naudet, *Industrial heavy water reactors*, 1979;
- D. Landel, *Industrial heavy water reactors*, 1989;
- S. Yu, J. Hopwood, D. Meneley, *Heavy water reactors*, 2000.

[3] *Status of advanced technology and design for water cooled reactors: heavy water reactors.* AIEA-TECDOC-510, July 1989.
[4] *Canada enters the nuclear age: a technical history of Atomic Energy of Canada Ltd. as seen from its research laboratories.* McGill-Queen's University Press. 1997.
[5] J. Whitlock, http://www.nuclearfaq.ca/; "The Canadian Nuclear FAQ".
[6] https://canteach.candu.org/aecl.html; Courses on the design and technology of CANDU reactors, McMaster University, Canada.
[7] L.M. Surhone, M.T. Tennoe, S.F. Henssonow, *Pressurised Heavy Water Reactor.* Betascript Publishing, 2010.

Appendix 1: Heavy-water production

All compounds containing hydrogen also contain deuterium. Heavy-water, for example, is a natural substance, present in "ordinary" water in a proportion of approximately 1 part of heavy-water for 7000 parts light water. Deuterium (H^2) and oxygen combine to form heavy-water (D_2O). A comparison of the physical properties of heavy-water and light water are given in § 4.2.1 – Table 4.4.

Heavy-water is produced through the extraction of deuterium contained in water. Different processes have been developed, most of which use the differences in physical properties between "ordinary" and "heavy" water (see Table 4.A1 below).

The first process implemented for the production of heavy-water was electrolysis of water (remember the Battle of heavy-water, in Norway, during the Second World War). It is by far the most efficient process, but it also uses the highest energy. The current process,

Table 4.A1. Heavy-water production processes.

Process:	Separation factor
1. Elevated energy need	
• Gaseous diffusion	1.20
• Thermal diffusion	1.05
• Electrolysis of water	10.00
2. Large volumes necessary	
• Distillation of water	1.05
• Distillation of ammoniac	1.03
• Freezing of water	1.02
3. Low volume and energy need	
• Distillation of hydrogen	1.40
• Isotopic exchange	
• Water/hydrogen	
• Water/sulfurized hydrogen	3.00
• Ammoniac/hydrogen	2.00
• Amines	6.00

used in Canada, is the GS (Girdler-Sulphide) process. It exploits exchanges between water and sulfurized hydrogen, the sense varying in function of the temperature:

$$H_2O_{(liq)} + HDS_{(gas)} \underset{T\ high}{\overset{T\ low}{\leftrightarrows}} HDO_{(liq)} + H_2S_{(gas)}$$

T low is at room temperature, and T high is around 130 °C. This process uses three successive stages of exchanges columns (plate columns), to a title of 30%. It is finished by a distillation stage to a purity of 99.75%, (see Figure 4.A1).

Canada has constructed or planned the following installations, using the GS process (see Table 4.A2)

Appendix 2: A Heavy-water reactor with a reactor pressure vessel

A German company, KWU, has developed a reactor concept of small power, cooled with heavy-water, but having a vessel analogous to light water reactors. A reactor was delivered to Argentina, ATUCHA 1 (Figure 4.A2).

The reactor has a pressure vessel, but of a much greater size. Figure 4.A3 below gives a comparison of the sizes of the heavy-water reactors and a vessel of a 1200 MWe PWR.

The pressure vessel (see Figure 4.A4) contains the moderator calandria that is made up of 252 vertical fuel channels, arranged in a hexagonal geometry. The moderator

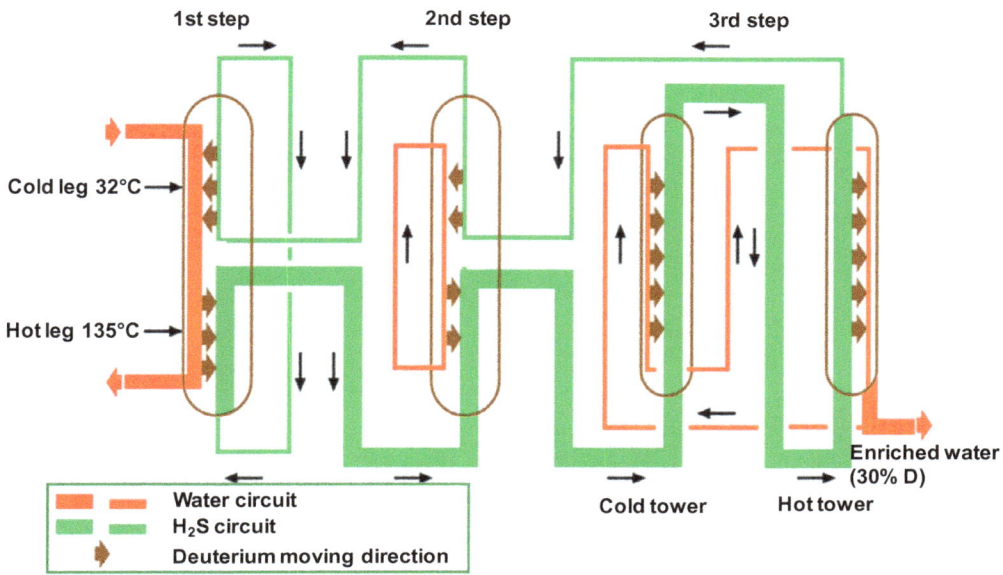

Figure 4.A1. Diagram of the principle of heavy water production.

Table 4.A2. Heavy-water production facilities.

Installation	Capacity	(Situation in 2006)
Port-Hawkesbury	400 t/year	Decommissioned and dismantled: 1985 Implementation, 1970
Glace Bay	400 t/year	Implementation, 1976 Decommissioned and dismantled: 1985
La Prade	800 t/year	Stop construction: 1978
Bruce A	800 t/year	Implementation, 1973 Decommissioned and dismantled: 1984
Bruce B	800 t/year	Implementation: 1981
Bruce C	800 t/year	Cancelled in 1976
Bruce D	800 t/year	Stop construction: 1973

and the coolant, both heavy-water, are at very similar pressure, which allows a small channel thickness and no pressure tube. Considering the size of the shield, the absorber control rods cannot be lodged between the channel heads; they are implanted in the periphery of the cover and the rods move obliquely between the channels.

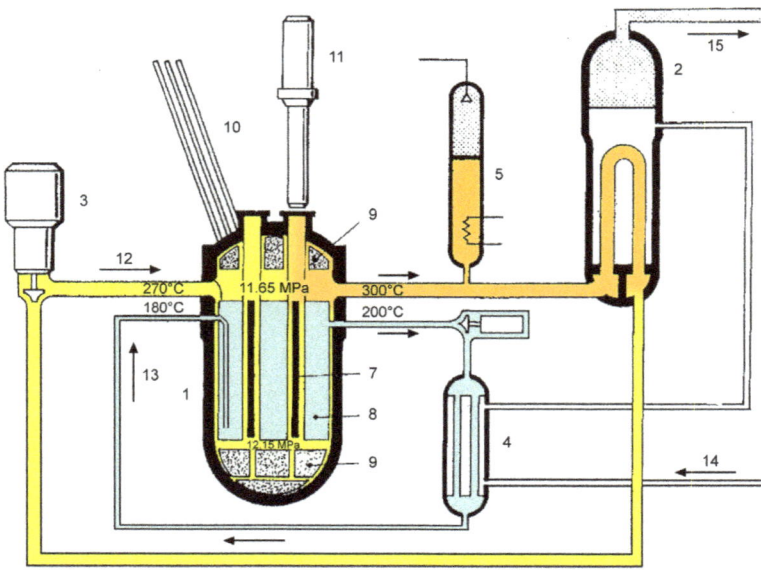

1 Reactor
2 Steam Generator
3 Primary pump
4 Moderator heat exchanger
5 Pressuriser
6 Moderator pump
7 Fuel element
8 Moderator
9 Internal structures
10 Control rods
11 Handling machine
12 Primary loop
13 Moderator loop
14 Secondary water supply
15 Steam

Figure 4.A2. The ATUCHA 1 reactor principle.

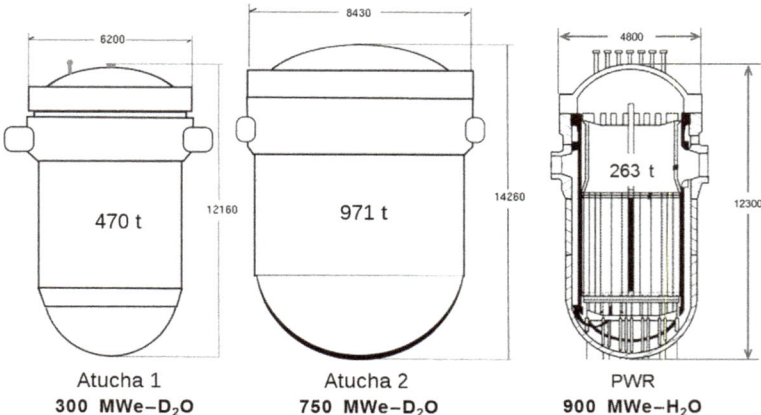

Atucha 1
300 MWe–D$_2$O

Atucha 2
750 MWe–D$_2$O

PWR
900 MWe–H$_2$O

Figure 4.A3. Comparison of pressure vessel size for various power size of pressurized heavy-water reactors (left) and a PWR (right).

4 – Heavy water moderated nuclear reactors

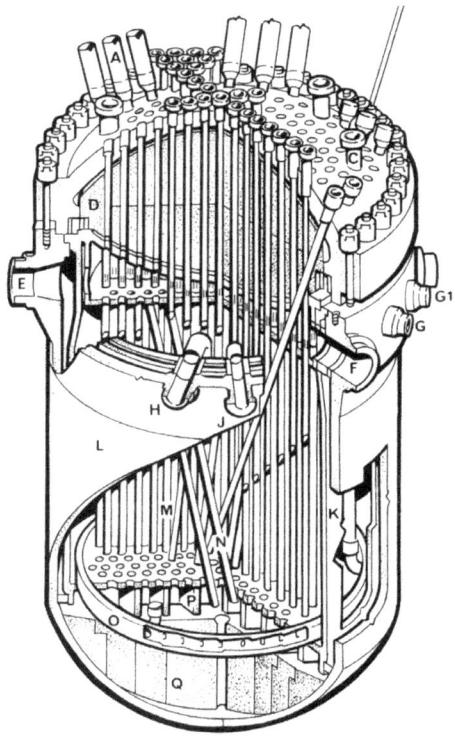

Figure 4.A4. The ATUCHA 1 pressure vessel.

5. Nuclear marine propulsion

R. Lenain

5.1. Introduction

As of 2010, more than 140 ships were powered by nuclear reactors, most of them submarines but also cruisers, aircraft carriers and icebreakers. Interest for these applications started early in the 1940's. The technology was mainly developed in the 1950's, and the number of nuclear ships peaked around 1990 (Figure 5.1) with 180 Russian and 150 US ships, UK, France and China accounting for about 35 units. At that time in the US, more nuclear reactors were used to power ships than to produce electricity.

The general principle of nuclear propulsion is as follows: a nuclear reactor feeds vapour to a turbine generating mechanical energy which is then either directly transmitted to the propeller shaft or transformed into electricity later used to provide fresh water and purified air.

The main point is that nuclear-powered submarines, unlike conventional engines, do not need oxygen for propulsion and that the long autonomy they provide allows long missions without refuelling. The main limitation, however, comes from crew capacities and the availability of loaded goods and supplies.

The major type of nuclear reactor technology used is the water pressurised reactor. Numerous other solutions have been studied, some have been tested, the Pb-Bi design was developed in the USSR but now abandoned. A nuclear naval fleet needs land-based test reactors for design validation, component tests and also for training activities.

This work gave a crucial thrust to land-based reactors dedicated to electricity production.

5.2. Main properties required for propulsion

This section gives an overview of the properties expected from nuclear designs for electricity production and for naval propulsion.

Electricity production on land-based reactors

Economy: (investment & cost/KWh) is the main motivation; preference is given to high unit power and load factor.

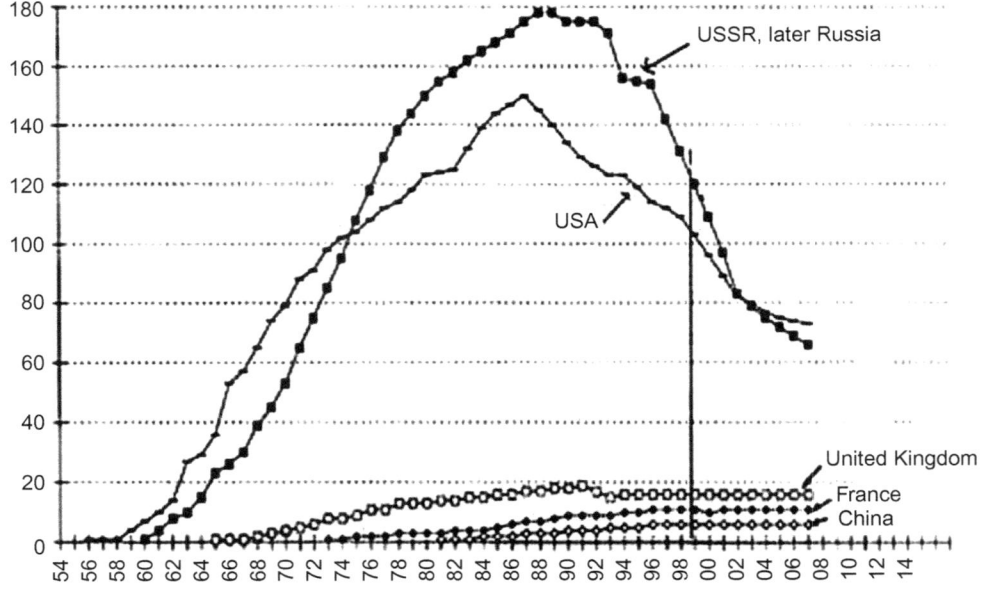

Figure 5.1.

Safety-security: external resources (human and material) are available in emergency situations, the design should ensure that no external radioactivity release can exceed the plant limit.

Cold source: its availability is an essential criterion to be respected in any situation; it should obey environmental constraints.

Operability: load following capability and adaptability to grid demand could be important, nevertheless slow transients are acceptable but their characteristics depend on grid limitations (some %/mn) and on the features of other connected energy sources.

Fissile material and fuel: optimisation of the cycle requires to optimise the design to economise uranium. The constraints of limited enrichment (<<20%), reducing fuel cost, back-end of the fuel cycle and optimised plant operating costs are major concerns.

Protection against external hazards: design considerations for water-flood, earthquake, airplane crashes, fires, as well as common reactor mode failures should be properly addressed.

Navy Applications

Operability: this is the key objective. Priority is given to propulsion over fuel or operating costs, weak reactor-propulsion system coupling, and avoiding shutdowns for maintenance or refuelling: long fuel cycle management capability is of the utmost importance, possibility to operate with natural circulation.

Safety-security: the crew has to be able to maintain a good overview of the reactor and ship. In accidental situations, no external resources (neither human nor material) are available and no internal radioactivity release can be tolerated.

Overall dimension and mass constraints: high specific energy, optimised shielding, limited space for operation require optimisation of operating and maintenance devices such as the control room and instrumentation. Such optimisation involves integrating and sharing components or functionality: confinement, shielding, water tanks…

Cleanness and radiation shielding: clean circuits and the associated shielding are required to limit the release of activated isotopes (e.g. 16N) or fission products in case of fuel leakage.

Core: no clad failure, capacity for fast power changes (some %/s), resistance to thermal shocks.

Autonomy: requires to limit the need for land support and the frequency of maintenance task.

Mobility: because of pitching and rolling, stable mechanical and physical properties are expected of the reactor.

Specific requirements: resistance to shocks (military shocks or, in the case of icebreakers, shocks caused by ice) and acoustic discretion, fresh water production…

5.3. History and development

USA

The activity was officially launched in 1948 by the creation of the Nuclear Power Division of the Bureau of Ships with Admiral Rickover at its head. General Electric (GE) and Westinghouse were engaged in programs developing different solutions: GE was working on liquid metal cooling (project Submarine Intermediate Reactor or SIR) and Westinghouse on pressurised water concepts (project Submarine Thermal Reactor or STR).

In water cooled reactors, Westinghouse find out that zirconium fuel cladding is optimal for corrosion point of view and low neutron absorption. Reactivity control was partially obtained with gadolinium or boron. By the 1950's, Westinghouse built prototypes of pressurised water reactors (STR). It was reported [1] that the fuel was sandwich plates made of high enriched uranium and zirconium, and clad in zirconium. Based on this PWR technology, the USS Nautilus (submarine, S1W reactor) was launched in 1955 and the USS Enterprise (aircraft carrier, A1W reactor) in 1960: the Enterprise remains in service today. The Seawolf, based on the SIR project was launched in 1957 but, after problems with the sodium circuit, the Seawolf reactor was replaced by a SW water cooled type reactor. The PWR technology was deployed for submarine and surface ships (cruisers and aircraft carriers) reactor.

USSR

The USSR developed both PWR and lead-bismuth cooled reactor concepts. The largest submarines were Typhoon class (ballistic missile submarine) powered by PWR, while Alfa class submarines were powered by fast neutron reactors cooled by Pb-Bi. Between 1950 and 2000, the USSR and now Russia built more than 200 submarines, as well as a fleet of icebreakers.

UK

The United Kingdom developed a nuclear propulsion naval fleet in the framework of a cooperation agreement with the US (US-UK Mutual Defence Agreement of 1958). The agreement covered the export of one complete US submarine nuclear propulsion reactor and its enriched uranium fuel (S5W type) which was installed in HMS (Her Majesty's Ship) Dreadnought. This became the Royal Navy's first nuclear submarine and it was launched in the 1960's.

FRANCE

In 1954, France launched a dedicated program on its own resources. The first program, named Q227, was a natural uranium fuel concept moderated by heavy water. The project was abandoned before fuel loading. The French program continued, focusing on LWR technology: in 1964 the PAT (Prototype à Terre) was operational. The submarine Le Redoutable (ballistic class submarine) was launched in 1971. Afterwards, the attack submarine class was developed with a compact reactor technology that produces the smallest nuclear submarine concept. The French nuclear fleet includes one aircraft carrier.

China

China was the last to join the small club of countries using nuclear energy for military navy propulsion.

5.4. Naval reactor development [2]

This section aims at illustrating the scope of the research launched at the early stages of nuclear propulsion: only a small number of these designs remain in present nuclear fleets. Major US industries were involved in these R&D projects. Feedback and knowledge gain gave a strong impetus for land-based power plant development. Consequently, we can say that more reactor designs were investigated by manufacturers and laboratories in the area of naval propulsion than for electricity production. In addition to military nuclear propulsion, some civilian ships have been constructed, all of them also equipped with PWR type reactors.

Pressurised Water Reactors

In the early 1950's, Westinghouse designed and built a PWR prototype for submarines; the STR was the test reactor (Figure 5.2) and the S1W became the first generation of submarine reactor. Large ships (aircraft carriers) and the associated A1W reactors were physically based on the S1W.

Figure 5.2. STR, Submarine Thermal Reactor (INEL).

Combustion Engineering developed the S1C reactor using an electrical drive system instead of a steam turbine one.

General electric developed a series of PWRs up to the S8G types that can be operated at high power by natural circulation or with coolant pumps. The S8G generation is claimed to have enough energy to last for the lifetime of the ship.

Intermediate flux beryllium sodium cooled reactors

General Electric designed and built a reactor with UO_2 fuel and stainless steel cladding. Beryllium was used as moderator and reflector in SIRs and S1Gs. The US Seawolf submarine was initially equipped with a sodium cooled reactor which was replaced in 1959 by a PWR. This design allows high power density. A prototype (EBOR) was also designed at the National Reactor Testing Station, but the project was cancelled before construction was completed.

SCWR supercritical water reactors

The supercritical water reactor (SCWR) was considered with intermediate energy neutron spectrum. The fuel, UO_2 distributed in a stainless steel matrix, consists of parallelepipeds gathered on a plate assembly with stainless steel cladding. This design enables to reach higher temperatures and thermodynamic yields.

Organically moderated reactor experiment

An organically cooled and moderated reactor was investigated for thermalised neutron spectra: the coolant could work at low pressure and was less corrosive than water. Initially intended to be loaded with rectangular aluminium fuel plates, the project was cancelled before any fuel could be loaded.

Lead-bismuth cooled fast reactors

The Russian Alfa class submarines used Pb-Li (45-50) coolant for their fast neutron spectrum reactors. This design allows high power density, but requires heating during shutdown phases. It was the fastest class of submarines but all of them have been decommissioned.

Gas-cooled beryllium oxide reactors

The Experimental Beryllium Oxide Reactor (EBOR) was built at the National Reactor Testing Facility. It was a helium-cooled reactor moderated by beryllium oxide directly coupled to a closed gas cycle or steam cycle. The goal was the availability of a nuclear reactor for merchant ship propulsion or for small/intermediate size power plant. Westinghouse, General Atomic, General Dynamics were involved in this project. The fuel was made of Hastelloy X tubes housing ceramic pellets of UO_2 and beryllium oxide. Pins were separated from each other by a helical spacer. Dysprosium oxide (Dy_2O_3) was chosen for the reactor control blades.

5.5. Civilian fleet

Some civilian ships using nuclear propulsion were built: the major advantages of nuclear propulsion are energy reserve, availability and, in certain conditions, cost. Only Russian icebreakers remain in operation. Some of the major concerns are the lack of international laws and the lack of adapted installations in harbours.

In Japan, the cargo ship Mutsu (Mitsubishi PWR) was built in 1968 and decommissioned in 1991.

In Germany, the cargo Otto Hahn was built by Howaldtswerke Deutsche Werft AG in 1963. The nuclear propulsion reactor was replaced by a conventional propulsion engine in 1979.

In Panama, the MH-1A reactor (a PWR) was built for the U.S. Army by Martin Marietta and placed in a Liberty ship, operating between 1968 and 1975.

In Russia, 10 icebreakers were been launched between 1959 and 1993. In 2007, the last Russian icebreaker, the NS 50 Let Pobedy (whose construction began in 1993), completed the testing phase before officially beginning operating in March 2007.

In the US, the cargo-passenger ship NS Savannah (now a museum ship) was successively ordered in 1955, launched in 1959, in service in 1964 and de-fuelled in 1975. The installed power was a 74 MW Babcock and Wilcox nuclear reactor, 5.2m high and 160m in diameter, housing a core 1.7m high and 1.6m in diameter, with 32 fuel elements of enriched uranium (4.2 to 4.6%) and 21 control rods (Figures 5.3 and 5.4).

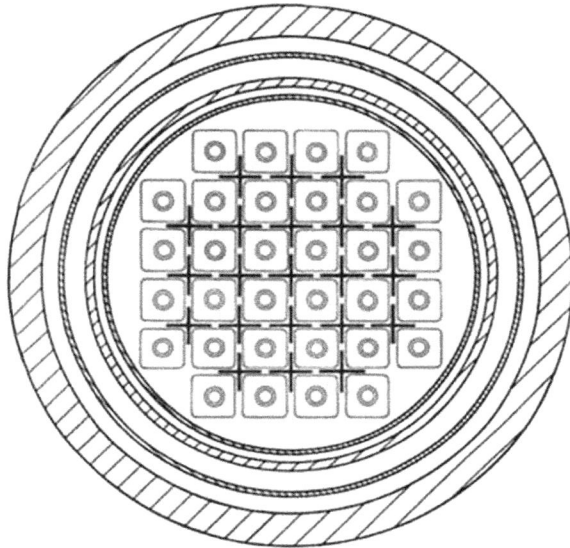

Figure 5.3. Core layout of the Savannah reactor.

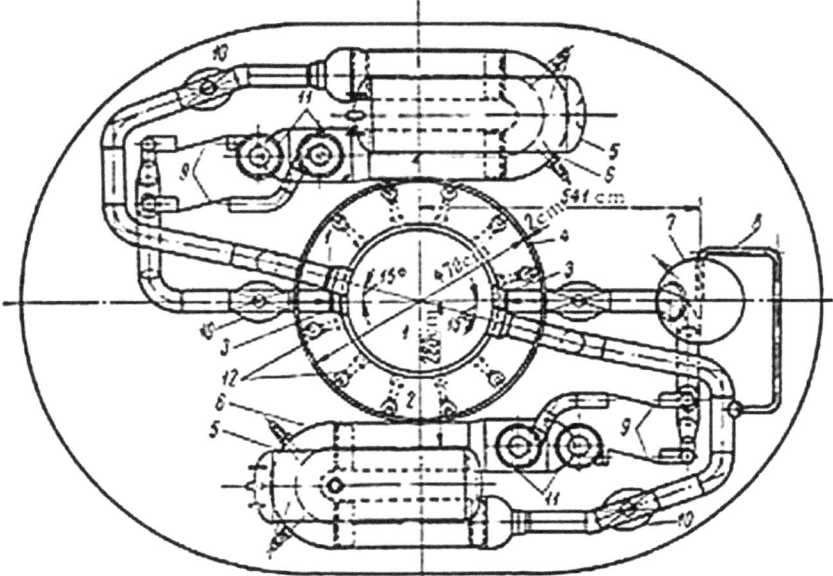

Figure 5.4. Overview of the Savannah reactor. © 2011, Historic Naval Ships Association.

References

[1] *Nuclear Power - Deployment, Operation and Sustainability*, edited by Pavel Tsvetkov, 510 pages. InTech, September 09, 2011, CC BY-NC-SA 3.0 license, ISBN 978-953-307-474-0 DOI: 10.5772/704.
[2] M. Ragheb, *Nuclear Naval Propulsion; Department of Nuclear, Plasma and Radiological Engineering*, University of Illinois at Urbana-Champaign 216 Talbot Laboratory, Urbana, Illinois, USA.
[3] Techniques de l'ingénieur, Référence BN3140, 2001, C. Fribourg.

6. Experimental reactors

B. Barré

6.1. Different types of experimental or research reactors

Power reactors generate electricity, naval propulsion reactors spin propellers, urban heating reactors heat water, but the category of "experimental" or "research" reactors covers a wide range of types as well as uses for these reactors. Here are some examples:

- critical mock-ups, which are very low power reactors, well instrumented and dedicated to reactor physics (flux mapping, macroscopic cross sections, ßeff, reactivity coefficients, etc.);
- reactors dedicated to teaching and training;
- small size "demos" of new reactor concepts;
- neutron sources for physicists, chemists and (increasingly) biologists to investigate solid state structures;
- radionuclide production for medicine, industry and agriculture;
- irradiation of "technological" materials, mostly in support of power reactor developments;
- safety or criticality experiments to be tested on smaller scales (fuel meltdown, reactivity excursions, LOCA, etc.);
- miscellaneous: activation analysis, dating, neutronography, homogeneous silicon doping for microelectronics…

Experimental reactors are often multi-purpose: this critical mock-up is used for teaching and a little neutronography, that materials irradiation facility dopes some silicon ingots on the side, etc.

There are many research reactors in the world: over 200, a third of which are located in developing countries (Figure 6.1). Among those, there are only few reactors whose power exceeds 15 MWt, shared between two categories: materials irradiation facilities and neutron sources for physics.

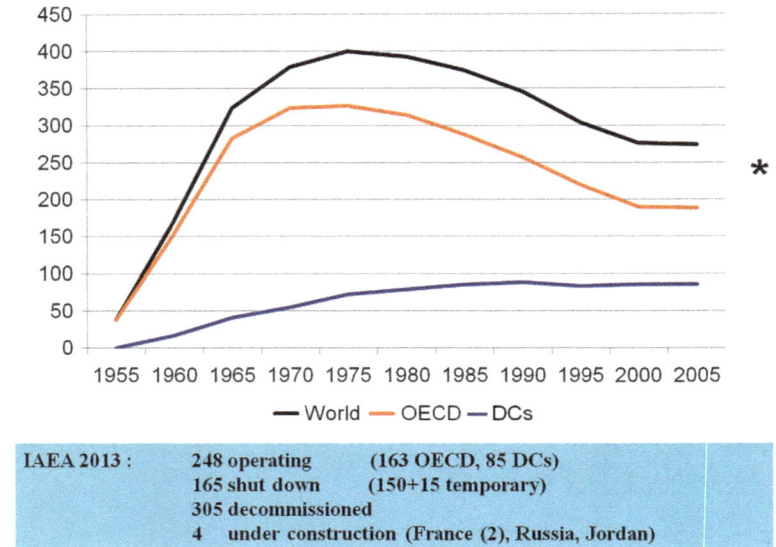

Figure 6.1. Research reactors in the world (source IAEA).

6.2. Materials irradiation reactors (MTR, TRIGA...)

Most irradiation facilities derive from the Materials Testing Reactor (MTR) which operated between 1952 and 1970 at Idaho Falls. From the MTR, they inherit the general "pool-type" open-core architecture and the plate-type bundle fuel elements.

The core is a square assembly of bundles, whose bottom end is inserted in a support structure made of aluminium alloy. The fuel is cooled by ordinary water and its support structure sits at the bottom of an open pool which shields operators from the radiation while allowing free access, from above, to and around the core. The core layout can be modified to suit specific experimental requirements, opening irradiation positions in the center or in an empty bundle slot at the periphery.

6.2.1. OSIRIS, in Saclay

The OSIRIS reactor, located at the CEA Saclay Research Centre, is an open-core MTR rating 70 MWt, cooled by ordinary water flowing upwards and sucked from the core into an open chimney located within the pool. It was designed to test and qualify the behaviour of structural materials and fuel elements under irradiation, but it is also used marginally for silicon doping and radionuclide production as an emergency alternative to other European production facilities. Its main characteristics are listed in Table 6.1.

The core is shaped like a compact cube of side length 60 cm. It comprises 44 MTR-type bundles, 38 standard fuel bundles and 6 control bundles (hafnium) driven from below (Figure 6.2). The whole core is encased in an open zircaloy crate which channels the cooling water. The top of the core is covered by water over a depth of around 10 meters.

Table 6.1. A typical MTR: OSIRIS.

Thermal Power (MWt)	70
Fuel	U3Si2 plates
Enrichment U (%)	19.75
Moderator	H_2O
Reflectors	H_2O, beryllium
Coolant	H_2O
Thermal flux (n.cm^{-2}.s^{-1})	3×10^{14}
Fast flux (n.cm^{-2}.s^{-1})	2.5×10^{14}
Inlet/Outlet temperatures (°C)	38/47
Water velocity between plates (m/s)	8.3
Maximum thermal flux (W/cm^2)	310

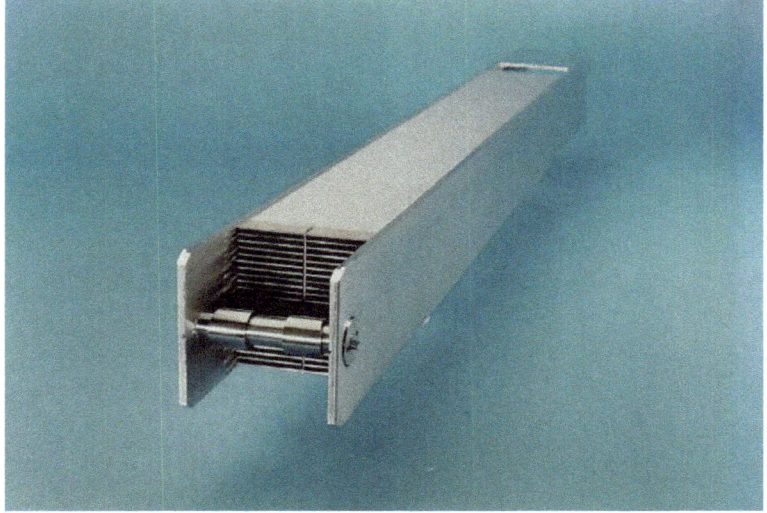

Figure 6.2. MTR fuel bundle (or sub-assembly). © Cea.

Experimental irradiations are carried out in instrumented devices: the "capsules", whose coolant is static, and the water, gas or sodium "loops", within which a physico-chemical environment representative of a power reactor is created around the sample (usually a fuel sample) to be tested.

Loops and capsules can be located in the core or at its periphery, in the reflector. Some loops are equipped with devices which can simulate power ramping by moving the loop closer to or farther away from the core.

OSIRIS stands among the most powerful pool-type reactors, but it is not the largest of the materials irradiation facilities: in Europe, the champion is the 100 MW Belgian (pressurized) reactor BR 2, much less powerful though than the 250 MW American ATR.

6.2.2. TRIGA

The TRIGA family counts as many, or maybe even more members than the MTR family. A few of units (TRIGA Mark 3) exceed 10 MW.

They use homogeneous fuel: 20% ^{235}U enriched uranium is dispersed with an 8% concentration in a zirconium hydride moderator. Compared with water, ZrH has the same number of hydrogen nuclei per unit volume and it remains solid at high temperatures. This mixture, obtained by hydriding a (U, Zr) alloy, is clad in aluminium or zirconium to constitute rods of approximately 4 cm in diameter. Packed together, those rods constitute the reactor core inside a pool whose water cools down the core by natural convection, for the smaller units (Figure 6.3).

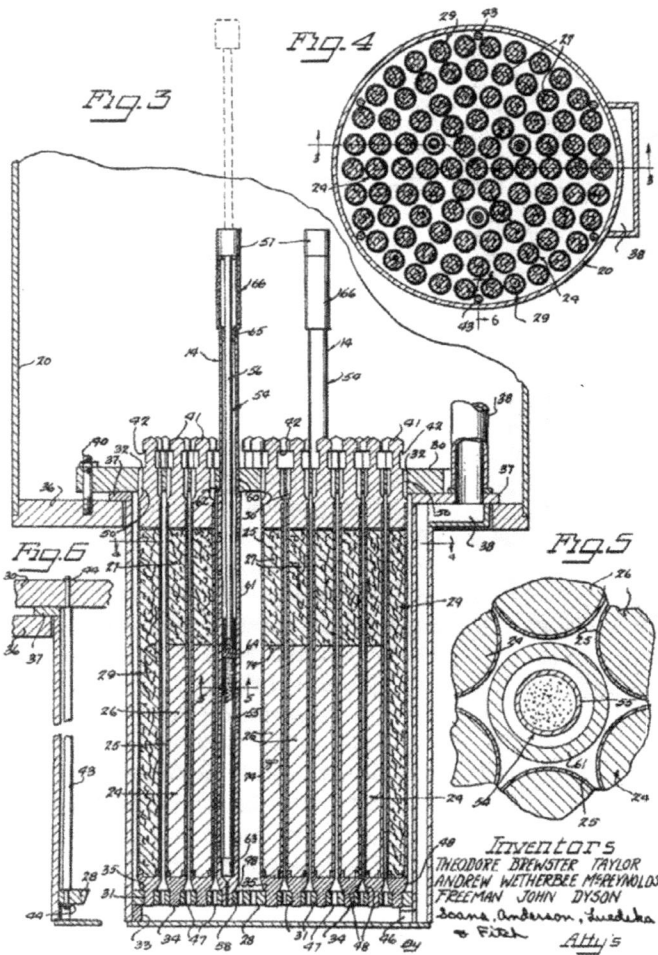

Figure 6.3. The original TRIGA patent.

This special fuel exhibits a very large negative temperature coefficient: in case of a power excursion, it shuts itself down instantly. This ultra-safe design makes this reactor well suited for teaching.

6.3. MTR Fuel, RERTR Programme

The plates of a conventional MTR subassembly are formed by co-laminating a sandwich in which the "meat" of uranium/aluminium alloy is inserted between two aluminium sheets. This alloy contains 40% uranium with a ^{235}U enrichment ranging between 20 and 93%. The plates are typically 1.5 mm thick (three times 0.5 mm), separated by a 3 mm water channel. A standard subassembly comprises 10 to 20 plates inserted in "comb" side plates.

In the middle of the 1970's, when the Indian nuclear explosion triggered a reinforcement of the non-proliferation policy, the USA (which used to be the sole Western suppliers of highly enriched uranium or HEU) declared that they would soon stop exporting uranium enriched above 20% in ^{235}U. Subsequently, they launched the RERTR Programme to study and promote conversion solutions for research reactors to use less enriched fuel without losing too much performance (in terms of neutron fluxes).

The flux performance being roughly proportional to the density of ^{235}U nuclei in the fuel, it was necessary to compensate for the lower enrichment by increasing the uranium concentration in the meat, which thus shifted from U Al to 10% enriched UO_2 "caramels", and then to U_3Si_2 silicide with 20% enriched uranium. R&D is still ongoing to qualify even denser fuel (Figure 6.4).

Only US reactors have been exempted from this conversion, with a temporary dispensation for a very limited number of neutron sources (see below).

6.4. Neutron source reactors

With their short wavelength close to the inter-atomic distance and their low absorption by heavy nuclei, neutrons are very complementary to hard X photons as probes for the exploration of the structure of solid matter. Specialized reactors can produce high-intensity (the equivalent of luminance with photons), well-collimated neutron beams which can be carried by neutron guides over dozens of meters to the experimental areas housing spectrometers, scattering devices, time-of-flight measurements, etc.

Varying the neutrons' velocity and, therefore, their wavelength, can be achieved by bringing them to thermal equilibrium with hot (graphite heated to 1000 °C or more) or cold (liquid deuterium) sources before they enter the guides.

To produce neutron fluxes equal to or greater than 10^{15} n.cm^{-2}.s^{-1}, these neutron sources usually have a small and very compact core fuelled by HEU and moderated by heavy water. The core is located inside a heavy water tank. This tank is penetrated by channels, whose "nose" is as close as possible to the core or the hot/cold sources. Those channels are connected to the previously mentioned neutron guides and can be either radial or tangential to the core.

Today, the highest-performance neutron source is the *Réacteur à Haut Flux* (RHF), belonging to the Lauë-Langevin Institute in Grenoble, close to the European synchrotron light

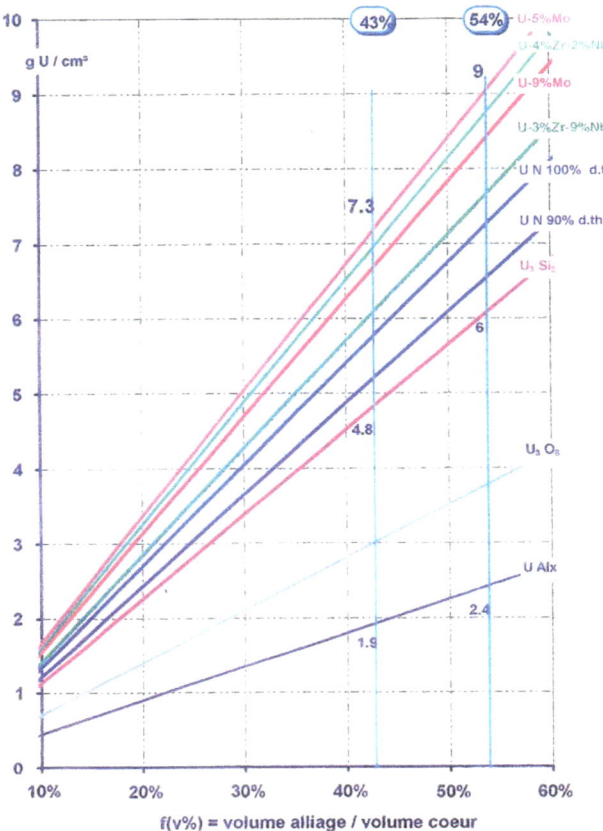

Figure 6.4. Uranium concentration in the "meat".

source ESRF. This geographic proximity is highly symbolic of the complementarity between neutrons and photons in the study of living or inanimate matter. Thoroughly renovated in the 1990's, the RHF is both cooled and moderated by slightly pressurized heavy water and its core is composed of a single fuel element. In this fuel element, U Al plates are curved following an anti-clothoid curve and set between two concentric cylinders (see Figure 6.5). The control rod can move inside the inner cylinder.

In Saclay, the smaller reactor Orphée, used for preliminary tests for many RHF experiments, is also used for neutron activation. FRM2, put into operation in Munich in 2004, rates 20 MWt and supplies a neutron flux of 8×10^{14} n.cm^{-2}.s^{-1} (Figure 6.6). The fact that it is fuelled with HEU raised a severe controversy between the USA and Germany.

6.5. Spallation sources

Spallation is the reaction in which a high energy particle, proton or deuteron, smashes a heavy nucleus, thereby releasing many high energy neutrons and a number of fragments

6 – Experimental reactors

Table 6.2. Two typical neutron sources.

	ORPHÉE	RHF
Thermal Power MW$_{th}$	14	57
Fuel Element(s)	8 MTR U/Al	1 Special U/Al
Enrichment U	93%	93%
Moderator	H$_2$O	D$_2$O
Reflectors	D$_2$O, Beryllium	D$_2$O
Coolant	H$_2$O	D$_2$O
Thermal Flux (n.cm^{-2}.s^{-1})	3×10^{14}	1.5×10^{15}
Inlet/Outlet Temperatures	35/49 °C	39/47 °C
Inlet/Outlet Pressures (bars)	4/2	14/3.2
Water Velocity (m/s)	7.5	15.5
Max. Heat Flux (W/cm^2)	172	500

Figure 6.5. RHF's single fuel element. © Areva.

not yet well identified. A 1 GeV proton colliding with a heavy nucleus (Hg, Pb, Au, W, etc.) can thus generate around thirty spallation neutrons whose energy can reach several hundreds of MeV!

As accelerators are becoming increasingly powerful, the performances of spallation sources improve and they become potential competitors of experimental reactors (Table 3).

Pulsed sources may offer advantages for physics measurements (time-of-flight) but they have no technological usefulness. Continuous sources still have relatively low fluxes, but those which would be needed for efficient ADS[1] would become quite competitive (for example, see the Belgian plan to replace their BR2 reactor by the Myrrha ADS below).

---

[1] Accelerator Driven Systems, or "hybrid" reactors.

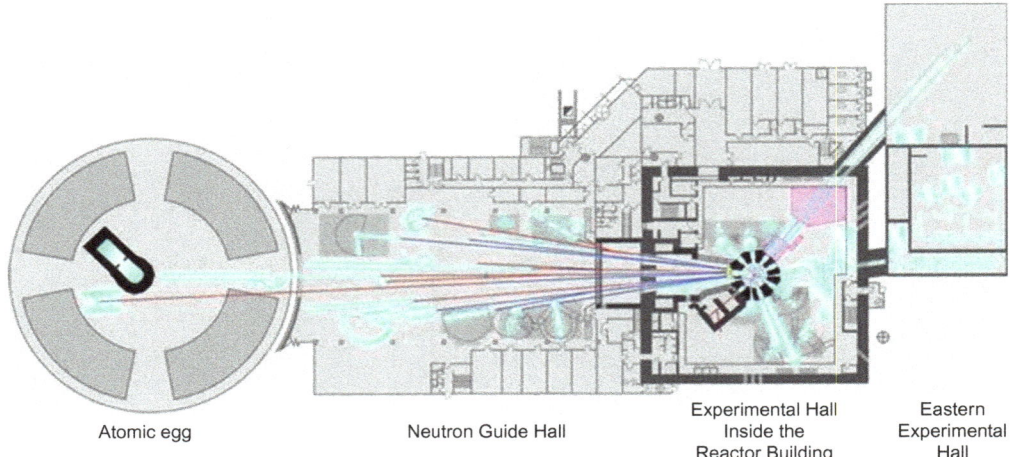

| | Atomic egg | Neutron Guide Hall | Experimental Hall Inside the Reactor Building | Eastern Experimental Hall |

Figure 6.6. Layout of FRM2 and its experimental halls. © FRM 2/Technishe Universität München.

Table 6.3. Main spallation sources.

	SINQ	ISIS	MLNSC
Country	Switzerland	UK	USA
Accelerator Type	Cyclotron	Synchrotron	LINAC
Energy (GeV)	0.57	0.8	0.8
Frequency (Hz)	Continuous	50	20
Power (MW)	0.6	6.16	0.08
Target	Zircaloy	Tantalum	Tungsten
Peak Th Flux	3.2×10^{13}	2.3×10^{15}	2.3×10^{15}
Time-average	3.2×10^{13}	2×10^{12}	1×10^{12}
(with Pb/Bi target)	(8.5×10^{13})		

The advantages of reactors are the volume available for high-fluence irradiation and the flux stability. Spallation sources are simpler Basic Nuclear Facilities (INB) without the problems of fuel supply and management, but their availability is still low and access is limited by the very thick shielding needed against 100 MeV neutrons. Furthermore, some threshold reactions can blur activation measurements.

6.6. Materials irradiation facilities in Europe, the JHR project

After reaching a peak in 1970, the number of research reactors operating in the OECD has been steadily decreasing. Irradiation facilities exhibit the same tendency but the decrease in their number remains less steep than that of their main customers, nuclear R&D programs.

In Europe, the situation is rather peculiar: there is a plethora of irradiation facilities, which led the CEA to close Siloé in 1997, but all these facilities are ageing. OSIRIS, the third youngest of the list, turned 48 in 2014! (Table 6.4).

Table 6.4. European Research Reactors.

Country	Reactor	Power (MW)	Initial Operation
Germany	BER-2	5–10	1974
Belgium	BR 2	100	1961
Norway	HBWR	25	1961
Netherlands	HFR	45	1960
Czech Republic	LVR-15	10	1057
France	OSIRIS	70	1966
Hungary	Budapest RR	10–20	1959
Romania	Pitesti	14	1980

In order to replace OSIRIS, and most of the others as well, the CEA is currently building the Jules Horowitz Reactor (JHR) in Cadarache, where it should operate for the first half of this century. It has already been mentioned that SCK*CEN in Mol (Belgium) are studying the replacement of BR-2 by the Myrrha ADS.

The JHR, a multi-purpose facility, should be twice as performing as OSIRIS and would be mainly dedicated to R&D in support of European nuclear power programmes:

- Support of existing plants in operation (EdF and other electricity producers);
- Support of the replacement fleet of ALWRs;
- R&D on Generation 4 systems (HTR, GFR, SFR, Thorium cycle, etc.).

The main characteristics of the JHR are summarized in Table 6.5.

To reach a fast neutron flux of 5×10^{14} in the core and a thermal neutron flux of 8×10^{14} in the reflector, this pool-type reactor must be slightly pressurized. Its 20% enriched uranium fuel will have subassemblies with concentric cylindrical plates to withstand a water flow running faster than with standard MTR bundles (Figure 6.8). Irradiation positions will be located in the center of the subassemblies and in the beryllium reflector. Mobile devices will allow simulations of power ramping, and a loop in the center of the core will house cadmium and uranium screens to adjust the neutron spectrum (Figure 6.7).

6.7. Myrrha, Pallas

MYRRHA, **M**ulti-purpose h**y**brid **r**esearch **r**eactor for **h**igh-tech **a**pplications, is a multi-purpose irradiation facility designed to replace the ageing BR2 reactor, MTR, which has been in operation since 1962.

Table 6.5. Technical characteristics of the JHR.

Power	100 MW
Type	Pool, Pressurized Primary Circuit (1.5 MPa max)
Active Height	600 mm
Fuel	UMo_7, 8 gU/cm^3
Reflector	Beryllium and Water
Power Density	600 kW/l
Neutron Fluxes	
In Core	6.5×10^{14} n.cm^{-2}.s^{-1} (E > 0.907 MeV)
In Reflector	7.0×10^{14} n.cm^{-2}.s^{-1} (E < 0.635 eV)
Coolant Flow	Upwards

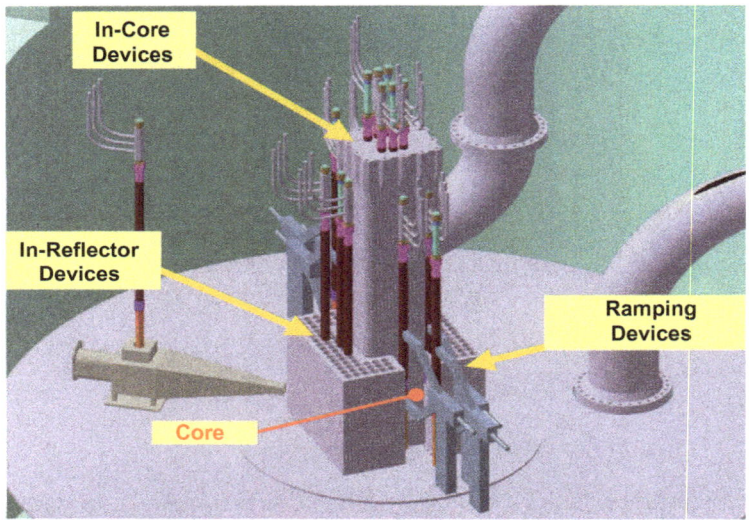

Figure 6.7. Inside the pool. © Cea.

MYRRHA, a flexible fast spectrum research reactor (50–100 MW$_{th}$) was conceived as an accelerator driven system (ADS), able to operate in sub-critical and critical modes. It contains a 600 MeV proton accelerator, a spallation target and a multiplying core with MOX fuel, cooled by liquid lead-bismuth (Pb-Bi), as shown in Figure 6.9.

MYRRHA will be operational at full power around 2023. Construction of the facility and assembly of the components is foreseen in the 2015–2019 period. The total investment cost is currently estimated at 960 M€ (€ 2009).

To replace the ageing HFR, the Netherlands (NRG) promote a project called Pallas, a "tank-in-pool" reactor type in the 30 to 80 MW power range. Pallas would combine

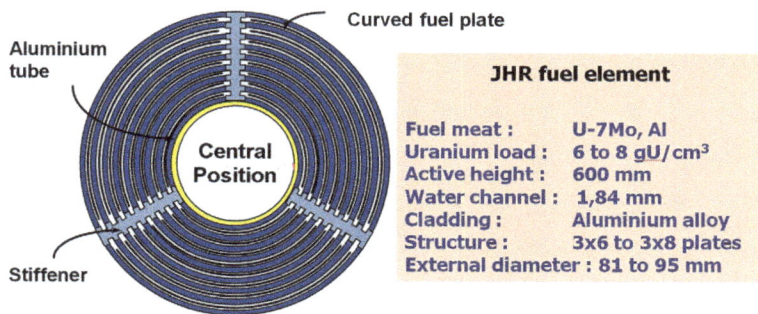

Figure 6.8. JHR fuel element. © Cea.

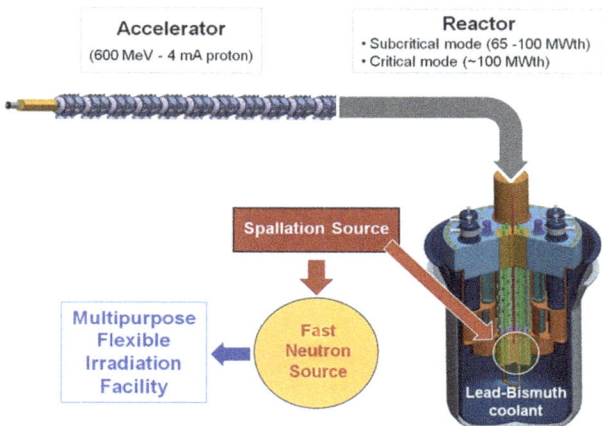

Figure 6.9. Myrrha diagram. ©SCK.CEN.

isotope production requirements (^{99}Mo) with nuclear research capabilities in the reflector zone and reactor core. It would use LEU fuel: uranium silicide or U/Mo.

Complete funding would allow the construction of Pallas to begin in 2016.

References

[1] F. Merchie, *Réacteurs de recherche et d'essais de matériaux*. Techniques de l'Ingénieur B3030, 1996.
[2] *Nuclear Research Reactors in the World*. IAEA Reference Data Series n°3, December 1995.
[3] *Research Facilities for the Future of Nuclear Energy*. Proceedings of ENS Conference, Brussels, June 1996.

7 Advanced "Generation III" reactors

B. Barré

7.1. Introduction: Genesis of "Generation III"

The accident which occurred on March 28, 1979 in Unit 2 of Three Mile Island, a Babcock & Wilcox PWR, had a deep and long lasting influence in the USA. Even though there was no impact on public health, many US citizens are still, more than 30 years later, haunted by the panic it triggered over a period of a few weeks. In retrospect, it was a close call: if the corium[1] had actually leaked through the pressure vessel, the accident could have been much more severe.

This accident did not reveal any basic flaw in the PWR system (it did in fact prove the sturdiness of its design) but it underlined the crucial need for decay heat removal, the failure of which is the main contributor to the total risk of core meltdown[2]. TMI 2 also demonstrated how important it is that plant operators have access to clear and unambiguous information in order to understand the status of the main reactor systems, without being overwhelmed by redundant signals. Awareness of the "human factors", data display in the control room, alarm hierarchy, *conduite par états*: all these concepts were born from TMI. Lessons learned were implemented on existing reactors and incorporated into new designs such as the N4. TMI was also the trigger for design simplification and "passive safety" which both characterize the "revolutionary" concepts which will be described below.

The April 1986 accident in Chernobyl Unit 4, whose severity was incommensurable, brought into question the very design of the RBMKs which was intrinsically unstable under certain operating conditions. It vividly underlined the vital role of ultimate barrier played by the containment building, especially in the early days after a core meltdown, when short-lived radionuclides have not yet decayed. All post-Chernobyl designs, such as the EPR, include in their design specifications the prevention of containment failure or early by-pass. The total "deterministic" exclusion of reactivity accidents was also the main motivation behind Accelerator Driven Systems (ADS) described in another chapter.

---

[1] Lava-like mixture of various molten metals and oxides resulting from the reactor core meltdown.
[2] More recently, the Fukushima accident reinforced this lesson.

7.2. Evolutionary or Revolutionary?

"Generation III"[3] Light Water Reactor designs consist of three Boiling Water reactors, GE-Hitachi's ABWR, General Electric's ESBWR and AREVA's KERENA, and six Pressurized Water reactors: AREVA's EPR, Westinghouse's AP 1000 (Toshiba), Rosatom's AES-92, Mitsubishi Heavy Industries' (MHI) APWR, AREVA-MHI's ATMEA and the APR 1400, South Korea's version of CE's System 80+.

All these reactors use "conventional" PWR or BWR fuel assemblies with their latest improvements.

These concepts belong to two distinct families:
    Considering the *return of experience* to be paramount for safety, the designers of "evolutionary" reactors intend to draw maximum benefit from the cumulative operating experience of past and present reactors of a similar type. Innovations are therefore evaluated by weighing the improvements that they would bring to the system against the loss of continuity which would ensue. The EPR, System 80+ and, to a lesser extent, ABWR follow such an approach.
    Present models have more or less been the result of an evolutionary process in which every significant safety issue that arose was solved by adding a new system or a new component. As a result of this "onion peel" process, present reactors are complex and sophisticated. "Revolutionary" reactor designers intend to go back to the drawing board and design "from scratch" a PWR or BWR meeting present day safety standards and requirements, integrating recent trends such as *passive* safety systems, while taking advantage of modern technologies. Of course, these breakthrough designs cannot benefit from the return of experience to the same degree as their evolutionary competitors. The AP 1000 and ESBWR are typical products of this approach (which was undoubtedly favoured in the USA by the very lengthy lack of any domestic order).
    It is important to underline that the controversy does not lie in the level of safety to be reached by Generation III LWRs, but only in the best way to reach this level, to guarantee it, and to demonstrate it both to the Safety authorities and the public at large.

7.3. EPR, the Evolutionary Power Reactor [1–6]

7.3.1. Genesis of the EPR

The EPR Model (Evolutionary Power Reactor) is the result of three parallel approaches.
    The first approach was of an industrial nature: it stemmed from the willingness to merge into a common product the French and German PWR designs, both derived from US models but subsequently "Frenchified" or "Germanized". Well within a general trend of intra-European (and especially Franco-German) consolidation of hi-tech industries, such a convergence was necessary to meet the nuclear reactor market's demands in the context of the late 80's: saturation in Europe, stagnation in the United States, limited growth in Asia, with

---

[3] Vendors use terms like Generation II+, III, III+ as marketing gadgets. The author considers as Generation III only those designs including a full core meltdown as Design Basis Accident.

too many vendors and too many reactor types. Framatome and Siemens, while retaining their *de facto* monopoly on their respective domestic markets, therefore established in 1989 a joint subsidiary, Nuclear Power International (NPI), in order to design and commercialize a *common product*, strictly for export.

This convergence progressively dampened the once raging controversy about the different safety approaches practiced on each side of the Rhine River, each having its internal consistency. The net result of this controversy, fuelled by commercial motivations, was to trouble the public and cause anti-nuclear organizations to rejoice.

In order to avoid a simple overlap of systems and procedures from both countries, which would have resulted in unreasonable increases in capital costs without any real benefit to the overall safety, the convergence of both reactor designs required the parallel convergence of the safety approaches in both countries.

Therefore the second move came from the Safety Authorities, based on lessons learned from the aftermath of the Chernobyl accident. Very schematically, the consequences of Chernobyl can be summarized as: fewer casualties than anticipated, but contamination spreading far beyond expectations and a quasi-universal diffidence toward nuclear power. The reaction of the Safety Authorities materialized in France as a 1991 letter from DSIN[4] to Électricité de France, and in Germany as an amendment to the Nuclear Law. But behind those purely national reactions, broad and deep discussions were started in order to reach, at last, a French-German consensus on Safety approaches, a consensus likely to spread across all of Europe. This process, thoroughly described in [7], involved the regulatory authorities themselves, DSIN and BMU, their expert advisory groups GPR and RSK as well as their technical support organizations IPSN and GRS, working in perfect unison.

Thus, in June 1993, the following three common objectives were defined:

1. reduce even further the probability of core meltdown,

2. "practically eliminate" accidents which could lead to an early and massive release of radioactivity,

3. in the case of a core meltdown (at low pressure), guarantee *per design* that the maximum radioactivity release would only require protection measures limited both in time and in space.

The first two objectives are in continuity with the previous safety philosophy. In contrast, the third objective (which could be summarized as: no permanent evacuation of people, temporary evacuation limited to the close neighborhood of the plant, and no lengthy lockout of agricultural areas) was a breakthrough, as it introduced full core meltdown into the design basis.

The last move was initiated by European electricity producers. Previously, in the mid-80's, US utilities had written down in the URD (Utilities Requirements Documents) the conditions to be met by any new reactor design if it were to revive their interest in nuclear technology. Following this example, in December 1991, five European companies – to be followed by many more – launched the project which led to the EUR (European Utilities Requirements) [8].

---

[4]DSIN later became ASN, Autorité de Sûreté Nucléaire, and IPSN became IRSN, Institut de Radioprotection et de Sûreté Nucléaire.

The EUR formulated common requirements to be met by light water reactors to in any future European orders – later dubbed "Generation III LWRs", be they PWRs or BWRs, evolutionary or largely passive. These requirements did not only address safety, but also included absolute or relative specifications for costs, margins, availability, lifetime, operating flexibility and recycling capability. Here are some of those requirements:

- total core damage frequency $< 10^{-5}$ *per annum*, and

- cumulative frequency of accidents leading to significant radioactivity releases $< 10^{-6}$ *per annum*,

- no evacuation beyond 800 meters from the reactor building,

- guaranteed lifetime of 60 years for all non-replaceable components (pressure vessel, containment, civil works),

- capability to load 50% of the core with MOX subassemblies,

- 12 to 24 month refuelling cycles, with possible extension by 60 equivalent full power days,

- economic competitiveness with the most modern large coal plants, for an operating load of 4500 to 5500 equivalent full power hours *per annum*.

The peak of the convergence was the transformation of the NPI *common product* into the EPR *European Pressurized Water Reactor* by merging Framatome's and Siemens' design activities with parallel efforts led by EdF and some German utilities. The French-German cooperation expanded to include R&D, mostly shared between CEA and Forschungszentrum Karlsruhe FZK. Later on, when called to compete outside the European market, the EPR was renamed **Evolutionary Power Reactor**.

7.3.2. EPR General Characteristics

Table 7.1 compares the EPR to its "parents" N4 and Konvoi.

Roughly speaking, EPR is a combination of both designs, where the most conservative option in terms of margins was systematically selected. It fits both the 1993 Safety Directives and the EUR: the design is "evolutionary" in that it maximises the benefit drawn from the return of experience of the French and German nuclear fleets, with the safety level reinforced against both internal incidents and external aggressions, and *ad hoc* devices and systems limiting releases in the unlikely event of a core meltdown.

7.3.3. Primary and secondary circuits

The general layout of the primary and secondary circuits closely resembles the N4's, with a few significant differences (Figure 7.1):

- larger volumes of primary and secondary water to dampen the transients,

Table 7.1.

Characteristics	EPR	N4	Konvoi
Thermal Power (MWth)	4250–4500	4250	3850
Net Electric Power (MWe)	1500–1600	1475	1360
Primary Circuit			
Number of Loops	4	4	4
Operating Pressure (MPa)	15.5	15.5	15.8
RPV inlet/outlet Temperature (°C)	291.3/326.3	292.1/329.1	291.3/326.1
Mass Flow (kg/s)	21 900	20 200	19 900
Pressurizer Volume (m^3)	75	60	65
SG Exchange Surface (m^2)	4×7300	4×7300	4×5400
Secondary Circuit			
Steam pressure at full power (MPa)	7.8	7.31	6.45
Steam mass flow (kg/s)	2400	2400	2050
Secondary water volume (m^3)	75	62	46
Core			
Subassembly type	17×17	17×17	18×18
Number of subassemblies	241	205	193
Active height (m)	4.20	4.27	3.90
Linear rating (W/cm)	155	179	163
"In-core" instrumentation penetration	top	bottom	top
Containment			
Type	Double cylinder concrete + liner/concrete	Double cylinder concrete/concrete	Double sphere metal/concrete
Internal Volume (m^3)	90 000	73 000	70 000
Pressure (bar)	6,5	5,3	6,3
Leakage Rate (%/day)	1	1	0,25
Pressure Peak Control (severe accident)	Aspersion, 2 trains	Filtered Release	Filtered Release

- larger RPV. Heavy neutron reflector (baffle) to reduce neutron fluence on the vessel (which, as a side effect, improves fuel economy). Suppression of RPV bottom penetrations,
- shutdown heat removal through the low pressure injection system.

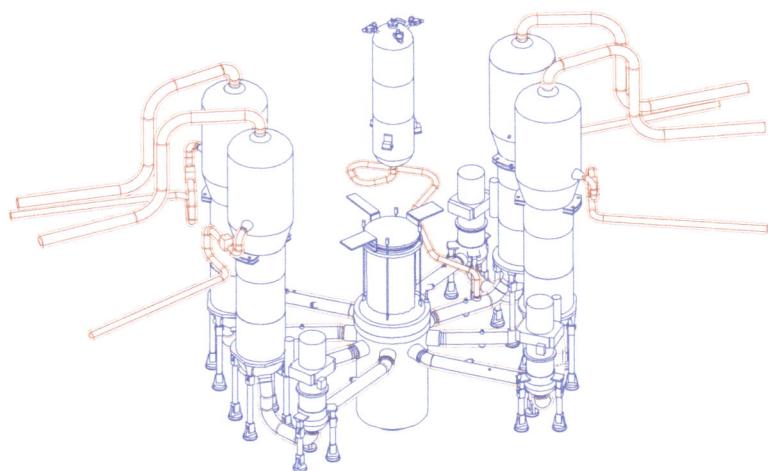

Figure 7.1. EPR primary and secondary circuits.

7.3.4. Systems architecture

The main safety systems, safety injection, SG emergency water feed and electrical distribution are designed and built along a 4-train architecture (4 times 100%). Such very high redundancy facilitates maintenance, at full power or during shutdown states, thereby improving availability through shorter periodic outages.

Each safety train is set up in a strictly separate building (Figure 7.2). Furthermore, two of these 4 buildings are "bunkerized" to strengthen them against external aggressions. Common mode failure is therefore excluded by design.

The EPR containment is of French design, with a double concrete hull (Figure 7.3). The internal hull is made of pre-stressed concrete with a steel inner liner. The external hull is made of reinforced concrete, much thicker than the N4's in order to withstand much more brutal aggressions (airplane crash, explosion, etc.). This design was hotly debated between the French and German teams. Last but not least, a single base mat allows the EPR to resist seismic accelerations of 0.3 g (0.15 g for N4).

7.3.5. Mitigation of severe accidents

The general design of the EPR already decreases total core meltdown frequency by a factor 10 as compared to plants presently in operation. In addition – and this is its major innovation – the EPR is equipped with special devices and systems to limit the consequences of such an accident if it should happen (Figure 7.4):

- a depressurization system to prevent pressurized meltdown,

- a "core-catcher" to spread the molten corium, if it were to escape the RPV, and cool it down through passive flooding,

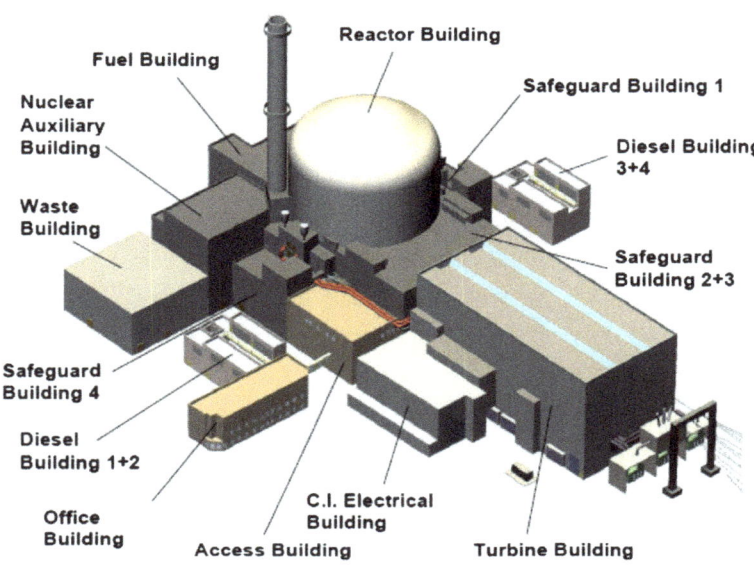

Figure 7.2. EPR general layout.

- implementation of some 40 catalytic recombiners inside the containment, to guarantee that in no location can hydrogen reach a potentially explosive concentration,

- an aspersion system dedicated to pressure control and cooling inside the containment. This system needs to be activated within 12 hours of the passive cooling on the molten corium.

In the event that AC power from the grid is lost and the switch to house load operation (i.e. to operate, the plant uses the electricity it produces) fails, 4 emergency diesel generators (housed in 2 separate concrete buildings together with their fuel tanks) supply 4 × 100% of the power needed to maintain the plant in safe shutdown. If a common mode failure were to affect all four generators, 2 additional diesel generators relying on a different technology, called "station blackout diesel generators", would still act as ultimate power sources.

All these systems allow the EPR to comply with the common safety criteria mentioned above. R&D carried out in France mostly focused on the mitigation of severe accident consequences, notably corium behaviour (Vulcano, Corinne) and hydrogen (Kali, Mistra).

7.3.6. *Future economics of the EPR*

Broader margins, more concrete, additional devices and systems: one could conclude that the EPR should be more expensive than the N4, even though, during its design phase, the competition with combined cycle gas turbines became increasingly fierce (and natural gas was predicted to remain durably inexpensive).

As a matter of fact, the overnight investment of EPR is indeed higher. But other factors influence the total *busbar cost*:

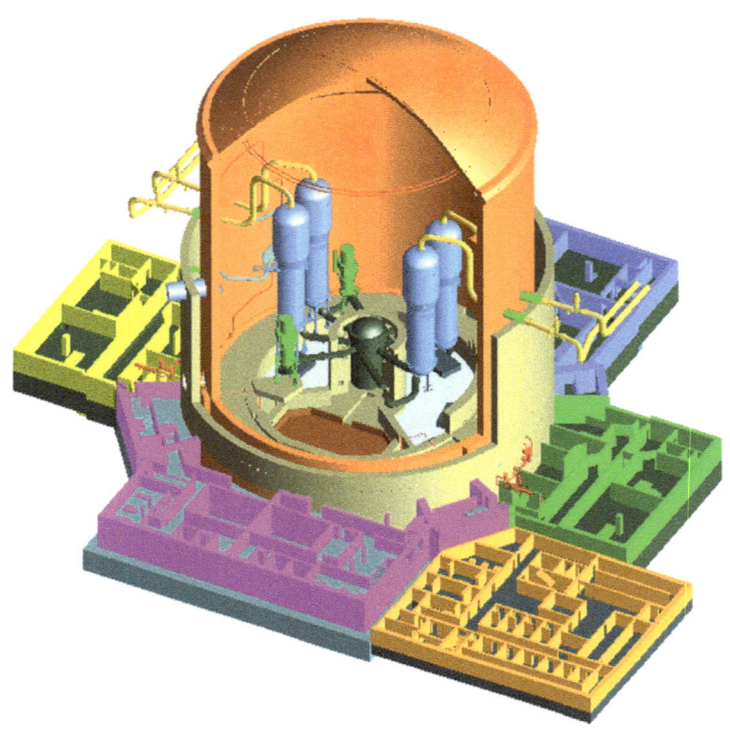

Figure 7.3. EPR nuclear island.

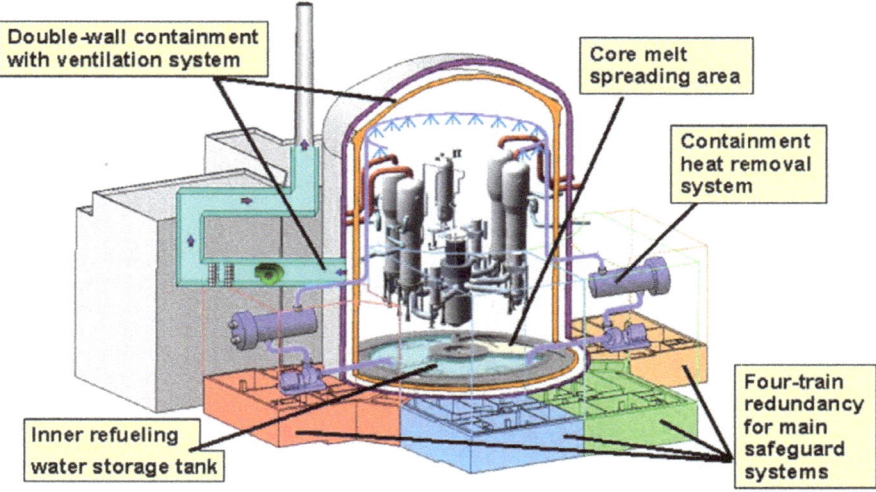

Figure 7.4. EPR Safety.

- longer technical lifetime of the plant, extended from 40 to 60 years;

- increased availability factor (at least 90%);

- longer power cycle up to 24 months;

- better neutron economy (heavy baffle, lower volumetric power).

All in all, the total MWh produced by the "nth" EPR, n>5, should cost around 70 euros, in 2010 currency. By comparison, both combined cycle gas turbines and fluidized bed coal plants would, in France, generate electricity at a cost of 70 to 100 euros/MWh[5].

7.3.7. EPR status in 2014

After a competitive bid against the ABWR, the KERENA and a VVR 1000, a 1600 MWe EPR was selected in December 2003 by the Finnish utility TVO. Presently under construction at the Olkiluoto site, it should be operational by the end of 2017.

In France, the first unit of an expected series of EPR's has been under construction since 2007 on the Flamanville site, where two 1300 MWe PWR units are already in operation. A procedure to authorize the construction of another EPR as unit 3 of the Penly site has been suspended. Two EPR units are under construction in Taishan, China. The EPR is also under review for certification by the US NRC, ready for construction in Great Britain and under consideration by India.

Both Olkiluoto and Flamanville have experienced very significant delays and overcosts, which is not unusual for prototypes in any heavy industry, but the Chinese reactors have benefited from the return of experience from both prototypes and might even be operational before them.

7.4. The Korean APR 1400

Before being bought by BNFL and merged with Westinghouse (before Westinghouse was sold to Toshiba), ABB/Combustion Engineering had developed the "System 80+" [9–10], an evolutionary reactor designed to be closer to the N4 than to the EPR. Its qualification as Generation III is questionable. But while the N4 NSSS was an economic "optimization" of the previous P4-P'4, reducing some margins to increase the output, System 80+ was designed with larger margins than its predecessor, the Palo Verde plant of Arizona Power & Light (all 3 units of which constitute the only "series" in the USA). This technology was transferred in its entirety to South Korea and, with some improvements, is now commercialized as the APR 1400.

The development of System 80+ started in 1985 by a joint design team from ABB-CENP and Duke Power Company, along with Stone & Webster and the Korea Atomic Energy Research Institute, to fit the previously mentioned URD [11]. Submitted to the NRC in 1991, System 80+ was the first to receive, in July 1994, the *Final Design Approval*.

---

[5] Figures supplied by EDF for the public enquiry held for the Penly 3 Project during the summer of 2010.

7.4.1. S 80+ basic options

System 80+ complies with the four basic URD requirements: safer, simpler, more reliable and easier to operate. It comprises several devices to limit the consequences of a core meltdown accident (Figure 7.5):

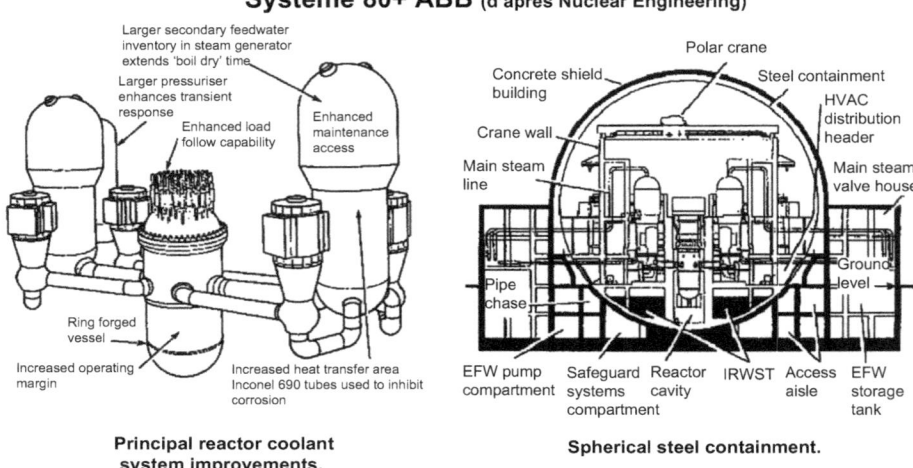

Figure 7.5. System 80+.

- large spherical steel containment vessel;

- gravity flooding of the reactor pit, from an in-reactor water storage tank (IRWST);

- primary circuit depressurization system (followed by *feed and bleed* cooling);

- limitation of the hydrogen concentration inside the containment through natural diffusion (no inside partitioning) and the presence of igniters.

Probabilistic assessments estimate core meltdown frequency to be decreased 100-fold compared to previous plants.

7.4.2. General characteristics

Table 7.2 compares System 80+ with the EPR.
 The emergency injection system comprises 4 trains, as for the EPR, but with only $4 \times 50\%$ capacity.

Table 7.2.

Characteristics	System 80+	EPR
Plant		
Thermal Power (MWt)	3800	4250
Net Electric Power (MWe)	1300	1500
Core		
Active Height (m)	3.81	4.20
Number of Fuel Assemblies	241	241
Type	16×16	17×17
Linear Rating (W/cm)	177	154
Number of Control Mechanisms	93 (boron-less operation)	81
Primary Circuit		
Number of Loops	2	4
Operating pressure (bars)	158	155
Vessel Inlet Temperature (°C)	292.2	291.3
Vessel Outlet Temperature (°C)	323.9	326.3
Mass Flow (kg/s)	20 890	21 050
Pressurizer Volume (m^3)	68	75
SG Exchange Surface (m^2)	2×13 605	4×7 308
Inner Containment	Spherical, Steel	Cylindrical, Concrete/liner
Volume (m^3)	96 000	90 000
Outer Containment	Cylindrical, Concrete	Cylindrical, Concrete

7.4.3. Primary circuit

As mentioned above, System 80+ was designed to increase margins: large pressurizer (+33%), SG tubes made of Inconel 690, large SG exchange area (+17%), feedwater inventory larger by 25% and direct in-vessel injection, which should prevent any core uncovery.

The ring-forged vessel should last 60 years and the reactor can operate in load following mode without adjusting borication with its many control rods, some of which are grey rods.

7.4.4. The APR 1400

Based on System 80+, South Korea developed the APR 1400, two units of which are under construction in Korea (Shin-Kori 3 & 4) while 4 units were ordered (against 2 EPRs) by the United Arab Emirates in December 2009. The APR 1400 is a twin unit plant, an option no longer used in Europe, which allows for lower construction costs but entails a significant risk of human error during maintenance.

The single containment offers less protection against external aggressions than does the EPR, and it is not proven for such a large vessel, that molten corium can be effectively

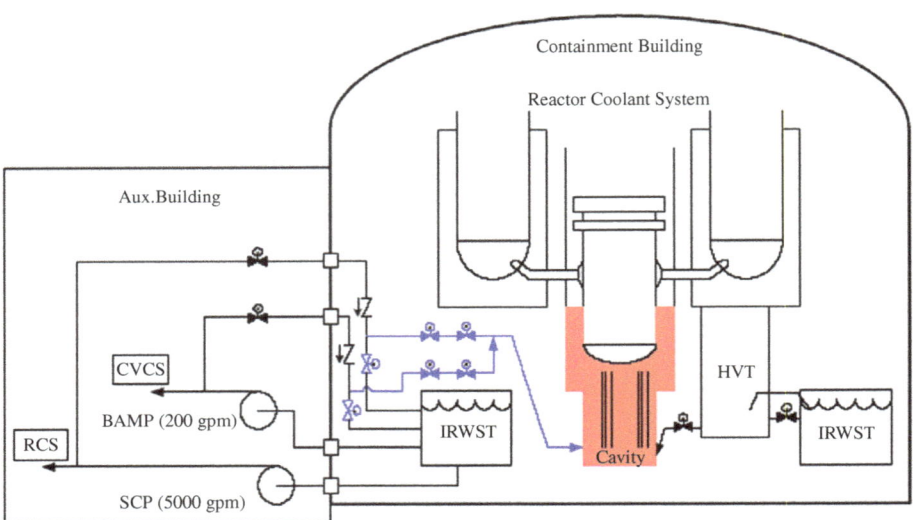

Figure 7.6. AP 1400 RPV flooding.

solidified within the pressure vessel by flooding its pit (Figure 7.6). The limiting factor is the surface/volume ratio as the decay heat is generated in the corium volume while the cooling occurs through the vessel surface.

On the other hand, one Korean improvement is the design of an integrated pressure vessel head (Figure 7.7) which incorporates all control rod mechanisms into a single set. This feature reduces refuelling downtime, as well as the corresponding doses to the operators.

7.5. The AP 600 and AP 1000 by Toshiba-Westinghouse [12–14]

During the 80's, Westinghouse developed two projects in parallel: the APWR, an evolutionary reactor in the 1000–1400 MWe range, and the AP600, a "revolutionary" or "passive" design with lower power. Westinghouse abandonned its APWR, but Mitsubishi designed a similar project to be built at the JAPC Tsuruga site, as will be described later in the chapter.

By contrast, the AP 600 is a really innovative concept, even though its promoter invokes the return of experience from many Westinghouse 2 loop PWRs. The development of the AP 600 started in 1985 on DOE + EPRI funding, with many partners: Westinghouse, Bechtel, Burns & Roe, as well as Ansaldo, the Chinese CNNC and other foreign companies.

The aim was to develop a medium size reactor (600 MWe) according to the URD specifications, using already qualified components, but implementing "passive" systems taking advantage of phenomena such as gravity, natural convection, evaporation and condensation, without calling on any external energy source or operator action.

7 – Advanced "Generation III" reactors

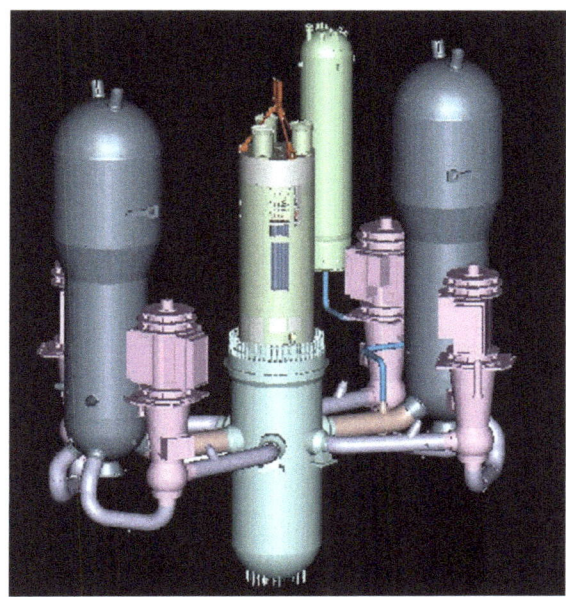

Figure 7.7. APR 1400 primary circuit.

7.5.1. General characteristics

The AP600 was advertised by Westinghouse as an advanced nuclear reactor suitable to be *accepted* in countries where the public refuses today's plants. This is why they underline:

- the simplicity of construction and maintenance;
- the use of "intrinsically safe" passive systems.

They claim, of course, that simplification allows for offsetting the additional costs due to passivity, which usually leads to a reduction in compactness.

Actually, the simplification is significant (Figure 7.8). When compared to a standard 600 MWe PWR, the AP 600 has:

The main emergency systems with passive characteristics are the following:

- gravity-driven emergency cooling water system,
- decay-heat removal through a natural circulation heat exchanger,
- final heat removal by natural air circulation around the metallic containment vessel.

Table 7.3 compares the characteristics of the AP 600 and a conventional Westinghouse 600 MWe PWR.

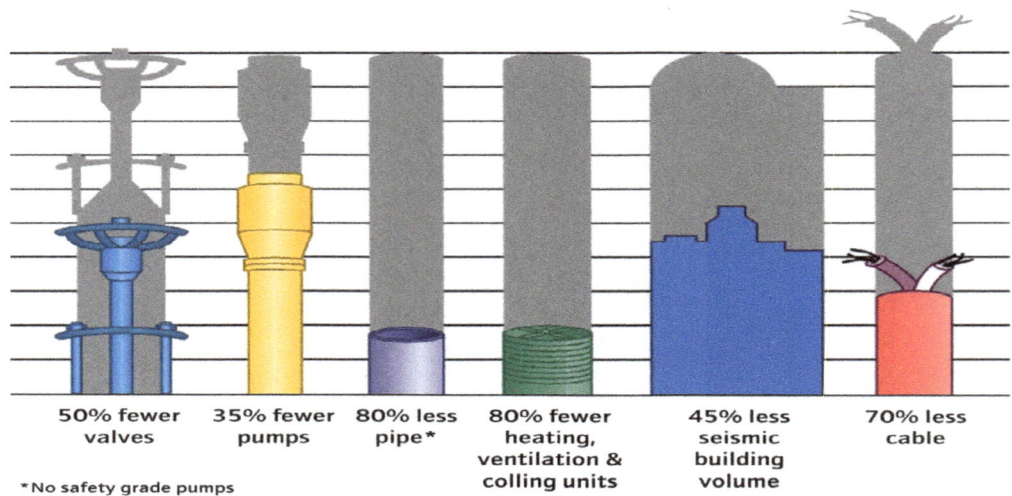

Figure 7.8. AP 600 design simplifications.

Table 7.3.

Characteristics	2 Loop PWR	AP 600
Technical Lifetime (years)	40	60
Thermal Power (MWt)	1876	1933
Net Electrical Power (MWe)	620	610-640
Core Outlet Temperature (°C)	324	316
RPV Inner Diameter (m)	3,17	3.99
End-of-life RPV Fluence (n.cm^{-2})	5×10^{19}	2×10^{19}
Cycle Length (months)	12	24
Core		
Number of Fuel Assemblies	121	145
Type	16 × 16	17 × 17
Active Height (m)	3,66	3,66
Linear Rating (W/cm)	165	126
Number of Black (Grey) Control Rods Grapples	33	45 (16)

7.5.2. Core and primary circuit

The AP600 is a PWR with two primary loops. Each loop comprises a conventional U-tube steam generator, one hot leg and two cold legs connected to the outlet of wet rotor primary pumps located at the bottom of the SG (Figure 7.9). This design has no inlet pump duct and eliminates the risk of leaking pump seals. In addition, there is no pump cooling auxiliary circuit. The previous NP 300 French project had a similar layout.

Reactor Coolant Loop

- **Larger Reactor Vessel**
  - Based on Standard 3 Loop
- **Two SGs / Four RCPs**
  - Existing size RCP design
  - Model Delta 75 SGs
- **Simplified Main Loop**
  - 50% fewer welds
  - 80% fewer supports
- **Larger Pressurizer**
  - 60% increase in volume

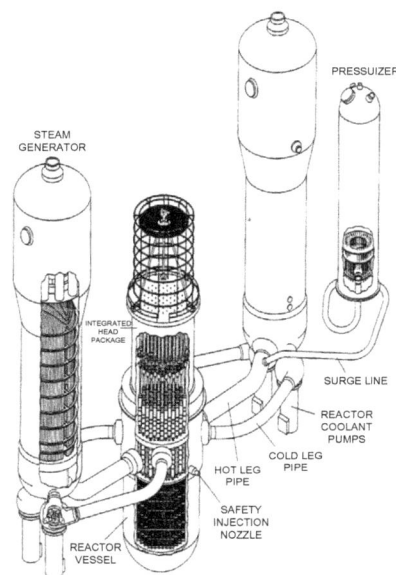

Figure 7.9.

The loop architecture reduces pressure losses, which helps initiate natural circulation in case of unanticipated primary pumps trips. The lowest point of the primary ducts lies above the top of the core.

The pressurizer is quite large so as to be able to dampen transients and supply a large initial water inventory in case of an accident.

The reactor pressure vessel is the same size as Westinghouse's 900 MWe PWR RPV; the *downcomer* is wider and the bottom is free from instrumentation penetrations.

The AP600 has roughly the same core as 3-loop 900 MWe PWRs, its power density being reduced by 27%. Very little soluble boron is used for reactivity adjustment: grey rods are used for load following. The use of Gadolinium oxide as burnable poison allows for long cycles of up to 24 months. A heavy baffle around the core reflects neutrons to reduce RPV embrittlement.

7.5.3. Emergency systems

Entirely passive emergency systems are located inside the containment. There is no penetration of emergency circuitry in the metallic containment vessel, which acts as the cold source after a LOCA.

A gravity driven emergency injection system (Figure 7.10) uses water from a large in-reactor water storage tank (IRWST) located within the containment, above the top of the RPV. A heat exchanger, which operates like a primary loop, is immersed in the IRWST; it

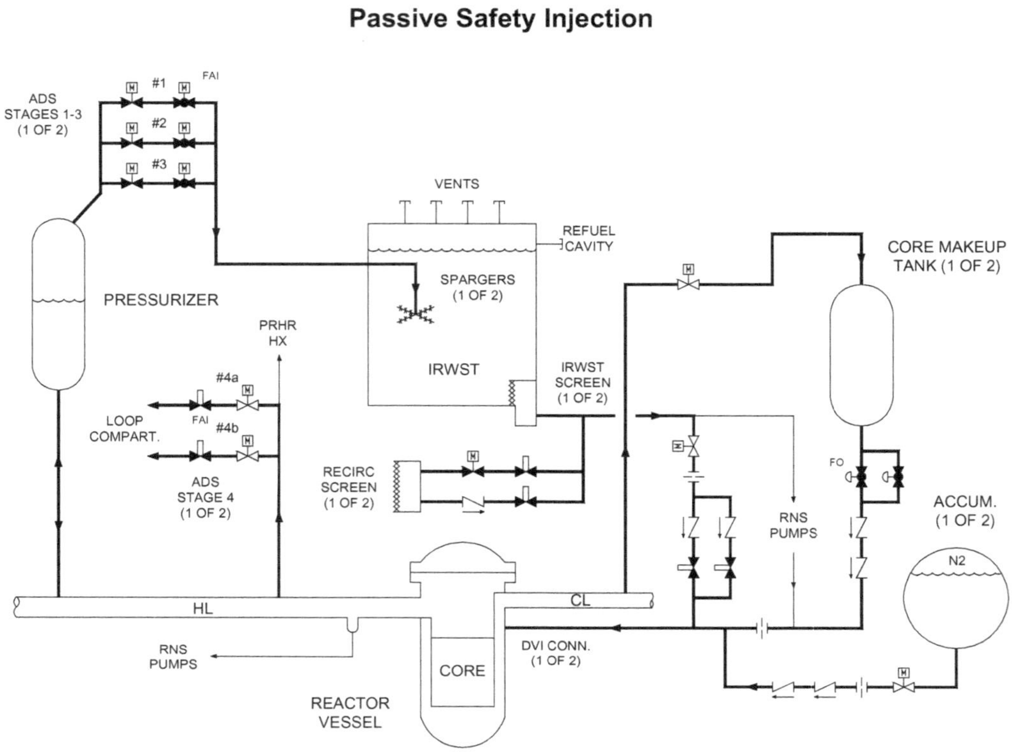

Passive Safety Injection

Figure 7.10.

is triggered by compressed air actuated valves which automatically open in case of loss of electric power. In this heat exchanger, the primary water circulates by thermosiphon effect.

Two boricated water tanks can be used in case of small primary leaks when the normal water supply fails. These tanks are located above the primary loops and their water flows into the downcomer under the force of gravity.

In case of larger breaches, accumulators re-flood the primary circuit as soon as the primary pressure falls below 48 bars. If water is still missing, core pressure is decreased by the automatic discharge system to the level at which water can flow from the IRWST to the core under gravity. This requires around ten hours. When the IRWST is empty, water in the containment vessel covers the primary loops, allowing the cooling of the core by natural convection within the RPV (Figure 7.11).

Cooling of the containment vessel is also passive: steam escaping from the breach condenses on the steel wall. The wall is cooled from the outside through natural air convection in the annular space between the two "containments". The steel containment vessel is in effect surrounded by a thick concrete "wrapping" with a wide chimney. A water tank located at the top of the chimney is used to sprinkle the steel vessel as soon as a given pressure or temperature threshold is reached inside. When this upper tank is empty, the decay heat has decreased to the point where it can be evacuated by natural air flow around the steel containment (hence the "chimney" around the containment vessel,

PCS Operation

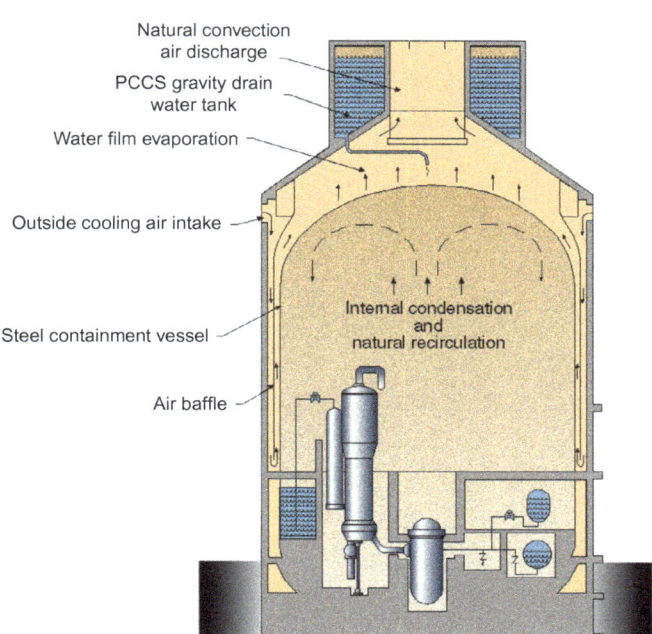

Figure 7.11.

which gives the AP 1000 its characteristic shape but offers poor protection against aircraft crashes).

In case of a full meltdown, the corium would be solidified *in situ* by flooding the pressure vessel pit.

7.5.4. From the AP 600 to the AP 1000

The AP 600 was granted NRC's *Final Design Approval* at the end of 1999, after a 6 year study and with the support of an impressive international R&D programme. The safety analysis of such a passive reactor really was a first [15]. It should be noted that China did continue with its "AC 600" [16].

The AP 600's innovative design is indeed attractive, but its economic competitiveness remains to be proven. Despite Westinghouse's claims that the overnight cost would be lower by 27% and the fuel cost lower by 14% (as compared with a conventional 600 MWe PWR), no customer was interested. No nuclear plant whatsoever was ordered in the USA and, in Europe and Asia, 600 MWe units are considered too small to be competitive.

Westinghouse therefore worked with other partners, notably EdF and JAPC, to design a more powerful passive reactor, in the 900 to 1000 MWe range. This led to the Toshiba-Westinghouse[6] AP 1000 Project: a 2 loop PWR with the W 1300 reactor vessel and the CE System 80+ steam generators (Figure 7.12).

---

[6] At the beginning of 2006, BNFL sold Westinghouse to Toshiba, better known in the BWR market.

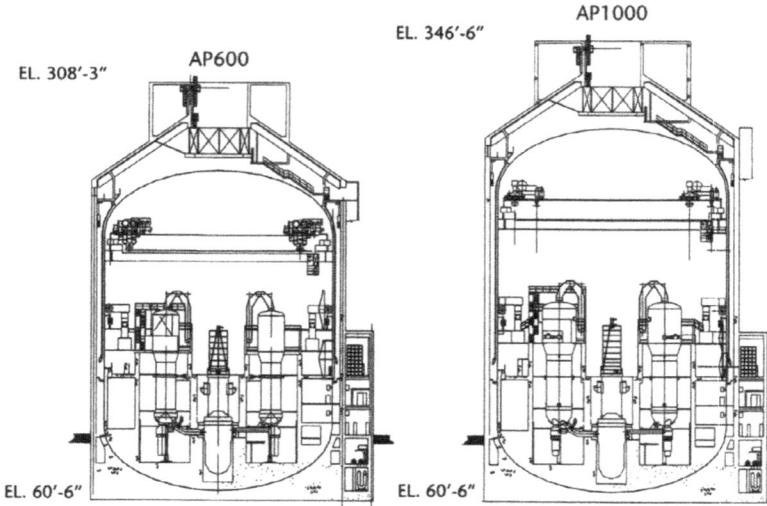

Figure 7.12.

According to a study by Bechtel, the total cost of an AP 1000 would only be 12% higher than the cost of AP 600. But increasing the size without significantly altering the design raised some difficulties in the design of the actual steam generator, for instance. This design was given its final certification by the NRC in 2005, but many amendments had to be submitted, most notably to answer the post 9/11 constraints on anti-airplanes crashes.

Four 1200 MWe AP 1000 units are under construction in China (Haiyang and Sanmem sites) and the construction of four units has begun in Vogtle and Summer in the USA, the first new projects in the USA since 1974. The construction of the AP 1000 is characterized by the "open-top" assembly of big prefabricated modules, which should reduce construction time.

Critics of the AP 1000 underline the following:

- it only offers one containment when most modern plants have two;
- from a seismic point of view, the large tank on top of the chimney is weak;
- the whole design is vulnerable to plane crashes (such concern was already expressed before September 11, 2001).

7.6. Other generation III PWRs

7.6.1. The ATMEA

As many countries do not have an electric grid large enough or offering sufficient interconnection with its neighbours to accommodate a 1600 MWe unit, AREVA and Mitsubishi Heavy Industries are jointly developing the ATMEA, a 3 loop PWR in the 1100 MWe range. Table 7.4 compares the main parameters of the ATMEA and EPR reactors. The top part of the containment building is a single-walled, 1.8m thick, reinforced concrete cupola, but the lower part is double-walled along the full height of the primary circuit inside. The ATMEA has 3 × 100% safeguard trains and a core catcher similar to the EPR's. The ATMEA also uses Mitsubishi advanced accumulators (Figure 7.13).

Table 7.4.

	EPR (OL3)	ATMEA 1
Core thermal power (MWth)	4590	2860–3150
Electrical output (MWe)	1600	1000–1150
Number of fuel assemblies	241	157
Type of fuel assembly	17 × 17	17 × 17
Active fuel length (m)	4.2	4.2
Rod linear heat rate (W/cm)	167	175
Number of control rod clusters	89	60
Total flow rate (m^3/h)	108 720	74 376
Core outlet temperature (°C)	330	326
Core inlet temperature (°C)	295	291
Sg heat exchange surface (m^2)	7960	7960
Steam pressure (bar)	78	73

As of 2014, the ATMEA is supported by GDF-Suez and considered for bids to Turkey and Jordan.

7.6.2. The APWR

MHI developed a 1530 MWe twin unit PWR, called APWR, jointly with Westinghouse. Like the ATMEA, it has a cylindrical single containment (pre-stressed concrete + steel liner) and an Inside Refuelling Water Storage Tank as well as four SIS trains with Reactor Vessel Direct Injection and Advanced Accumulators. The core envelop is surrounded by a neutron reflector which is not as heavy as the EPR's. The core is made of 257 17 × 17 fuel assemblies.

Two APWRs are under construction in Japan on the Tsuruga peninsula and a more powerful unit (1700 MWe) is being offered in the USA (although the potential customer has withdrawn its request).

7.6.3. The AES 92

Russia also has its Generation III VVR model, the AES 92, a 1060 MWe PWR with horizontal steam generators and hexagonal fuel assemblies (as is the case with all VVRs). Developed partly with Finland, it has the following characteristics:

- single unit concept,
- mitigation system for severe accidents (core catcher),

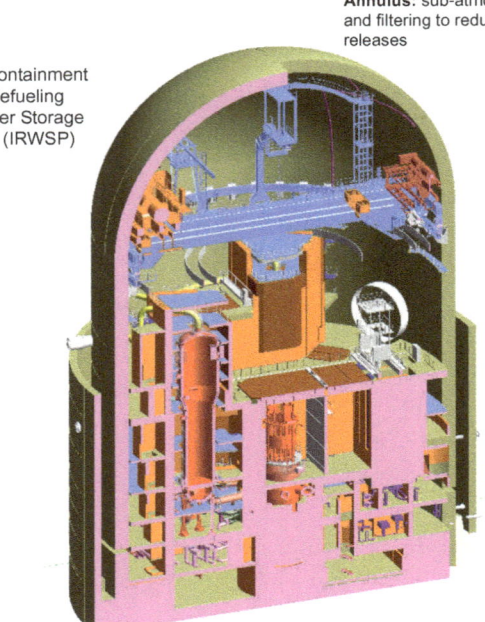

Figure 7.13. ATMEA Nuclear Island.

- large double containment with liner (internal diameter: 44 m),
- large free volume of internal containment (69 170 m^3),
- containment wall thicker than the previous VVR 1000 (1.2 m vs. 0.9),
- four independent safety system trains, direct RPV safety injection (Figure 7.14).

Two AES 92 units were completed in 2011 in the Tamil Nadu province of India, but the Fukushima accident delayed their start-up until 2014.

7.7. Japanese and American ABWRs [17–22]

The ABWR Advanced Boiling Water Reactor which can boast a construction time of only 48 months, has been operating since November 1996 (at least in the case of the Japanese version[7] of the ABWR). It constitutes the next-to-last evolution stage of the BWR line (in Chapter 1, we mentioned to what extent this evolution was more significant than the PWR's in terms of primary circuit, containment shape and general layout).

---

[7] Or rather its Japanese versions, because even though the general design comes from General Electric, Hitachi and Toshiba built two different final versions on the TEPCO Kashiwazaki Kariwa site.

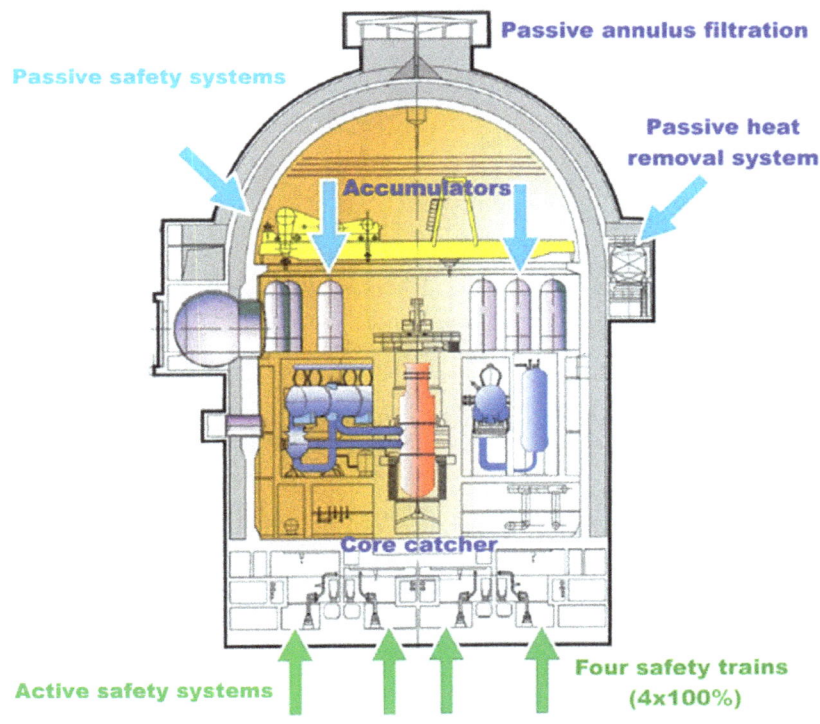

Figure 7.14. AES 92 safety systems.

The Japanese model does not include devices to mitigate the consequences of severe accidents, which excludes them from Generation III, but the American version, in compliance to the URD, does. The US ABWR, which received its *Final Design Approval* from the NRC, is close to completion on the Lungmen site in Taiwan.

7.7.1. General characteristics

According to both its designers and operator, when compared with previous BWRs, the ABWR improves on competitiveness, safety and operating flexibility. As a matter of fact, there is a real effort to simplify architecture and circuitry, and to reduce on-site construction time through shop pre-fabrication.

7.7.2. Architecture simplification

In comparison with relatively recent Japanese BWRs, it appears that the ABWR reactor building is more compact and the general layout significantly simpler (Figure 7.15). The turbine building is also more compact.

Table 7.5.

Characteristics	BWR-5	ABWR
Plant		
Thermal Power (MWt)	3293	3926
Net Electrical Power (MWe)	1100	1356
Steam Mass Flow (t/h)	6400	7500
Reactor Building Dimensions	85 m × 85 m × 75 m	55 m × 60 m × 65 m
Turbine Building Dimensions	120 m × 80 m × 50 m	80 m × 100 m × 50 m
Containment Type	Steel	Concrete + Liner
Primary Circuit		
Operating Pressure (bars)	70.7	72.1
Temperature (°C)	282	284
Mass Flow(t/h)	48 300	52 200
Vessel Height (m)	22	21
Vessel Inner Diameter (m)	6.4	7.1
Recirculation Pumps	20 + 2 ext.	10
Core		
Number of Fuel assemblies	764	872
Type	8 × 8	8 × 8
Power Density (MW/m^3)	50	50.6
Number of Control Rods	205	185

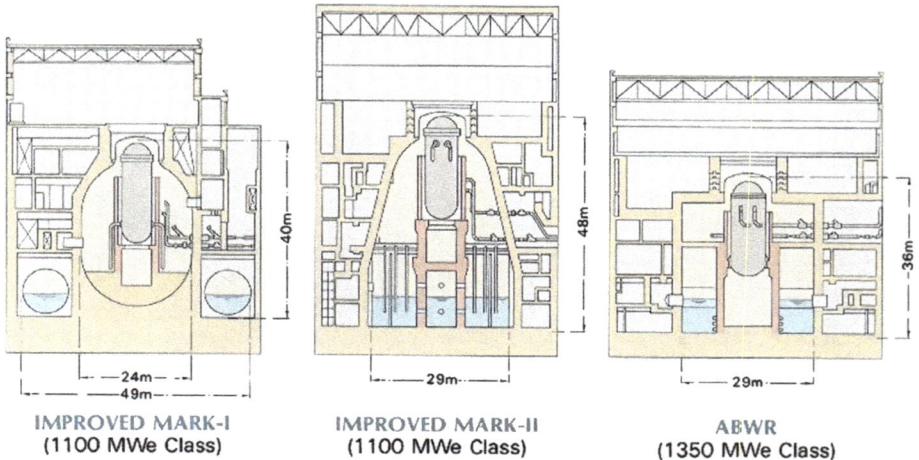

Figure 7.15. Evolution of BWR architecture.

Such compactness not only significantly reduces the volumes of steel and concrete, but also facilitates the on-site assembly of in-shop pre-fabricated elements. This pre-fabrication has proven to be essential in reducing the overall construction time of both Kashiwazaki Kariwa units. Reducing volumes and shortening construction time are key elements of competitiveness.

At the same time, smaller buildings and shorter ducts improve seismic resistance and therefore safety.

7.7.3. Simplification of the primary circuit

With the ABWR, GE has adopted design improvements already implemented in Europe by Sweden and Germany. Old external recirculation loops were removed (Figure 7.16): There are no longer ducts below the top of the core, and recirculation is driven by ten wet rotor pumps located at the bottom of the pressure vessel, around the "forest" of control rod mechanisms. Their motors can be maintained from the bottom and the impeller from the top without impacting circuit leak tightness. Nine of these pumps are sufficient for operation at full power.

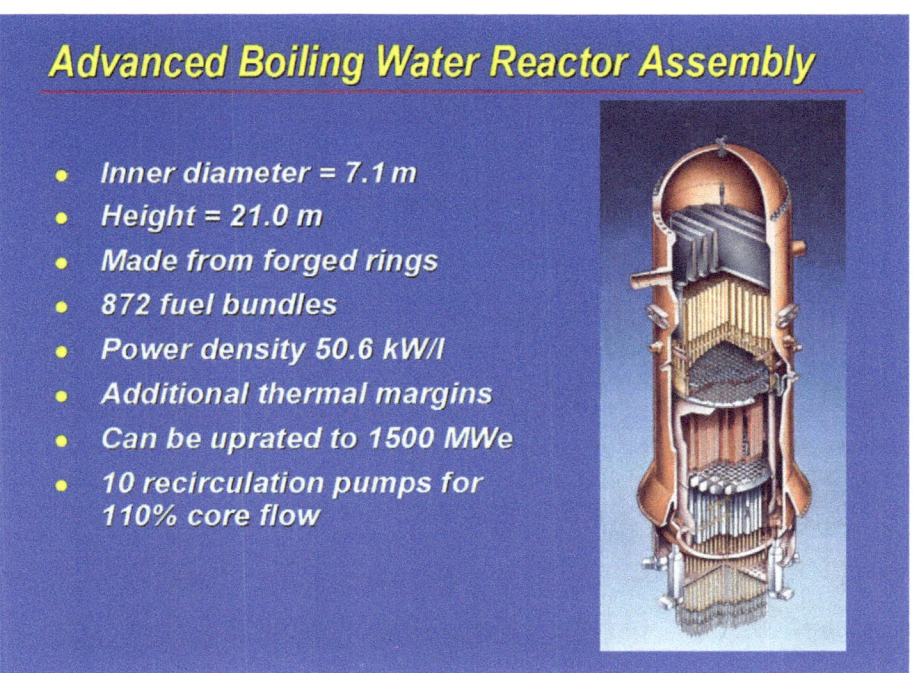

Figure 7.16.

Fine-motion electro-mechanical control rods with hydraulic scram capability were also imported from Europe (Figure 7.17). The mechanism redundancy increases reliability and fine motion improves load-following capability (which is already high thanks to adjustment of the recirculated flow).

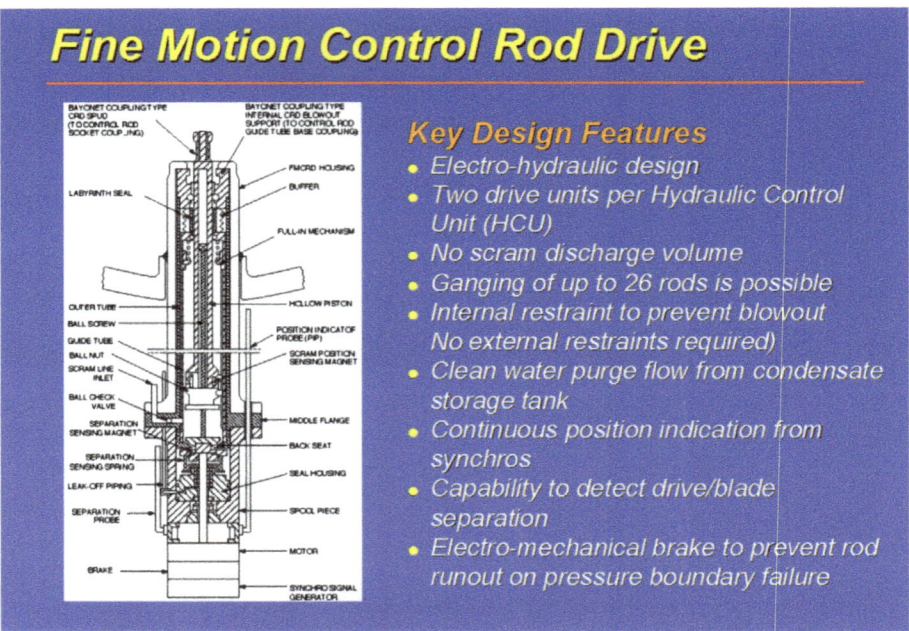

Figure 7.17.

7.7.4. Additional improvements

The digital control room is more modern than in previous GE designs, but rather less modern than in N4's. The fuel assemblies are of the same type used in modern BWRs, with sandwich zircaloy/zirconium cladding to reduce PCI, and pellet enrichment is axially zoned to reduce fuel costs (decrease in enrichment). As a result, the core is complex, with 9 different enrichment levels. The emergency core cooling system (ECCS) has three trains, any of which would be sufficient. Each train comprises a high pressure system, a low pressure system and a shutdown heat removal system. There is, in addition, one single depressurization system.

In contrast with the Japanese design, the American ABWR has severe accident mitigation features summarized in Figure 7.18.

7.8. General Electric Simplified BWRs [24–29]

While Westinghouse was developing both the APWR and the AP 600 in parallel, General Electric was designing both the evolutionary ABWR and the Simplified BWR (SBWR), a "passive" 600 MWe reactor complying with the same URD (Figure 7.19).

During the mid-90's, as the US market was still dormant while European and Japanese utilities were not interested in any 600 MWe plant, General Electric and its partners designed a larger 1190 MWe reactor along the same principles and in compliance with the EUR. This ESBWR, standing for *European* SBWR, is even more compact so that, despite increased

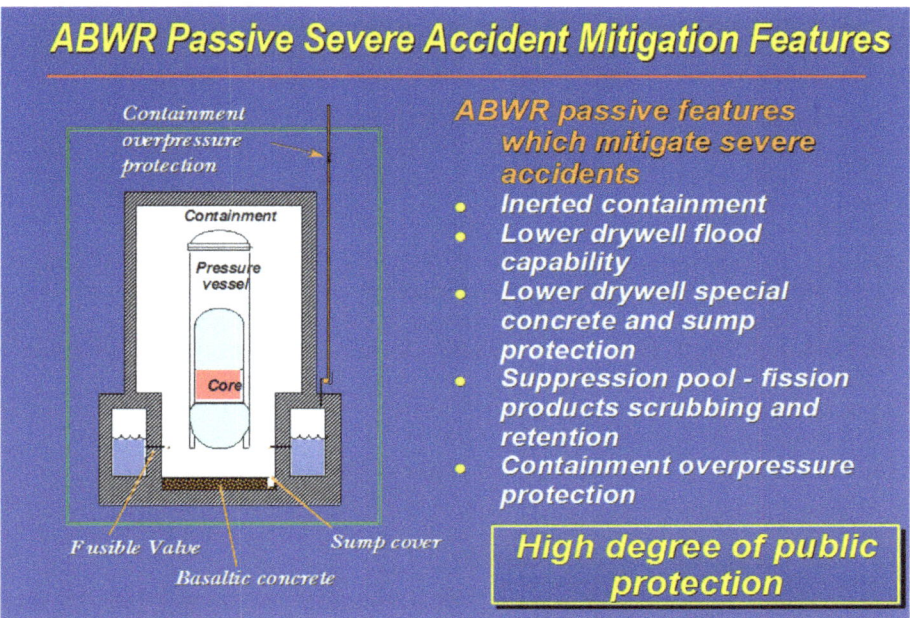

Figure 7.18.

Evolution of the BWR Reactor Design

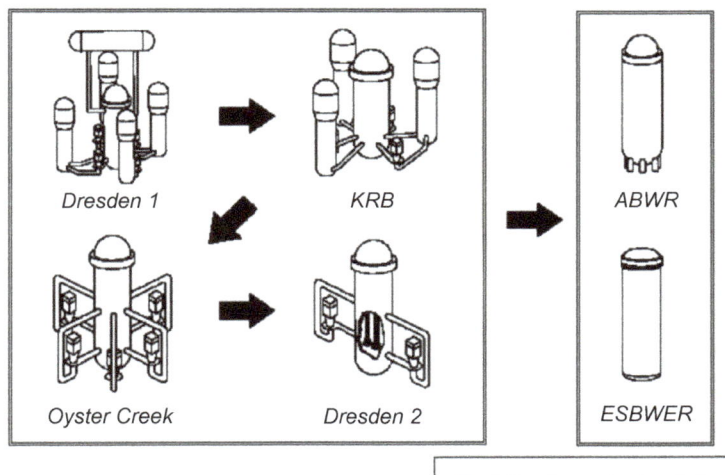

Figure 7.19.

power, its footprint is smaller. After 2000, and after the ABWR lost the competition with the EPR in Finland, GE again increased the rating of the ESBW, now dubbed *Economic Simplified BWR*.

The basic features common to the SBWR family are the following:

- no forced recirculation at all: the reactor operates at full power with natural convection;

- emergency core cooling and decay-heat removal are driven by gravity or by previously stored energy, without requiring any operator action.

To make natural convection efficient, all pressure losses are reduced in the very large vessel. Use of natural circulation is somehow a return to the roots of the design since it was how the now defunct 180 MWe Dodewaard BWR used to operate in the Netherlands. It is important to note that *this suppression eliminates the flexibility given by the reactivity control of BWRs relying on recirculation flow rate adjustment.*

7.8.1. General characteristics

The SBWR and ESBWR both benefit from the main ABWR improvements: digital instrumentation, multiplexed cables, electro-hydraulic control rod mechanisms, and large turbines (Figure 7.20).

7.8.2. The SBWR (600-670 MWe)

To eliminate forced recirculation, the SBWR had lower power density and shorter core height together with a wider downcomer annulus than conventional BWRs. Its pressure vessel was bigger than the ABWR, which was twice as powerful. A large water volume above the top of the core replaced conventional accumulators, and to keep the chimney free of obstacles, steam dryers were annular and located in the vessel head.

Emergency core cooling in case of loss of water inventory was based upon the classical BWR principles: depressurization followed by low pressure water injection. In the SBWR, this injection is gravity driven from a tank located above the top of the core.

Long term decay heat removal is carried out by isolation condensers immersed in a pool located high enough to allow for natural convection. The pool itself is cooled by evaporation into the atmosphere. Six depressurisation valves discharge steam into the pressure suppression pool, where most of the caesium and iodine would remain trapped if need be.

7.8.3. The ESBWR (1300-1550 MWe)

As its designers openly admit [26], while the SBWR was successful in simplifying circuitry and operation, it was not economically competitive.

1300 to 1500 MWe plants such as the ABWR or EPR benefit from a size effect. Furthermore, such evolutionary designs present little uncertainty as to their feasibility and licensability. As a rule of thumb, to become serious competitors, passive reactors should offer a kWh cheaper by at least 10 to 15%. Such is the target of the ESBWR.

7 – Advanced "Generation III" reactors

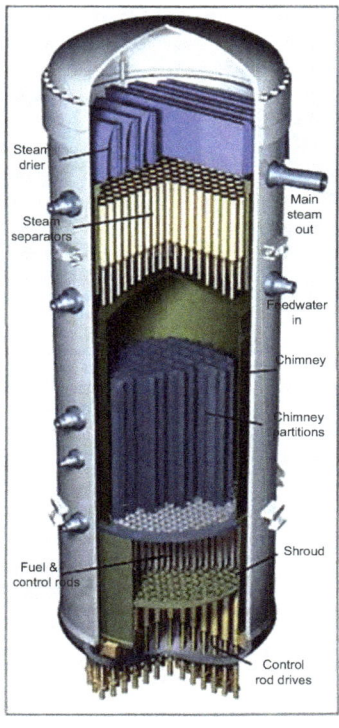 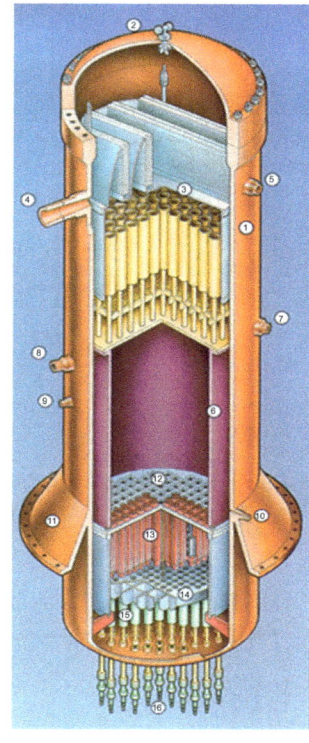

Simplified Boiling Water Reactor Assembly

1 Reactor Pressure Vessel
2 RPV Top Head
3 Integral Dryer-Separator Assembly
4 Main Steam Line Nozzle
5 Depressrization Valve Nozzle
6 Chimney
7 Feedwater Inlet Nozzle
8 Reactor water Cleanup/Shutdown Cooling Suction Noozzle
9 Isalation Condenser Return Zozzle
10 Gravity-Driven Cooling System Inlet Nozzle
11 RPV Support Skirt
12 Core Top Guide Plate
13 Fuel Assemblies
14 Core Plate
15 Control Rod Guide Tubes
16 Fine Motion Control Rod Drives

ESBWR   SBWR

Figure 7.20.

Table 7.6.

Characteristics	ABWR	ESBWR 2005	SBWR
In compliance with	URD	EUR/URD	URD
Thermal Power (MWt)	3920	4500	2000
Net Electrical Power (MWe)	1350	1550	670
Power Density (MW/m^3)	51	54	42
Active Fuel Height (m)	3.7	3.0	2.7
Number of Fuel Bundles	872	1132	872
Number of CRD	205	269	217
Feedwater Trains	3	3	1
Turbine	1 HP, 3 LP	1 HP, 3 LP	1 HP, 1 LP
Safety Systems	Active	Passive	Passive
Recirculation Pumps	10	0	0

It was therefore necessary to reduce the SBWR capital cost. This was made possible by the modular nature of its safety systems: power was increased by parallelising identical modules (isolation condensers and decay-heat removal exchangers). Even though the pressure vessel is larger, it affects the containment building only in a limited manner.

The whole architecture was redesigned so that the safety circuits only were left inside the containment and auxiliary systems were moved to cheaper and less seismically-resistant buildings (Figure 7.21). This reduced the overall footprint, and led to the expectation of shorter construction times.

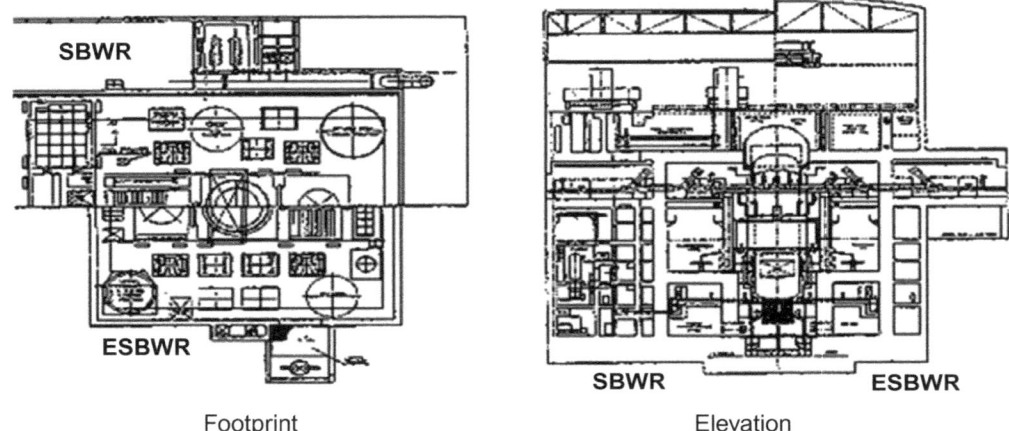

Figure 7.21.

In order to minimize financial as well as regulatory uncertainties, the ESBWR uses as many components and systems as possible from the ABWR. Specific components and systems have been tested since the inception of the SBWR project through a remarkably international validation program (on which GE's partners spent more than 250 million dollars over 10 years). Nobody has yet ordered an ESBWR, but the NRC just delivered its design certification in the fall 2014.

7.9. The KERENA [30, 31]

During the 90's, while developing the EPR in the French-German framework described above, and before merging its nuclear division with Framatome ANP, Siemens together with some German utilities developed a medium size (650, then 1000 MWe), rather innovative BWR. Originally named SWR 1000, this reactor is now offered under the trade name KERENA (Figure 7.22).

KERENA's main technical characteristics are shown in Table 7.7. Here are the salient features which should by now sound familiar:

- reduced core power density;
- large water inventory in the primary circuit;

7 – Advanced "Generation III" reactors

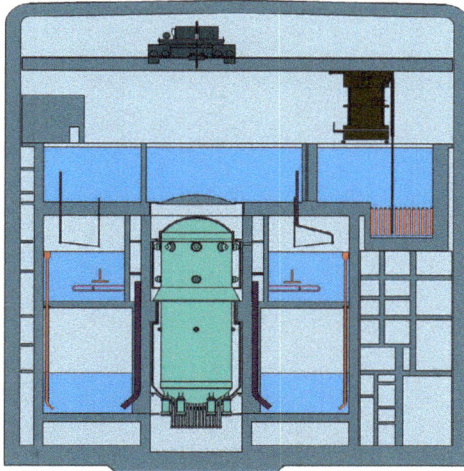

Thermal power	2,778 MW
Gross power output	~1,000 MW
Reactor Pressure	70.6 bar (1010 psig)
Type of fuel assemblies	ATRIUM 12
Number of fuel assemblies	624
Number of control rods	145
Length of active core	2.8 m. (9.2 ft.)
Max. accident pressure of containment	7.5 bar (95 psig)

Figure 7.22.

- transients management without additional water;

- large thermal capacity of the water tanks inside the reactor building (pressure suppression pool and core reflood tank);

- "passive" vessel and containment heat removal equipment (Figure 7.23);

- large water inventory available to reflood the core by gravity driven flow after depressurization;

- passive actuation of essential safety functions (scram, depressurization);

- post-accident management requiring no operator action for the first few days, followed by simple measures to remove the decay heat;

- containment inerting with nitrogen to avoid post-accident hydrogen risk;

- highly sturdy containment;

- passive cooling outside of the RPV after meltdown for *in situ* corium solidification;

- cycle length flexibility (12 to 24 months, 65 GWd/t);

- construction in 48 months and lifetime of 60 years.

The SWR 1000 was offered to Finland in parallel with the EPR. During the bid analysis, the project was upgraded to 1250 MWe, which appears to be the limit of such a design.

Table 7.7.

Characteristics		KERENA	Gundremingen B (REB 1300)
Plant			
Thermal Power	MW	2778	3840
Net Electrical Power	MW	977	1373
Efficiency	%	35.2	35.7
Core			
Number of Assemblies	–	568 (13 × 13)	784 (10 × 10)
Total Uranium Mass	t	121	138
Active Fuel height	m	2,80	3,71
Power Density	MW/m^3	47	56.8
Average discharge Burnup	GWd/t	65	50
Average Enrichment	% ^{235}U	5.45	3.63
Mass Flow	kg/s	12 000	14 300
Vessel			
Height	m	22.55	22.35
Inner Diameter	m	7.0	6.62
Nominal pressure	bar	88	87.3
Recirculation Pumps		6	8
Turbine			
Number		1	1
Speed	rpm	3 000	1 500
Stages HP/LP		1/3	1/2
Containment			
Inner Diameter	m	32,0	29
Inner Height	m	28,7	32,5
Nominal Pressure	bar (abs)	7,5	5,3
Volume of the Drywell	m^3	5 700	8 200
Water Volume in the PP Pool	m^3	2 900	3 100
Gas Volume " " " "	m^3	5 500	6 000
Water Volume in the Reflood Tank	m^3	3 100	–
Technical Lifetime	years	60	40
Construction Time	months	48	60

7.10. SMRs [32, 33]

The first nuclear "plant" (Obninsk, URSS, 1954) rated 5 MWe. Inaugurated in 1956, Calder Hall had 4 60 MWe units and Shippingport, the ancestor of the PWR family, also rated 60

7 – Advanced "Generation III" reactors

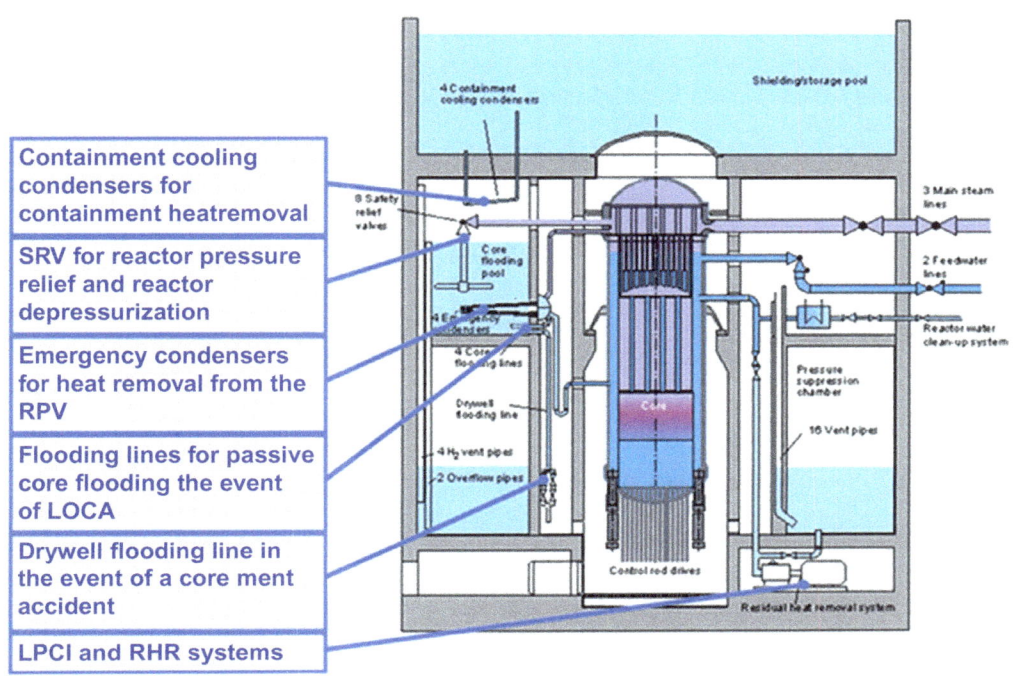

- Containment cooling condensers for containment heatremoval
- SRV for reactor pressure relief and reactor depressurization
- Emergency condensers for heat removal from the RPV
- Flooding lines for passive core flooding the event of LOCA
- Drywell flooding line in the event of a core ment accident
- LPCI and RHR systems

Figure 7.23.

MWe. From then on, unit ratings escalated very fast: 125 MWe, 300 MWe, 600 MWe, 900 MWe. In 1974, this led the newborn USNRC to set a provisional regulatory limit of 4000 thermal MW (around 1300 MWe) to NPPs in the USA. Today, AREVA's EPR (1650 MWe) and Mitsubishi's APWR (1700 MWe) significantly exceed this limit.

Reasons for this escalation were mostly economical, because nuclear reactors exhibit a significant "size effect", but considerations of site selection and licensing burden played a role too. With the exception of Chinese reactors sold to Pakistan, all recent LWR orders were at least 1000 MWe.

Periodically, however, studies and developments of SMRs appear; "small" if rated below 300 MWe or "medium" between 300 and 700 MWe. Additionally "M" tends to mean "modular".

During the early 80's, US utilities decided to no longer let vendors impose their catalogues but to produce their own sets of requirements, leading to the publication of the Utilities Requirements Documents or URD. In the URD, US electric companies requested NPPs in the 500-600 MWe range. One must realise that there were many utilities, most of them very small by European or Asian standards, and not well interconnected: to them, the SMR seemed to fit the size of their grids better and to be easier to finance. Based on the URD, Westinghouse developed the AP 600 and General Electric the SBWR, both largely resting their safety on "passive" systems rather than relying on operator actions (since operator errors were at the origin of the Three Mile Island accident).

Earlier in this chapter, we saw that neither SMR found any customer, either in the USA or abroad, so Westinghouse expanded the AP 600 into the AP 1000, rating 1200 MWe, while GE designed the ESBWR for up to 1500 MWe. And the first 4 NPPs ordered in the USA since 1974 were, indeed, 1150 MWe AP 1000.

Recently however, SMRs have come back in fashion in the USA where, in January 2012, the DOE (US Department of Energy) launched a special program with a 450 million dollar budget to finance 50% of the engineering, certification and licensing of 2 SMR demonstrators. Both models were selected in 2013: Babcock & Wilcox's "m-Power" and Fluor's "NuScale". As expressed in the DOE's announcement, "small modular reactors represent a new generation of safe, reliable, low-carbon nuclear energy technology and provide a strong opportunity for America to lead this emerging global industry."

7.10.1. SMRs' potential advantages and drawbacks

Advantages expected from SMRs are usually as follows:

- adaptation to small and poorly interconnected grids, or to grids where slow growth is expected,
- more "passive" safety (which also presents drawbacks),
- ability to load long lifetime cores,
- better adaptation to co-generation of heat and power,
- whole or partial plant fabrication and control, with on-site assembly,
- short construction (or assembly) time, which means reduced interests.

These advantages are balanced by a higher overnight cost per installed kWe and less efficient fuel use (smaller cores have higher neutron leakage and therefore require higher enrichment level). Furthermore, there is as yet no return of experience on SMR licensing (hence the DOE's support): it remains to be seen whether modularity can expedite or simplify procedures - and there would be many licensing procedures for an SMR fleet!

In September 2012, the IAEA published a "Status of Small and Medium Sized Reactor Designs" listing 31 SMRs, from old obsolete designs still in operation to advanced "paper" designs. They include a majority of LWRs but also HWRs, FBRs and HTRS. Within this series, the most promising designs could reach the prototype stage around 2025-2030.

7.10.2. Short description of four SMRs

Based on the IAEA report, we shall describe the two models supported by the USDOE, as well as the French "Flexblue" and the Korean "Smart" (all PWRs).

Babcock & Wilcox mPower

(Figure 7.24)

Reactor type:	Integral pressurised water reactor
Electrical capacity:	180 MW(e)
Thermal capacity:	530 MW(th)
Coolant/moderator	Light water
Primary circulation:	Forced circulation
System pressure:	14.1 MPa
Core outlet temperature:	320 °C
Thermodynamic cycle:	Indirect Rankine cycle
Fuel material:	UO2
Fuel enrichment:	< 5.0%
Fuel cycle:	48 months
Reactivity control:	Rod insertion
Emergency safety systems:	Passive
Residual heat removal systems:	Passive
Design life:	60 years
Planned deployment:	2020
Distinguishing features:	Internal once-through steam generator, pressurizer and control rod drive mechanism

Figure 7.24.

The reactor core consists of 69 fuel assemblies (FAs) whose enrichment is less than 5%, Gd_2O_3 spiked rods, Ag In–Cd (AIC) and B_4C control rods, and a 3% shutdown margin. No soluble boron is present in the reactor coolant for reactivity control. The FAs, of a conventional 17 × 17 design, were shortened to an active length of 241.3 cm with a fixed grid structural cage.

The reactor uses eight internal coolant pumps with external motors. The integrated pressurizer at the top of the reactor is electrically heated.

The inherent safety features of the reactor design include a low core linear heat rate (which reduces fuel and cladding temperatures during accidents), a large reactor coolant system volume (which gives more time for safety system responses in the event of an accident), and small penetrations at high elevations (increasing the amount of coolant available to mitigate a small break LOCA).

The emergency core cooling system is connected with the reactor coolant inventory purification system and removes heat from the reactor core after anticipated transients in a passive manner, while also passively reducing containment pressure and temperature.

The mPower reactor deploys a decay heat removal strategy based on a passive heat exchanger connected to the final heat sink, an auxiliary steam condenser on the secondary system, water injection or cavity flooding using the reactor water storage tank, and passive containment cooling.

Fluor NuScale

(Figure 7.25)

Reactor type:	Integral pressurized water reactor
Electrical capacity:	45 MW(e)
Thermal capacity:	160 MW(t)
Coolant/moderator:	Light water
Primary circulation:	Natural circulation
System pressure:	8.72 MPa
Core outlet temperature:	329 °C
Thermodynamic cycle:	Indirect Rankine
Fuel enrichment:	<4.95%
Fuel cycle:	24 months
Reactivity control:	Rod insertion
No. of safety trains:	Two trains
Emergency safety systems:	Passive
Residual heat removal systems:	Passive
Design life:	60 years
Planned deployment:	2020
Distinguishing features:	Synergy through plant simplicity; reliance on existing light water technology and availability of an integral test facility

Figure 7.25.

The NuScale Power concept can consist of 1 to 12 independent modules, each capable of producing a net electric power of 45 MW(e). Each module includes a PWR operated under natural circulation. Each reactor is housed within its own high pressure containment vessel, which is submerged in a stainless steel lined concrete pool.

The NuScale plant safety features include a high pressure containment vessel, two systems for passive decay heat removal and containment heat removal, a shutdown accumulator and severe accident mitigation.

The NuScale reactor module operates solely on natural convection and resides in a high strength stainless steel containment vessel. The decay heat removal system consists of two independent trains operating under two-phase natural circulation in a closed loop. The designers claim that the pool surrounding the reactor module provides three days of cooling supply for decay heat removal.

DCNS Flexblue

Flexblue is a 160 MW(e) transportable and submarine nuclear power unit, operating at a depth of up to 100 m. It is 140 m long (Figure 7.26).

The power production cycle lasts 3 years. At the end of a production cycle, the unit is taken back to its support facility. The reactor core is then refuelled and periodic maintenance

7 – Advanced "Generation III" reactors

Reactor type:	Pressurized water reactor
Electrical capacity:	160 MW(e)
Thermal capacity:	600 MW(th)
Coolant/moderator:	Light water
Primary circulation:	Forced
System pressure:	15.5 MPa
Core outlet temperature:	310 °C
Thermodynamic cycle:	Indirect Rankine cycle
Fuel enrichment:	5%
Fuel cycle:	36 months
Emergency safety systems:	Passive
Residual heat removal systems:	Passive
Design life:	60 years
Distinguishing features:	Transportable nuclear power plant; submerged operation

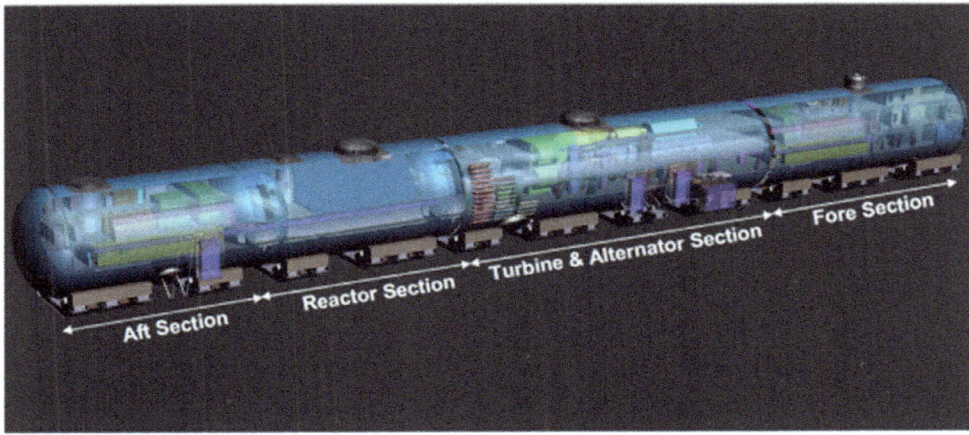

Flexblue system
The Sub Sea Module is the heart of the Flexblue Solution

- Power : 160 MW Electric
- Length ≈ 146 meters
- Hull Ø ≈ 14 meters
- Displacement ≈ 20 000 tons
- Moored up to a 100 meters depth
- Unmanned Operation, permanent accessibility

Figure 7.26.

is carried out. A major overhaul is scheduled every 10 years. At the end of its life, the power unit is transported back to a dismantling facility, which results in a quick, easy and full recovery of the natural site.

Reactivity is controlled without soluble boron. This simplifies primary chemistry management and reduces radioactive effluent and waste to the environment.

DCNS claims that Flexblue offers an extended level of nuclear safety that is enhanced by the submarine environment and based exclusively on proven technologies from the nuclear, naval and offshore industries. Water offers natural protection against most of the possible external hazards and guarantees a permanent and indefinitely available heat sink.

Within this framework, the use of passive safety systems brings the reactor to a safe and stable state without external intervention for an indefinite period of time.

KAERI Smart

The SMART (System Integrated Modular Advanced Reactor) is a small integral PWR with a rated thermal power of 330 MW(th). It could co-generate electricity and desalinate water. Its aims are enhanced safety and improved economics. To enhance safety and reliability, the design organization has incorporated inherent safety features and reliable passive safety systems. The aim is to improve economics through system simplification, component modularization, reduction of construction time and high plant availability.

By introducing a passive RHRS removal system and an advanced mitigation system for loss of coolant accidents, significant safety enhancement is expected by the design organization. The low power density design, with about a 5wt% UO2 fuelled core, will provide a thermal margin of more than 15% to accommodate any design basis transients with regard to the critical heat flux. This feature ensures core thermal reliability under normal operation and any design basis events (Figure 7.27).

Reactor type:	Integral pressurized water reactor
Electrical capacity:	100 MW(e)
Thermal capacity:	330 MW(th)
Coolant/moderator:	Light water
Primary circulation:	Forced circulation
System pressure:	15 MPa
Core outlet temperature:	323 °C
Thermodynamic cycle:	Indirect Rankine cycle
Fuel cycle:	36 months
Reactivity control:	Control rod drive mechanism, soluble boron and burnable poison
No. of safety trains:	4
Emergency safety systems:	Active and passive systems
Residual heat removal systems:	Passive systems
Design life:	60 years
Distinguishing features:	Coupling of the desalination system or process heat application

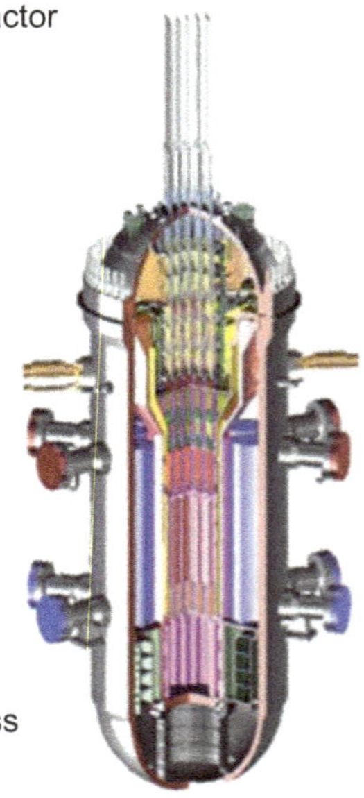

Figure 7.27.

The standard design of SMART was approved by the Korean Nuclear Safety and Security Commission in July 2012. Whether or not construction would take place was not yet decided by the end of 2014.

7.10.3. Prospects for SMRs?

SMRs were often developed in countries which did not need them, being already equipped with large, more competitive NPPs. There are, however, exceptions for specific usages: the Korean SMART targets the seawater desalination niche; some Russian designs would be a better fit for highly isolated Siberian locations; the small size of many US utilities faced with the need to replace outdated coal plants in the 300 MWe range also creates a special situation. Therefore, many models shall be offered without any operational reference in the vendor country, while most potential customers will be newcomer countries without previous nuclear experience.

On economic grounds, it remains to be seen whether the advantages of mass production and short assembly time can overcome the size effect.

The jury is still out. It is quite possible that the periodic wave of interest in SMRs will materialise this time, but not very quickly. However, it is doubtful that SMRs will be successful in countries where the electric grid can support large units above 1000 MWe.

References

[1] M. Yvon, *Réacteurs à eau ordinaire sous pression. Le projet EPR*. Techniques de l'Ingénieur BE 3 102, 1998.

[2] *Le Projet EPR*. Actes de la Conférence SFEN/KTG sur le projet EPR. Strasbourg, November 1995

[3] *The European Pressurized Water Reactor EPR*. Proceedings KTG/SFEN Conference, Köln, October 1997.

[4] F. Bouteille, H. Seidelberger, *The European Pressurized Water Reactor: A status Report*. Nuclear Engineering International, October 1997.

[5] *La sûreté des réacteurs du futur; Le projet EPR*. Dossier spécial. Contrôle n°105. June 1995.

[6] F. Bouteille, *L'évolution de la chaudière nucléaire EPR*. Conférence SFEN/SEE, Paris, December 1999.

[7] Dossier du débat public 2005-2006: http://cpdp.debatpublic.fr/cpdp-epr/documents/dossier-EDF.html

[8] *EUR European Utility Requirements for LWR Nuclear Power Plants*. Volumes 1 & 2 Rev.B, Volume 4 Rev.A CD-Rom EdF/SEPTEN, May 1997.

[9] R.S. Turk, R.A. Matzie, *System 80+: PWR technology takes a major step up the evolutionary ladder*. Nuclear Engineering International, November 1992.

[10] J.D. Crawford, R.A. Matzie, *System 80+, Evolution of a standard design*. Transactions ANS 61 (suppl. 1), March 1990.

[11] *Advanced Light Water Reactor Requirements Document*. EPRI, December 1993.

[12] H.J. Bruschi, *Commercializing the next generation: The AP 600 Advanced Simplified Nuclear Power Plant*. 9th Pacific Basin Nuclear Conference, Sydney, May 1994.

[13] H.J. Bruschi, *Les Centrales nucléaires de Westinghouse*. RGN N°6, November-December 1994.

[14] *AP 600. Ready for Commercialization*. Westinghouse Document, December 1999.

[15] B.A. Mc Intyre, *The challenges of Licensing the AP600 Passive Nuclear Power Plant Design*. ICONE-7, Tokyo, April 1999.

[16] Y. Qi, *The next generation reactor AC600/1000*. Nuclear Europe Worldscan **11–12**, 1999.
[17] J.R. Redding, *Advanced LWR technology for commercial application*. RGN N°6, 1994.
[18] A. Rao, *The ABWR Nuclear Plant*. Presentation, November 1999.
[19] *Kashiwazaki Kariwa Nuclear Power Station*. Tokyo Electric Power Company brochure, 1997.
[20] *Advanced Boiling Water Reactor*. TEPCO brochure, 1997.
[21] *Advanced Boiling Water Reactor*. Hitachi brochure, 1997.
[22] *Advanced Boiling Water Reactor ABWR*. Toshiba brochure, 1997.
[23] P. Bacher, *Les centrales nucléaires du futur*. Equipement 2000 n°16, March 1994.
[24] R.J. Mc Candless, J.R. Redding, *Simplicity: The key to improved safety performance and economics*. Nuclear Engineering International **34** (424), November 1989.
[25] A.S. Rao, *Simpler by design*. ATOM N° 430, September 1993.
[26] A.S. Rao et al., *ESBWR - An Economic Passive Plant*. Reprints from General Electric, September 1996.
[27] Y.K. Cheung et al., *Design evolution of natural circulation in ESBWR*. ICONE-6295, May 1998.
[28] A. Rao, *ESBWR Program and Design Overview*. Communication, September 1998.
[29] D. Hinds, *Next-generation nuclear Energy: The ESBWR*. Nuclear News, January 2006.
[30] *SWR 1000. An Advanced, Medium-Capacity Boiling Water Reactor with Passive safety Features*. Siemens, November 1998.
[31] E. von Staden, *The concept of the SWR 1000 Reactor*. Belgian Nuclear Society meeting, June 2001.
[32] *Status of Small and Medium Sized Reactor Designs*. IAEA Report, September 2012.
[33] http://www.world-nuclear.org/info/Nuclear-Fuel-Cycle/Power-Reactors/Small-Nuclear-Power-Reactors (updated December 2013).

8 High Temperature Reactor

B. Barrè

8.1. Obsolete or futuristic

In Chapter 1, we saw how ordinary water reactors, PWRs and BWRs, have progressively invaded the nuclear power plants' "ecological niche", leaving only a small breathing space for the CANDU. We also listed a few weaknesses of this species which, under the right circumstances, could constitute opportunities for other types of reactors.

Among the many possible designs, two seem to warrant deeper examination: High Temperature Reactors or HTRs (Chapter 8), and Molten Salt Reactors or MSRs (Chapter 9). Both have already been developed, albeit to different extents: HTRs almost became an actual family – or even two families as we shall see – while only a small MSR prototype was actually built and operated for a few years. Both are characterized by their fuel, which is quite different from "conventional" reactor fuel. This fuel gives them both a high degree of flexibility in the choice of fuel cycle. In addition, both can reach high thermal efficiency.

Apart from those similarities, these designs are quite different.

8.2. HTR fuel [1–3]

What constitutes the specificity of HTRs and gives them their qualities is their fuel. It was invented in Harwell, UK, in the mid-1950's. The core, wholly refractory and helium cooled, is made of tiny fissile particles, less than 1 mm in diameter, dispersed within a graphite moderator.

The kernel of each individual particle is coated (Figure 8.1), by catalytic cracking in a fluidized bed, with a number of concentric layers similar to the sugar coatings of the almond in a *dragée*: inner layers of pyrocarbon protect a layer of silicon carbide (SiC) from the hot kernel, and outer layers of dense pyrocarbon can withstand the pressure of fission gases up to very high burnups. The SiC layer is a leak-tight barrier to contain the fission products: it plays the role of the cladding in a conventional fuel pin. The outermost carbon layer facilitates the agglomeration of the particles inside "compacts" or pebbles (Figure 8.2).

This fuel is extremely divided and fully refractory, which enables the reactor to operate with very high coolant temperatures (we shall later see how high) and therefore

Figure 8.1. Scanning Electron Micrography of a coated particle.

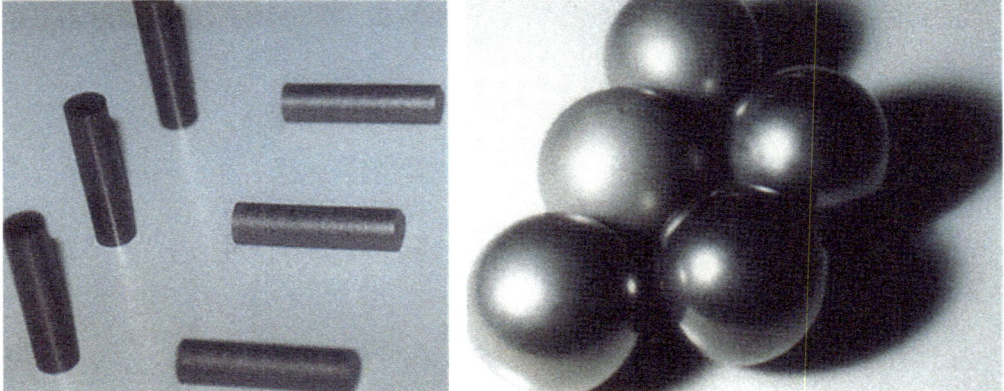

Figure 8.2. The two families of fuel elements (Compacts-in-Prism and Pebbles).

with excellent thermal efficiency. The coated particle is, indeed, a very special breed of fuel element:

- there are several tens of billions of particles in a reactor core: it is therefore a mass produced object, whose quality can only be assessed by statistical tools (no fewer than 10^{11} individual claddings constitute the first barrier against radioactivity dispersal, versus the 2×10^5 pin claddings of a PWR);
- there is an almost unlimited flexibility in the core composition. The nature (fissile, fertile, burnable poison, mixture) and dimension (i.e. self-protection) of the kernels can be freely selected. The particle concentration within the graphite matrix of the compact or pebble can be adjusted as well as their distribution by size (double heterogeneity). HTRs can therefore be adapted to any fuel cycle whatsoever.

The actual flexibility offered to the designer can be illustrated by the two types of fuel elements used in HTR prototypes, prisms in Fort Saint Vrain and pebbles in THTR, not to mention the many other types tested in Dragon (annular, teledial, etc.)

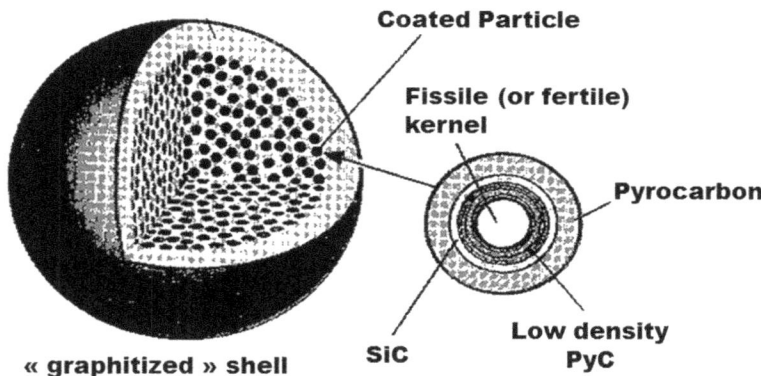

Figure 8.3. HTR pebble.

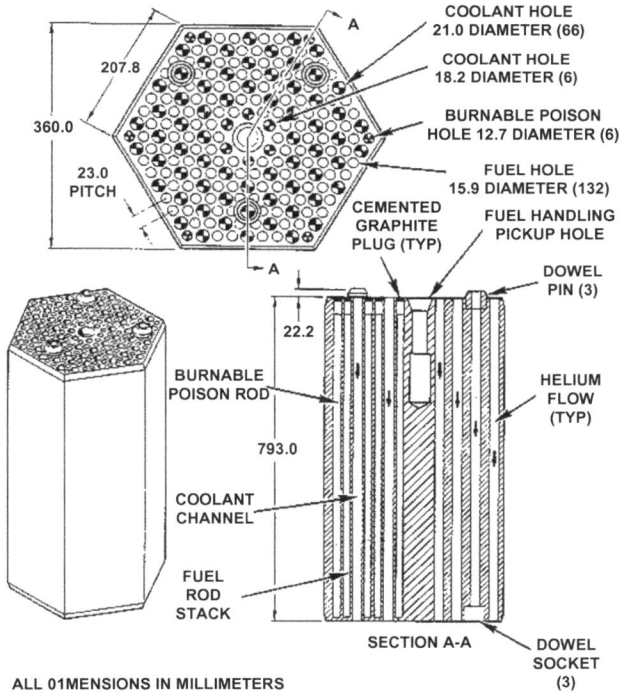

Figure 8.4. HTR prism.

8.3. HTR demos: Dragon, AVR, Peach bottom

8.3.1. Dragon

Around 1956, while the UK was launching its large Magnox programme, the Harwell discovery was developed within the Dragon Project, an *ad hoc* OECD enterprise located on the UKAEA Winfrith site. A demonstration facility was built and operated at Winfrith as soon as 1964, and it successfully established the HTR feasibility. In addition to building and operating the reactor, the 12-country Dragon team paved the way for future HTRs by exploring reactor designs and testing a number of fuel cycles (low enriched uranium LEU, thorium/^{235}U, and plutonium, both with oxide or carbide kernels).

Because of its international nature, the Dragon Project introduced the HTR to all European countries and triggered interest first in the USA, then in Japan.

Table 8.1.

Characteristics	Dragon	Peach Bottom	AVR
Criticality/shutdown	1964/1975	1966/1974	1966/1988
Thermal power (MWt)	20	115	46
Net electric power (MWe)	–	40	15
Helium pressure (absolute bars)	20	24	11
Core inlet temperature (°C)	350	340	260
Core outlet temperature (°C)	750	715	950
Core diameter (m)	1.1	2.8	3
Core height (m)	1.6	2.3	3
Power density MW/m^3	14	8.3	2.2
Fuel element	Prismatic	Prismatic	Pebble
Fuel cycle	Various	^{235}U/Th	^{235}U/Th

8.3.2. The AVR

Germany was a partner in the Dragon project through Euratom and developed HTRs as its first purely national design. As early as 1967, operation of the AVR, a very innovative demonstration reactor, began in Jülich, where it operated very successfully for more than 20 years.

Both the core and the steam generator were contained in a single steel double-walled pressure vessel. The helium coolant flowed upwards, and its outlet temperature was increased from 750 to 850 °C, and then to 950 °C during its last two years of operation. The temperature was even pushed to 1050 °C in the last days before shutdown.

The main innovation of the AVR was its spherical fuel element, the 6 cm diameter graphite "pebble" inside which coated particles were agglomerated. 100,000 pebbles were heaped inside a funnel shaped graphite cavity. The control rods moved in channels within the graphite reflector.

600 pebbles a day were continuously extracted from the bottom of the funnel, and tested for physical integrity and burnup. 90% were recycled on top of the heap, with the required complement of fresh pebbles: an intact pebble therefore travelled ten times through the core before disposal. Each pebble contained on average 1 g of HEU and 6 g of thorium,

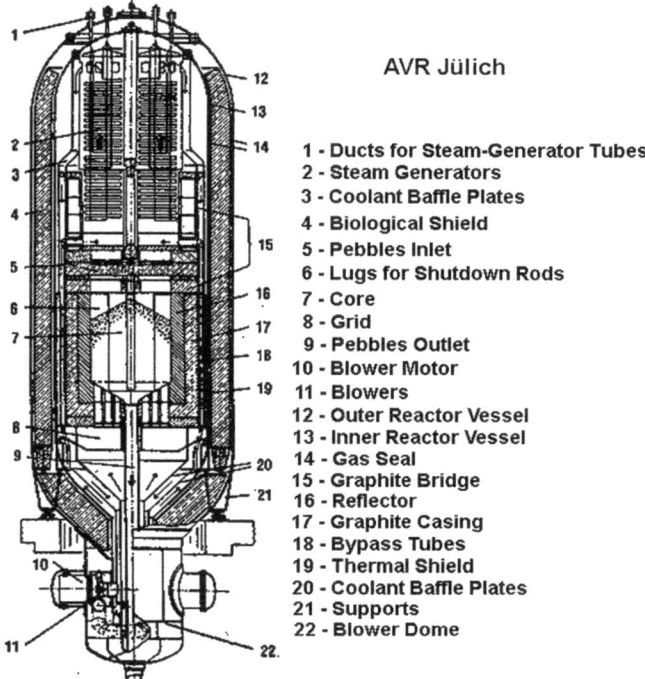

Figure 8.5. The Jülich AVR.

in particles with a "BISO" all pyrocarbon coating. Burnups as high as 150 GWd/t were routinely reached in the AVR.

8.3.3. Peach bottom

Very soon, 53 electricity producers, with the support of the US government, entered the HTR race and built a demonstration reactor at Peach bottom (Pennsylvania), which reached its nominal 40 MWe power in 1967. The Peach bottom fuel element is close to the Dragon design: a long hexagonal graphite prism housing a pile of annular "compacts". The core, surrounded by a graphite reflector, is located at the bottom of a steel pressure vessel. The first core was made using particles with a still imperfect coating. It was soon replaced by a core with a much improved retention of fission products. Peach bottom was decommissioned in 1974, just after the start-up of the Fort Saint Vrain prototype.

The very successful operation of these three demos gave great hopes concerning the future of the HTR families. Unfortunately, the performances of their immediate successors were less bright.

It should be noted that the Japanese 30 MWt demo HTTR, whose construction was started by JAERI in 1990, has been operating since 1998 [4], and a small Chinese demo has been running since 2000 [5].

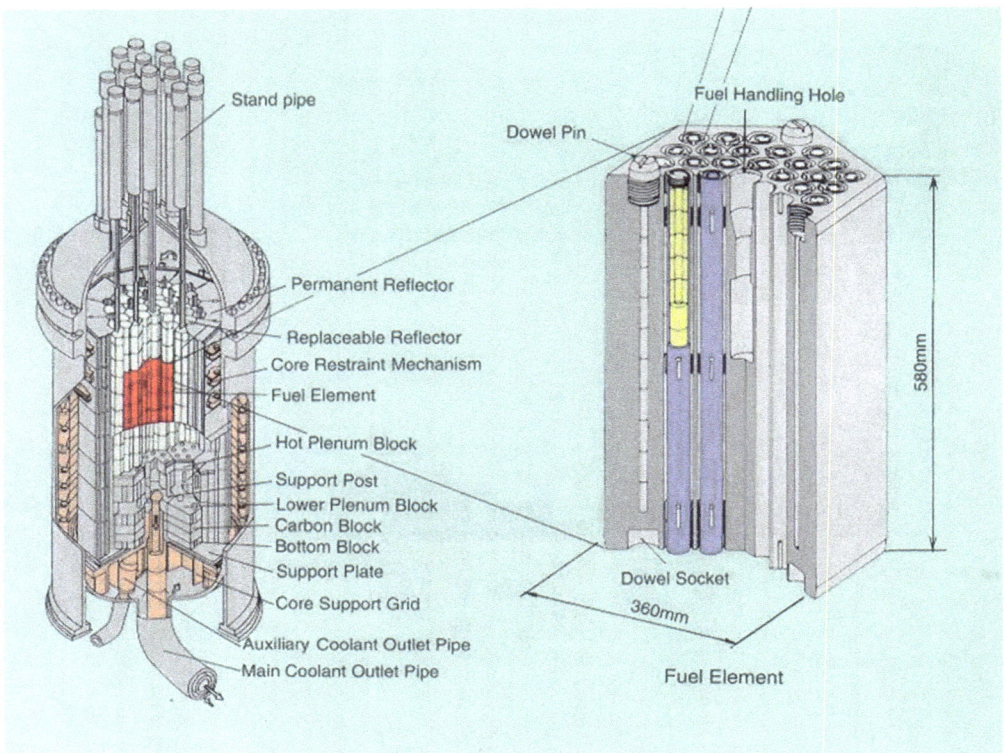

Figure 8.6. HTTR core and fuel element.

8.4. The "Astronuclear" Saga [6, 7]

Let us forget electricity for a while, and follow the HTR adventure in the wake of the famous APOLLO space programme

In parallel with their conventional program focusing on the development of chemical rockets, for 20 years or so, the US developed a project of nuclear propulsion rocket which, looking back, was incredibly ambitious. The goal was to heat liquid hydrogen at a temperature much above 2000 °C and for several dozens of minutes, using the heat generated by a nuclear reactor. For the sake of comparison, the Vulcain motors powering the Ariane V rocket are qualified to operate for no more than 10 minutes. Only the particle-based HTR fuel could allow dreams of such performances.

The most surprising is that, despite incredible technical difficulties, this project was a total success. From the tiny Kiwi to the big Phoebus, between 1959 and 1972, all records were consistently broken, as shown in the table below. The project was then completely dropped, as NASA never launched the APOLLO XVIII to XX rockets, although they had been completely built: 3 years after the first Moon landing, space was no longer a national priority. Over those years, the total cost of the Rover/NERVA programme amounted to

$ 1.5 billion. Since 2000, NASA and the DOE have somehow revived a few nuclear space propulsion projects. Whatever their future may be – and it is too early to guess – the early space attempt demonstrated the huge margins imbedded in the HTR fuel!

Table 8.2. A few Rover/NERVA tests.

Date	Test	Pmax (MWt)	Time at Pmax	Tmax outlet (°C)
7/1959	Kiwi 1	70	5 min	
7/1961	Kiwi B1A	300	30 s	
9/1962	Kiwi B1B	900	A few s	
5/1964	Kiwi B4D	1000	40 s	2220
7/1964	Kiwi B4E	900	8 min	2390
9/1964	NRX A2	1100	3.4 min	2300
5/1965	NRX A3	1122	13 min	2450
6/1965	PHOEBUS 1A	1090	10.5 min	2480
6/1966	NRX A5	1140	2 times 15 min	2450
2/1967	PHOEBUS 1B	1500	30 min	2445
12/1967	NRX A6	1100	62 min	2550
6/1968	PHOEBUS 2A	4300	12 min	2310
12/1968	PEEWEE	514	12 min	2750
7/1972	FURNACE	44	109 min, 4 tests	2450

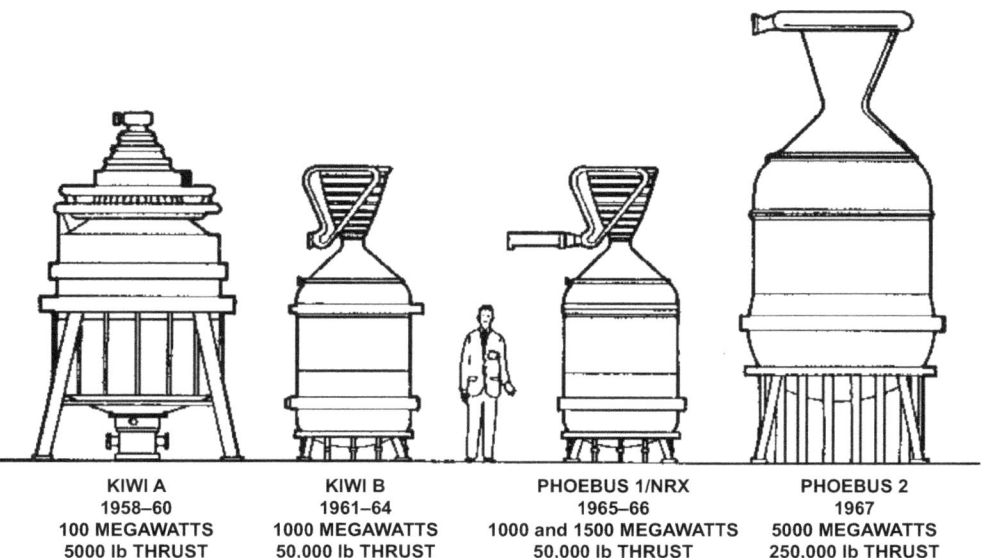

KIWI A
1958–60
100 MEGAWATTS
5000 lb THRUST

KIWI B
1961–64
1000 MEGAWATTS
50,000 lb THRUST

PHOEBUS 1/NRX
1965–66
1000 and 1500 MEGAWATTS
50,000 lb THRUST

PHOEBUS 2
1967
5000 MEGAWATTS
250,000 lb THRUST

Figure 8.7. The NERVA series of reactors.

Figure 8.8. The Phoebus 2 Test at Jackass Flats, Nevada.

8.5. Fort St Vrain and THTR Prototypes, the Thorium Cycle

8.5.1. Fort St Vrain

In 1968, only one year after the start-up of Peach Bottom, General Atomic started the construction of a 330 MWe HTR prototype on the Fort St Vrain site. The operator was to be

Table 8.3.

Characteristics	Fort St Vrain	THTR 300	1160 Project
Criticality/shutdown	1974/1989	1983/1989	-
Thermal power (MWt)	842	750	3000
Net electric power (MWe)	330	300	1160
Efficiency (%)	39	40	39
Active core height (m)	4.8	6	6.3
Equivalent diameter (m)	5.9	5.6	8.5
Power density (MW/m^3)	6.3	5.1	8.4
Inlet/outlet temperatures (°C)	405/780	260/750	320/740
Helium pressure (bars)	48	40	50
Particle	$UC_{2\,TRISO}/ThO_{2\,BISO}$	$UO_2 = ThO_{2\,BISO}$	$UC_{2\,TRISO}/ThO_{2\,BISO}$
Fuel element	Prism, 6 layers	Pebble	Prism, 8 layers
Number of fuel element	1462	6750000	3944
Average burnup (GWd/t)	100	<150	95
Fuel reload	Annual, per 1/6th	Continuous	Annual, per 1/4th

Public Service of Colorado, a small utility without any previous nuclear experience. Being a prototype, Fort Saint Vrain was built with federal support.

The general layout of the reactor (Figure 8.9) is strongly inspired by the 500 MWe St Laurent UNGG design, but with a reactor cavity six times smaller

The core is composed of 1483 prismatic fuel elements (Figure 8.4) superposed in six layers. Each fuel element is a hexagonal graphite prism. Cylindrical blind channels are

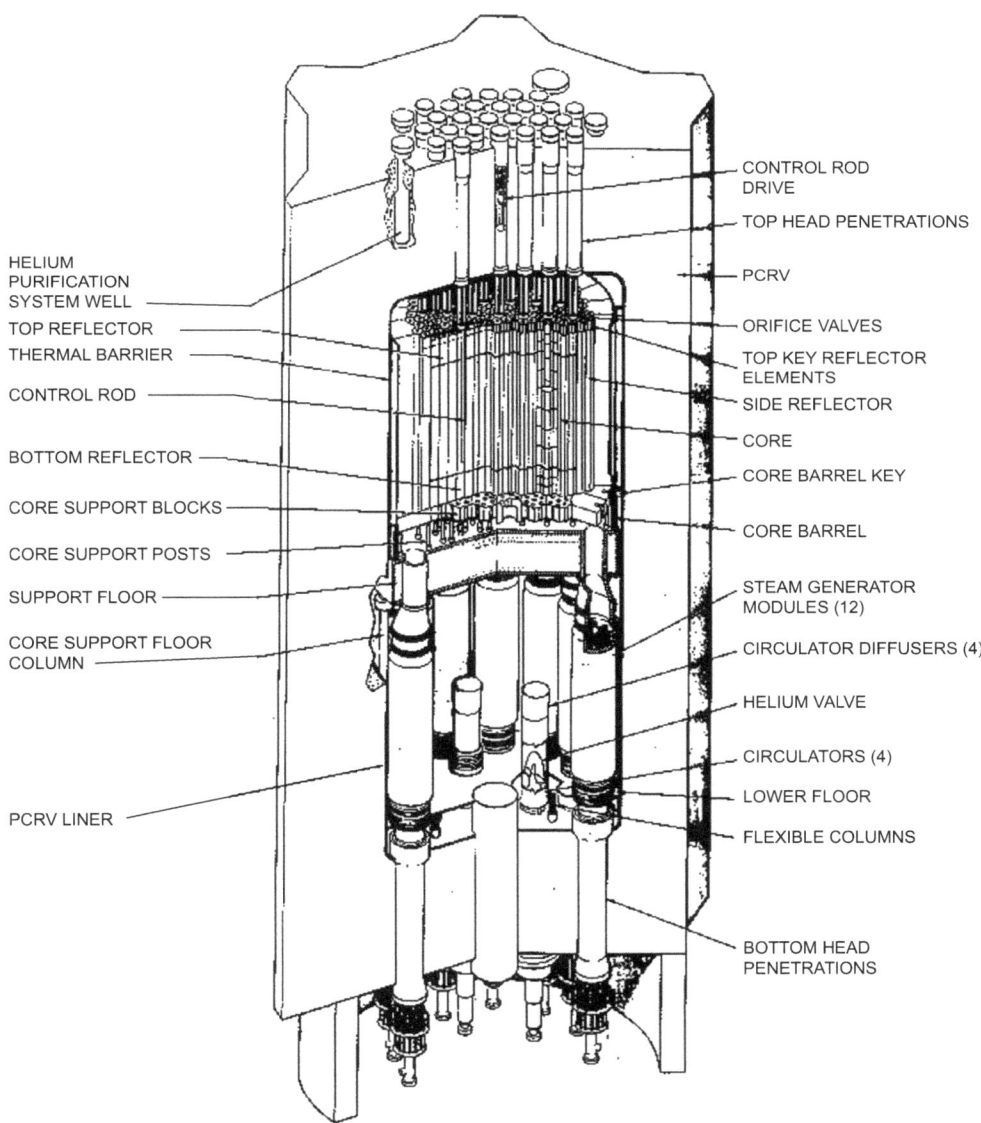

Figure 8.9. Fort Saint Vrain Reactor layout.

drilled in these prisms, filled with cylindrical compacts and surrounded by coolant channels in which helium flows downward under 48 bars of pressure.

The compacts are fabricated by mixing two types of particles: fissile particles with a kernel of HEU dicarbide $^{235}UC2$ and TRISO coating including one SiC layer, and fertile particles ThC2 with a BISO coating without silicon carbide. The core is axially and radially zoned, and it is reloaded by 1/6th at each annual outage

12 once-through helical steam generator modules are located below the core

Critical in 1974, Fort St Vrain was connected to the grid in 1976 and decommissioned in 1989 with a cumulative load factor of 30%. The fuel behaved successfully, but the reacto's overall design was rather a failure.

8.5.2. The Schmehausen (or Uentrop) THTR

In 1970, the German industry ordered a 300 MWe prototype based on the AVR model, to be built on the Schmehausen site. Extrapolation up 1200 MWe was expected of that design.

After a construction phase which lasted 14 years, the THTR operated for 4 years only. The decision to shut it down was taken for a combination of relatively minor technical difficulties and severe political difficulties, in the context of German public opinion becoming quite anti-nuclear and of bureaucratic feuds between the Land and Federal governments. A sad story, not unlike Superphénix's.

8.5.3. The thorium cycle [8–10]

The Fort St Vrain and THTR reactors being the largest ones having operated with thorium fuel, it is worth discussing this cycle, too often associated with ADSs alone.

Thorium, atomic number 90, is roughly three times more abundant in the Earth crust than uranium. Thorium only has one non-fissile isotope ^{232}Th. However, following one neutron capture and two β decays, it gives birth to ^{233}U, an excellent fissile material. With thermal neutrons, ^{233}U has $\eta = 2.28$, which means that strict control of neutron leakage (< 0.28) allows thermal neutron breeding (which was attempted in the last core of Shippingport, dubbed LWBR for Light Water Breeder Reactor).

^{233}U formation from ^{232}Th exactly parallels that of ^{239}Pu from ^{238}U, the main difference being the 27 days half-life of ^{233}Pa, much longer than ^{239}Np's 2 days. For good thorium utilization, one must avoid to submit protactinium to high neutron fluxes which would reduce ^{233}U formation through parasitic captures. The parallel between ^{233}U and ^{239}Pu extends to the fact that their daughter products induced by capture, ^{234}U and ^{240}Pu, are "fertile", which is not true of ^{236}U.

The main drawback of ^{233}U is that its production always generates traces of the ^{232}U isotope. ^{232}U gives birth by α decay and with a half-life of 72 years to ^{228}Th, α radioactive as well but with a relatively short half-life (1.9 y), and whose daughter products, ^{212}Bi and ^{208}Tl, emit very hard γ rays of several MeV. When ^{228}Th builds up, γ rays from its decay products and neutrons produced by (α, n) reactions with neighbouring light nuclei create serious problems in terms of radiation protection during the production of ^{233}U bearing fuel.

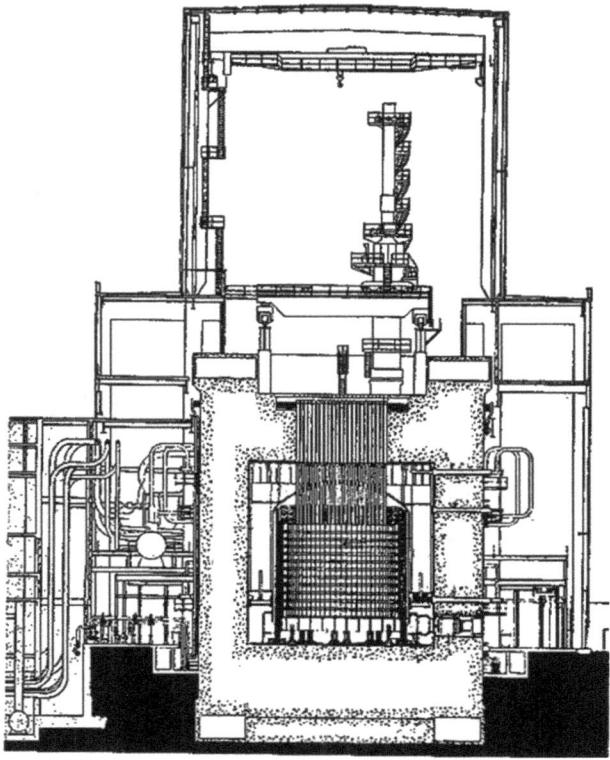

Figure 8.10. The Schmehausen THT.

In the 1960's and 1970's, the thorium cycle appeared attractive because it allowed good fertile material utilization. Breeding was hardly reachable but conversion factors greater than 0.8 were routine. Its comeback during the 1990's – on paper, that is – was mostly motivated by the fact that it does not produce plutonium nor, of course, higher actinides. It was also credited with producing a lot less long-lived radioactive waste, which is more debatable (it depends on the actual reprocessing flow-sheet).

8.6. False start in the USA

8.6.1. General atomic's 1160 and 770 project

At the very beginning of the 1970's, while Fort St Vrain was under construction and Peach Bottom was still operating, a few US utilities ordered from General Atomic, then a subsidiary of Gulf Oil (and soon of Shell as well), 8 large HTRs rating 1160 or 770 MWe, two very similar models with either 3 or 2 loops.

The general layout is derived from the pods-type AGR, but with downward coolant circulation to protect the upper structures and control rod mechanisms from the hot

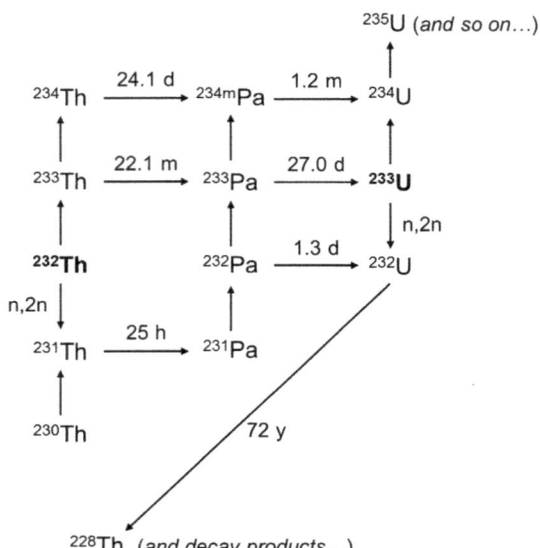

Figure 8.11. The thorium cycle.

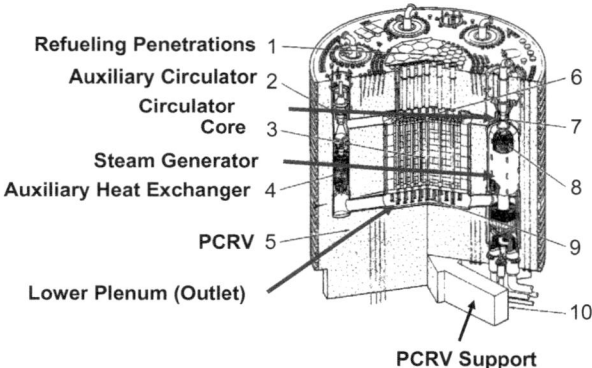

Figure 8.12. 1160 MWe HTR project (1973).

helium. The massive Pre-stressed Concrete Reactor Vessel PCRV, with vertical tendons and circumferential wire wrappings, is a very prominent feature of this design. In the centre of the PCRV, a main cavity contains the core while peripheral cavities (the "pods") contain the helical SGs and the helium circulators. The core, extrapolated from the Fort Saint Vrain design, was supported by a forest of graphite pillars above a lower plenum connected to the pods by hot ducts with thermal insulation.

The 1974 oil shock triggered an overnight rash of nuclear project cancellations in the USA: most HTR projects, being the latest ones ordered, were victims of this epidemic. In 1975, the vendor itself cancelled the last 2 survivors. And we should keep in mind that no order for a nuclear project took place in the USA between 1973 and 2010…

The mediocre performances of Fort St Vrain did not convince utilities from other countries to order HTR plants, but General Atomic, well introduced in the US Congress, managed to get, year after year, enough money from the US DOE budget to maintain a small but highly competent team of engineers and scientists. However, the thorium cycle was abandoned because it needed HEU to start the cycle and as a complement to ^{233}U because HTRs are not breeders. After 1974 and the Indian explosion, civilian use of HEU became taboo for non-proliferation reasons. Today, plutonium would more likely be used to start a thorium cycle.

8.6.2. The French HTR programme (first period)

As a member of EURATOM, France did join the Dragon Project, but its involvement remained very marginal until 1970. When France switched from UNGGs to PWRs (see Chapter 1), it appeared prudent to have an emergency alternative in case the "*filière américaine*" did not live up to its expectations. There was a short visit to Canada, resulting in the EL 600 HWR project, but HTRs seemed more in line with the French tradition of gas cooling. Such was the CEA's position, while some EDF teams were more attracted by the MSR.

After a late but significant involvement in the Dragon project – while the UK was completing the preliminary design of a large HTR to be built on the Oldbury site–the CEA signed a licensing agreement with General Atomic in 1972, to be paid in R&D services carried out in Saclay and Grenoble. Together with Technicatome, then Novatome when this company was established in 1974 to develop both FBRs and HTRs, the CEA and EdF made an offer to the Swiss utility EOS in 1976 to build a plant adapted from the GA 1160 project described above on the Verbois site.

After the US collapse, France looked toward Germany. Significant agreements were signed in 1977 to jointly develop both FBRs and HTRs. As the "window of opportunity" between PWRs and FBRs was closing – Fessenheim 1 was then critical and the construction of Superphénix was decided – the HTR programme was in search of a "niche". Two paths seemed promising: the direct cycle reactor on the one hand and, on the other, the process heat reactor for use in petrochemical industries (notably methane reforming for hydrogen production).

Because it is competitive in large units only, nuclear power has not yet found its way to the heat market, where local needs rarely exceed 100 MWt.

In preparation of the direct cycle, Germany built a 50 MWe helium turbine, using fuel oil burners as heat source, in Oberhausen in 1975. This demo facility which co-generated heat and electricity was a total success.

In 1979, to demonstrate to the government that it was able to terminate R&D programmes, the CEA put an end to all HTR developments, keeping only some irradiation contracts in the SILOÉ MTR.

8.6.3. An assessment of HTR programmes, as seen from 1980

Despite the US and German programmes' premature end, the results of this first stage of the HTR saga were far from negligible.

On the plus side

- This type of reactor can reach high thermal efficiency, as high as the best gas turbines.
- Cold fuel, refractory core, high thermal inertia, one-phase chemically inert coolant: all these elements result in a high level of safety and forgiveness of operator error.
- The particle-based fuel can accommodate any possible fuel cycle.
- The first small demos proved the feasibility of the concept (and the rocket programme demonstrated the existence of huge margins).
- The HTR is one of the very few concepts to offer real prospects of non-electrical uses of fission (together with the gas-cooled breeder... which exists only on paper).

On the minus side:

- The low core power density translates into a large vessel and therefore high capital costs.
- The GA 1160 Project did not include secondary containment.
- If a core meltdown is beyond credibility (though the SiC layer begins to deteriorate when the particle temperature exceeds 1600 °C), a massive water ingress in the hot core might provoke a dangerous weakening of the core support pillars by corrosion.
- The core itself is quite refractory... but the long-term behaviour of the materials outside the core exposed to very hot helium is of concern. This includes the concrete PCRV.
- Neither prototype was highly successful.

On the "plus or minus" side...

For all its great qualities, HTR fuel is not easily reprocessed – and reprocessing is indispensable if the thorium cycle is to be fully taken advantage of. The process which was originally developed at the laboratory scale at Idaho Falls involved the following stages:

- Crushing of the prismatic blocks to free the compacts. Graphite fragments would have been incinerated, thereby releasing ^{14}C, which would no longer be acceptable today.
- Burning of the compact matrix as well as the outer pyrocarbon layers. This operation bares the BISO kernels, while the TRISO keep their SiC shell.
- Dissolution of the BISO kernels in nitric acid: the remaining thorium is separated from ^{233}U by PUREX-type solvent extraction, while TRISO particles remain solid.
- Crushing of the TRISO SiC shell and burning of the inner pyrocarbon.
- Dissolution of the uranium kernel to recover the residual ^{235}U, contaminated with ^{236}U but still valuable.

The difficulty of reprocessing was considered a weakness in 1970, within a thorium cycle vision. After switching to LEU (about 8% ^{235}U enriched), and taking into account the very high burnups that could be achieved, the residual value of the remaining fissile materials is very low. Reprocessing is also less necessary because the spent fuel, whose graphite resists corrosion quite well at moderate temperatures, appears to constitute a rather acceptable waste form for direct disposal. The low power density of the core then becomes an asset. The fact that reprocessing is difficult and not very attractive is now advertised as an advantage in terms of non-proliferation.

In retrospect, HTRs have suffered above all from a bad timing of introduction. Their story might have been quite different if a few of the – better designed – large reactors ordered in the 1970's had actually been built. Their performance might have erased the bad memory of Fort St Vrain.

8.7. Why a renewed interest for HTRs?

8.7.1. A changing environment

Since the 1980's, a number of evolutions have triggered a renewed interest in HTRs, even from some PWR champions like AREVA-NP (formerly Framatome) or Westinghouse (at least when it was affiliated to BNFL).

1. To overcome the formidable "size effect" and attempt to remain competitive with plants adapted to the needs of emerging countries, i.e. much smaller than 1500+ MWe EPRs, many reactor vendors have considered "*modular*" concepts in the 1980's. A modular power plant would be built progressively by adding identical units, each small enough to have a high degree of prefabrication in factory and, therefore, requiring minimal construction – or rather assembly – time on-site. The progressiveness allows the first unit to start generating power, hence cash-flow, while later units are still under construction. Taking into account the weight of IDC[1] in the capital cost of a plant, both prefabrication and stepwise assembly offer the prospect of balancing the size penalty.

2. Below a certain size threshold, it becomes possible to remove the decay-heat after shutdown through purely passive means: thermal conduction and convection. Of course, the pressure vessel must be metallic. The threshold comes from the surface/volume ratio for a given power density. If the Safety Authorities accept the demonstration, expensive emergency core cooling systems could be dispensed with – there again, to overcome the size penalty.

3. Taking advantage of the aerospace development fallouts, gas turbines have spectacularly improved both in size and efficiency. This underlined the importance of thermal efficiency, one strong point of HTRs (Figure 8.13), and lent credibility to the feasibility of high power helium turbines, hence the direct cycle HTR.

---

[1] Interests during construction.

4. Strategic Arms Reduction agreements between the USA and the Russian Federation after the end of the Cold War have "freed" huge quantities of weapons-grade plutonium, which must be "disposed of". The best way is to burn this plutonium in power plants. With their wide fuel cycle flexibility, HTRs can be optimal plutonium burners.

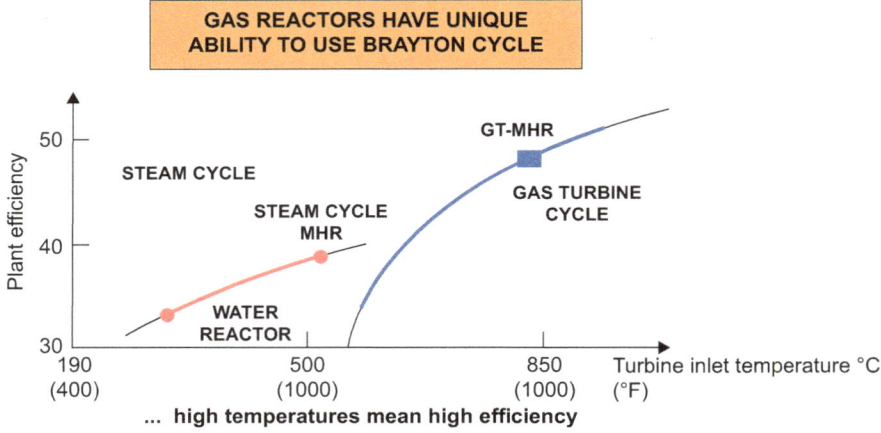

Figure 8.13. Achievable thermal efficiencies.

8.7.2. The GT-MHR, Gas turbine modular high temperature reactor [11–14]

As mentioned above, during the belt-tightening years, General Atomic still carried out some federally funded HTR development. In 1993, this development was boosted by an American-Russian decision to jointly design an HTR optimized to burn weapons-grade plutonium. Framatome (now AREVA-NP) joined the project in 1995, followed by Fuji Electric.

There were several motivations behind this (surprise) participation of Framatome: an opportunity to produce large steel components while the reactor is under construction, the possibility to introduce in its catalogue a unit smaller than the EPR but still competitive, the possibility to offer developing countries a plant whose safety is less "sophisticated", an opportunity for its design teams to work on an innovative project, an image of creativity for the company.

The GT-MHR nuclear island has the following characteristics:

- steel pressure vessel without thermal insulation to allow heat removal by radiation;

- prismatic "GA type" fuel elements constituting an annular core in order to maximize the surface-to-volume ratio (the center columns of the core, made of pure graphite, add to the overall thermal inertia);

- cooling panels inside the walls of the underground reactor containment building, operating in natural convection.

The conventional island features:

- a direct cycle helium turbine, allowing for excellent thermal efficiency and eliminating the risk of water ingress in the core;
- oil-free magnetic bearings (water ingress through the bearings was an endemic plague in Fort St Vrain).

The core and the single shaft turbo-alternator-compressors are located side by side in two separate steel vessels connected by a short co-axial duct. The whole layout is quite compact.

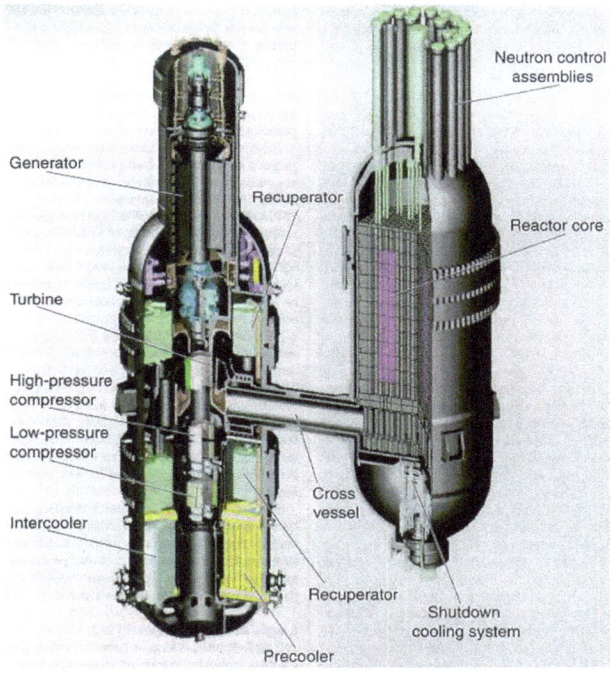

Figure 8.14. GT-MHR's two "bottles".

8.7.3. ESKOM PBMR pebble bed modular reactor [15]

While the GT-MHR remained directly in line with the General Atomic family, the PBMR derives from the Jülich AVR, revisited by the South African utility ESKOM. Table 8.4 summarises its original characteristics:

The PBMR is very simple and the fuel has been thoroughly qualified (but the German production facilities have disappeared). On the other hand, construction time and cost as quoted by ESKOM were not quite credible. Over the years, the PBMR increased its rating and adopted an annular core design, and South Africa stood ready to order the prototype… when the project was aborted in 2009.

Table 8.4.

Characteristic	PBMR	GT-MHR
Thermal Power (MWt)	265	600
Net electric power (MWe)	116	285
Efficiency (%)	44	47.5
Core inlet/outlet temperatures (°C)	536/900	490/850
Helium mass flow (kg/s)	140	320
Turbine inlet/outlet temperatures (°C)	751/554	850/510
Turbine inlet/outlet pressures (MPa)	7.0/4.3/2.6	7.0/2.6
Fuel element	Pebbles	Prismatic

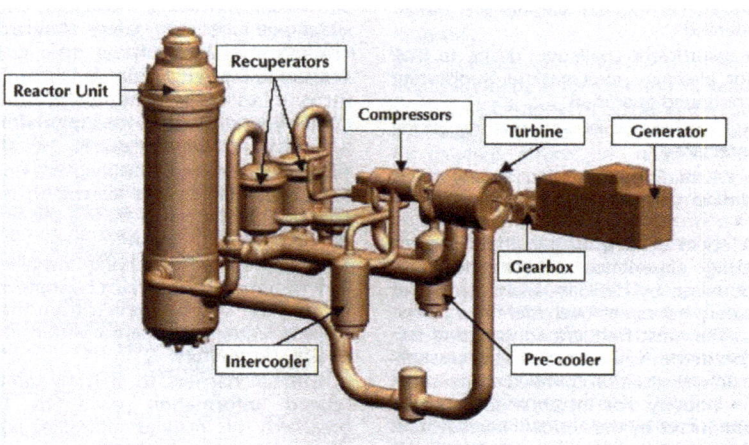

Figure 8.15. PBMR nuclear island.

8.7.4. The VHTR and ANTARES

The Very High Temperature Reactor VHTR is one of six concepts selected by the GIF (see Chapter 1) as a candidate model for Generation IV. Compared to the other concepts, the VHTR stands out: it offers very little improvement in terms of sustainability (fissile material utilization and waste management), but it opens nuclear fission to a wide range of new applications, the most promising of which appears to be hydrogen production in co-generation with electricity.

The first VHTR demonstration might be built in Idaho Falls with DOE financing[2], but the request for proposal was not yet issued at the beginning of 2015. This demonstration reactor would test both the reactor system and the various possible hydrogen production systems: high temperature electrolysis as well as a number of thermochemical water splitting cycles.

---

[2] As part of the NGNP, Next Generation Nuclear Plant program.

To be in a position to answer a potential DOE bid, AREVA-NC did develop the Antarès conceptual design of a VHTR co-generation demonstration reactor. Leaving out the direct cycle helium turbine of the GT-MHR, still considered a technological uncertainty, the Antarès reactor is coupled to the applications through an Intermediate gas-to-gas Heat Exchanger IHX. The high temperature calories can be used either to produce hydrogen or to generate electricity in a gas turbine (using an air-like mix of helium and nitrogen) and a bottoming conventional steam cycle generates additional electricity, like in a Combined Cycle Gas Turbine.

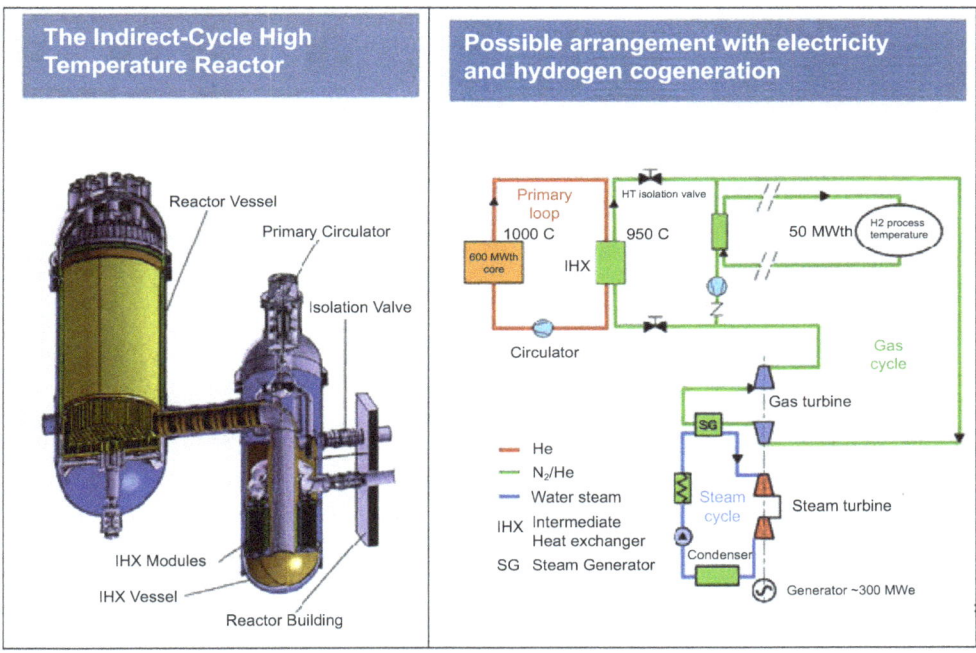

Figure 8.16. ANTARES conceptual design.

8.7.5. The Chinese HTR-PM

In 2014, the Chinese have started construction at Shidaowan of a prototype pebble-bed HTR with two 250 MWth modules connected to a single 210 MWe turbine.

References

[1] D. Bastien, *Réacteurs à haute température*. Techniques de l'ingénieur B 3190, 1993.
[2] G. Mélèse, R. Katz, *Thermal and Flow Design of Helium-Cooled Reactors*. ANS, 1984.
[3] W.K.H. Dent, *Gas-cooled reactors*, Nuclear Technology, Vol. 1. Clarendon Press Oxford, 1983.

[4] *Present Status of HTGR Research & Development.* Japan Atomic Energy Research Institute, February 1996.
[5] X. Yuanhui, *HTR-10: First step toward HTGRs.* Nuclear Europe Worldscan **11-12,** 1999.
[6] R.E. Malenfant, *The Solid-Core Heat-Exchanger Nuclear Rocket Program.* ANS Trans. **70**, June 1994.
[7] J.W. Simpson. *Nuclear Power from Underseas to Outer Space.* (Chapitre 14: The Astronuclear Technical Story) ANS 1995.
[8] Y. Cassagnou, *Comparaison des cycles Uranium et Thorium,* Electronucléaire, une présentation par des physiciens. Publication du CESN, 1999.
[9] A. Puill, A. Tsilanizara, *Nuclear energy in the long term. Thorium cycles.* Advances in Nuclear Fuel Management, March 1997.
[10] A. Puill, A. Tsilanizara, Y. K. Lee, *Cycle Thorium en REP : Etudes neutroniques – Radiotoxicité.* Rapport DMT 97/314.
[11] *Le GT-MHR, une nouvelle génération de centrale électrique.* Plaquette Framatome, 1999.
[12] M. Lecomte, *Pourquoi un regain d'intérêt pour la filière HTR ?* Communication Framatome, January 1999.
[13] *Le projet GT-MHR décrit par Framatome,* Documents ENERPRESSE, 16-17 août 1999.
[14] R. Kovan, *No showstoppers found for a GT-MHR route to plutonium disposition.* Nuclear Engineering International, August 1999.
[15] *PBMR Technical Description.* ESKOM Document PB-000000-25, May 1999.
[16] D. Whittal, A. Boudard, *Les sels fondus dans les systèmes nucléaires,* Electronucléaire, une présentation par des physiciens. Publication du CESN, 1999.

Molten Salt Reactors

B. Barré

9.1. Liquid fuel reactors [1–6]

Even more "exotic" than HTRs, Molten Salt Reactors or MSRs use fuel in liquid form and use this liquid itself as their primary coolant. It has been joked that PWRs were designed by engineers (and Navy men) while HTRs were designed by physicists. By comparison, MSRs were undoubtedly designed by chemists.

The concept of liquid fuel is quite attractive: no fuel fabrication, no question of heat transfer between fuel and primary coolant (they are the same fluid), no limitation of the burnup by the cladding integrity (there is no cladding), etc. It even includes an additional intrinsic safety feature, since coolant overheating decreases the fissile inventory in the core through thermal expansion.

On the other hand, a secondary fluid is still needed to transfer the calories to the electricity generating system (nobody, to my knowledge, proposed to directly "turbine" the primary fluid), the whole primary circuit is full of fission products (in terms of safety analysis, there are only two barriers) and its materials must withstand both radiation and corrosion, liquid leaks can create unusual risks of local criticality, etc.

Several types of liquid fuel reactors have been designed (and a few, built), with the main objective to breed in a thermal neutron spectrum. We have seen that such breeding is only possible in the ^{233}U-^{232}Th cycle, if all neutron losses have been minimized. Liquid fuel makes it possible to continuously eliminate fission products on a by-pass circuit, and to separate protactinium to let it decay out of flux.

Reactors were considered with ^{233}U dissolved in molten bismuth and graphite moderation. One reactor using a molten plutonium-bismuth eutectic as fuel was actually built. HR2, a 5 MWt prototype, was operated for a short time before corrosion levels became unacceptable: its core was a homogeneous solution of highly enriched uranium sulphate in heavy water, surrounded by a fertile blanket made of thorine (ThO2) suspended in heavy water, and the circuit was pressurized to 14 Mpa for reasonable thermal efficiency. Nowadays, the only concept still under consideration is the MSR.

9.2. MSRE, Molten Salt Reactor Experiment

Between 1965 and 1969, the 7.5 MWt MSRE demonstration reactor operated in Oak Ridge. To date, it remains the only MSR to have been built and operated. Compared with aqueous reactors, MSRs offer 3 advantages:

- uranium and thorium fluorides (or chlorides) are very stable and resistant to radiolysis. Therefore, they do not corrode stainless steel;

- far from their boiling temperature, molten salts – like liquid sodium – do not require pressurization;

- thorium is as soluble in molten fluorides as uranium: no need for thorine suspension in the blanket.

In the MSRE, the salt (1% ^{235}UF$_4$, 1% ThF$_4$, 5% ZrF$_4$, 70% ^{7}LiF, 34% BeF) flowed in nickel alloy channels across a cylindrical core made of graphite logs arranged in a compact hexagonal mesh. The fluid temperature was 635 °C at the inlet and 663 °C at the outlet, and calories were transferred to a secondary circuit filled with a non-fissile molten salt (66% LiF, 34% BeF).

During its last year of operation, the MSRE was fuelled with ^{233}U. 218 kg of ^{233}U had been extracted from the core using the fluoride volatilization process: fluorhydric acid attack followed by UF6 volatilization and condensation as UF$_4$. It even operated for some time with plutonium/thorium. After a very satisfactory period of operation, it was discovered during a shutdown phase that the Hastelloy N circuits had been severely corroded. The reactor is now decommissioned [7].

9.3. The Breeder MSR Projects

The success of the MSRE led the Oak Ridge teams to design a breeder MSR, without a blanket, illustrated in Figure 9.1. Helium bubbling would eliminate gaseous fission products (krypton and xenon) from the liquid salt. This programme was terminated in the mid-1970's in favour of the Clinch River LMFBR prototype, which was also dropped in 1983. Later studies determined that positive feedback coefficients can occur in the MSBR, leading to a potentially unstable reactor.

Between 1976 and 1982, preliminary studies of various non breeding MSRs were carried out, notably by EdF teams. The thermal spectrum concepts relied on fluoride fuel while the fast spectrum concepts used chlorides. Lead was considered as the secondary coolant.

9.4. Generation IV MSRs

The MSR concept resurfaced with the study of minor actinides transmutation in ADSs and projects such as the Los Alamos ATW (Accelerator Transmutation of Wastes) studies or the OMEGA programme of JAERI (now JAEA).

The MSR concept was also selected by GIF as one of the six Generation IV models, considered to be the most futuristic. In France, studies of MSRs in view of Generation IV are mostly carried out by the CNRS with some support by EdF. New concepts which appear interesting have been developed [5,6], using a thorium-^{233}U cycle that may be started with plutonium. There are two main motivations behind this renewed interest: fissile inventory is very low for a breeder, and no plutonium, neptunium nor americium is formed in the pure thorium cycle (Figure 9.2).

9 – Molten Salt Reactors

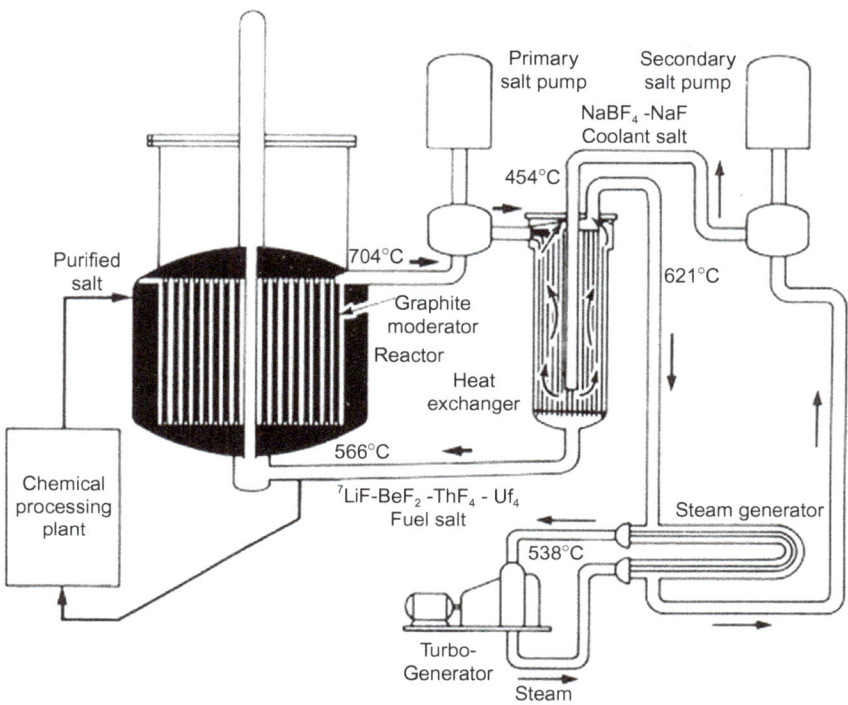

Figure 9.1. Molten salt breeder reactor.

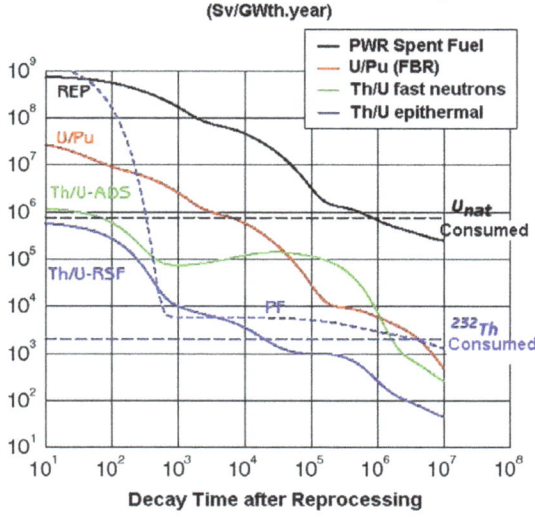

Figure 9.2. Actinides decay in various cycles.

The latest CNRS design is a fast neutron reactor, which does not have to deal with the problem of damage in the graphite structure (Figure 9.3).

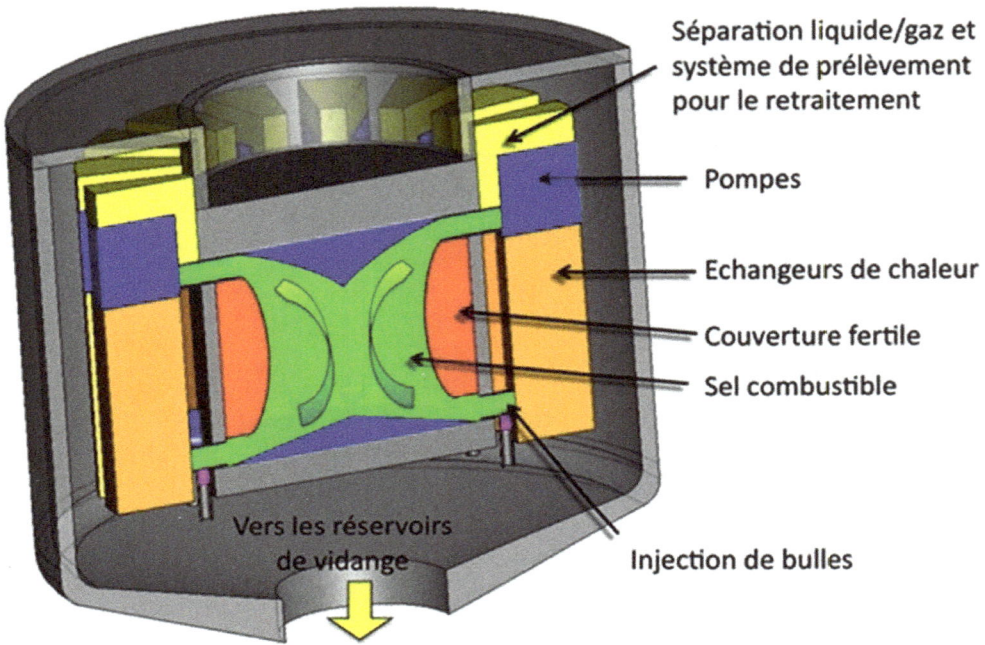

Figure 9.3. MSFR Molten Salt Fast Reactor.

9.5. AHTR

Before concluding this chapter, it is interesting to mention studies carried out in various US national laboratories on a mixed concept which is not included in the official Generation IV selection, the Advanced High Temperature Reactor AHTR.

It is a "mixed" concept because it uses the general layout of the Prism LMFBR project, the fuel element of the prismatic HTR, and a (non-fissile) molten salt as coolant.

It is interesting because it combines the attractive qualities of the particle-based HTR fuel with the use of a low pressure transparent liquid coolant, far from its boiling point, within a design of high thermal inertia (Figure 9.4).

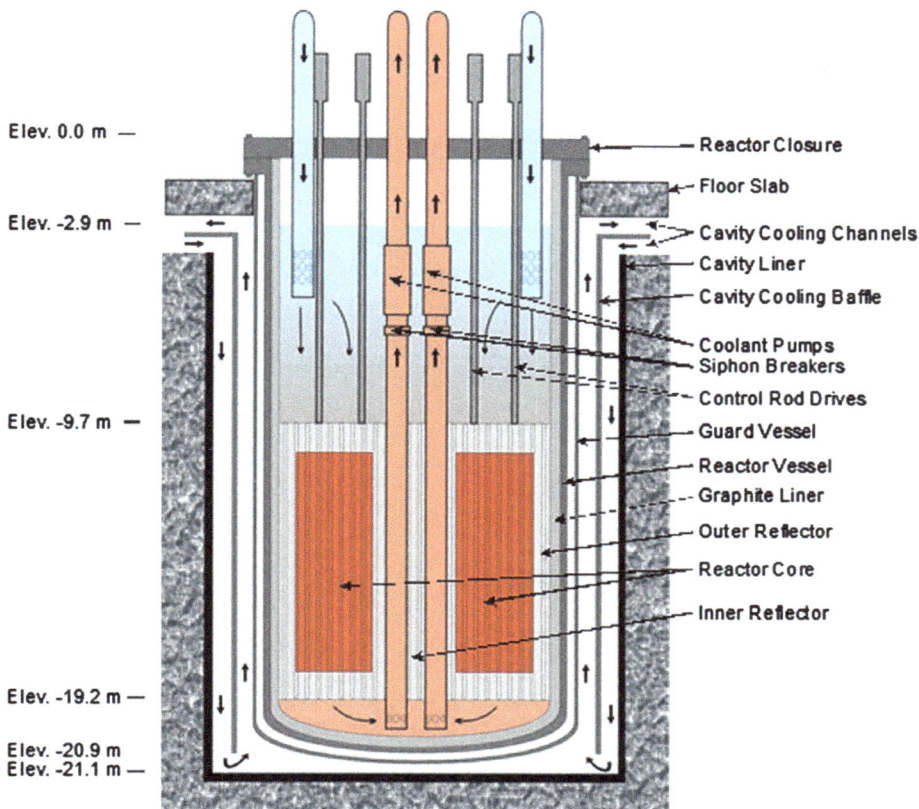

Figure 9.4. A 2000 MWt AHTR concept.

References

[1] *ANTARES, the AREVA HTR-VHTR Design.* AREVA publications, 2005.
[2] *Filière "Sels Fondus".* Dossier Technique CEA-EdF, 1977.
[3] A.M. Weinberg, *The First Nuclear Era*, (Chapter 6: The Fluid Fuel Breeder). AIP Press, 1994.
[4] J. Smith, *Novel reactor concepts, in* Nuclear Technology, Vol. 1. Clarendon Press Oxford, 1983.
[5] L. Mathieu et al., *Proposal for a Simplified Thorium Molten Salt Reactor.* GLOBAL 2005, Tsukuba, Japan.
[6] E. Merle-Lucotte et al., *Fast Thorium Molten salt Reactors started with Plutonium.* ICAPP'06, Reno, USA.
[7] M.R. Jugan et al., *Defueling the ORNL Molten Salt Reactor Experiment Facility.* Radwaste Magazine, Nov-Dec. 1999.

10 Liquid metal cooled fast neutron reactors

P. Anzieu

10.1. Introduction

The primary characteristic of fast neutron reactors, FRs, is that they operate with a non-slowed neutron spectrum. They do not use any moderator. This choice, which gives rise to a certain number of technical challenges, has resulted primarily from the desire to have a breeder reactor capable of optimising the use of natural resources by burning not only ^{235}U but, via its conversion into plutonium, a significant portion of ^{238}U.

Development of the industry relies essentially on sodium-cooled FRs, which will now be described. A typical sodium-cooled fast reactor is shown in Figure 10.1. In what follows, the European Superphenix reactor, SPX, is taken as an example of this industry.

10.1.1. Breeding

In any reactor core using uranium as a resource, fission energy comes essentially from ^{235}U or from ^{239}Pu; the excess neutrons convert ^{238}U into plutonium ^{239}Pu via the reactions[1]:

$$^{238}U \xrightarrow{(n,\gamma)} {}^{239}U \xrightarrow{(\beta^-)} {}^{239}Np \xrightarrow{(\beta^-)} {}^{239}Pu$$

^{239}U and ^{239}Np have short half-lives and everything happens as if ^{238}U were to generate ^{239}Pu directly. Fissile material is therefore consumed (^{235}U) and produced (^{239}Pu) simultaneously. If N_P is the number of fissile nuclei produced and N_C the number of fissile nuclei consumed, the **conversion rate** is CR = N_P/N_C. The **breeding gain** is defined as BG = CR − 1.

With the uranium plutonium cycle, only FRs can be breeders. CR can reach, depending on the core design, 1.3 or 1.4 although, whereas in a PWR CR barely exceeds 0.6.

---
[1] There is a similar schema for the thorium uranium system:

$$^{232}Th \xrightarrow{(n,\gamma)} {}^{233}Th \xrightarrow{(\beta^-)} {}^{233}Pa \xrightarrow{(\beta^-)} {}^{233}U$$

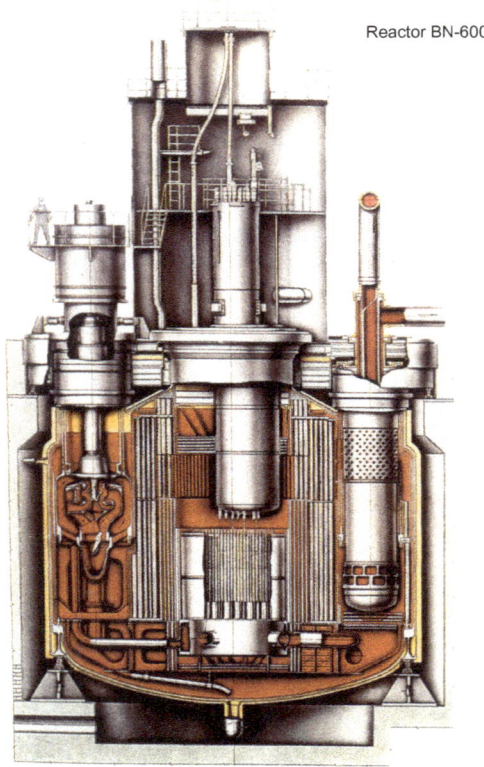

Figure 10.1. The Russian sodium-cooled reactor BN-600 (from B.A. Vasiliev, OKBM, 2007).

This property results from the physical characteristics of the nuclei, as shown by the following reasoning:

- ν being the number of neutrons produced by fission
- α being the ratio probability of capture/probability of fission (σ_f/σ_c)
- η the number of neutrons produced per neutron absorbed, is deduced from these two nuclei characteristics by $\eta = \nu\sigma_f/(\sigma_f + \sigma_c) = \nu/(1 + \alpha)$. For the main fissile nucle, η varies as shown in Figure 10.2.

A simplified breeding criterion is related to η: in order to provide, for each neutron absorbed, one fission neutron to sustain the chain reaction and one neutron for conversion, it is essential that η be greater than 2 to take losses into account. This is possible in particular for high energy neutron and Pu material.

If we now apply this reasoning at the scale of a reactor core and compare neutron economy for a light water reactor and a fast reactor respectively, it appears that in an FR,

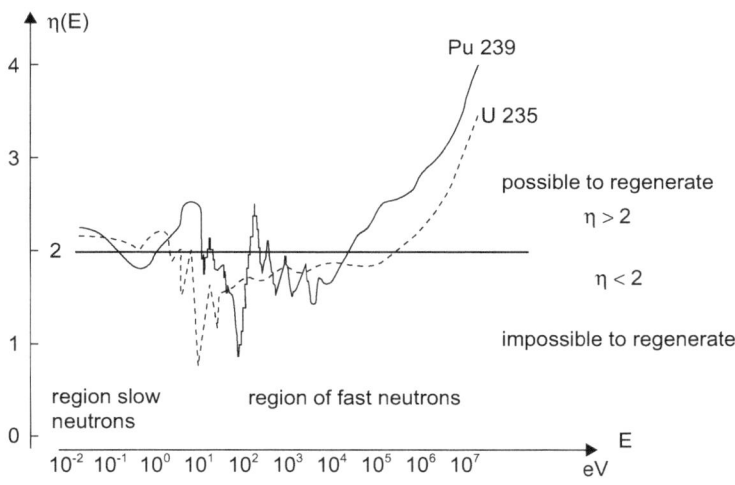

Figure 10.2. Variation of η, the number of neutrons produced per neutron absorbed, as a function of this initial neutron energy [from Ertaud, Revue générale de Thermique, May 1990].

there are actually far more neutrons for the conversion; firstly inside the core, where the economy is already greater than that for PWRs and second, outside. And this is actually the conversion in the blankets surrounding the core which leads to a surplus budget; this is not the case for PWRs, where very few neutrons escape from the core.

An established fleet of FRs is therefore fed with natural or depleted U, and with Pu. Breeding allows the maintenance of a Pu inventory corresponding to the monitoring of the energy demand using spent fuel re-processing facilities; Pu recovered by re-processing goes into the cycle and finally, the fleet of reactors operates by consuming ^{238}U.

The breeding capability, recognised very early on, results in a multiplication by a factor of approximately 70 of the energy drawn from a given reserve of natural uranium compared to that which light water reactors allow, which multiplies the natural usable reserves accordingly and leads to a possible use of nuclear energy for several thousand years. This is the only way to place nuclear power as a sustainable energy source at a level significantly greater than that for petroleum and gas, and even for coal.

10.1.2. Waste incineration

Most transuranium isotopes have a higher probability of fission than capture, if the neutron spectrum is fast rather than thermal (see chart of Table 10.1, in which the fission/capture ratio is shown). This, added to the surplus neutrons available by fission, allows the consideration of incineration of the minor actinides, americium, neptunium and curium, by transmutation. Heavy atoms are converted into lighter isotopes with much shorter half-lives. FRs therefore can achieve a reduction of the ultimate quantity of radioactive waste.

Table 10.1. Fission cross-section σ_f /capture cross-section σ_c ratio depending on the neutron spectrum.

Isotope	Thermal neutron reactor LWR			Fast neutron reactor FR		
	σ_f	σ_c	$10*\sigma_f/\sigma_c$	σ_f	σ_c	$10*\sigma_f/\sigma_c$
^{235}U	38,8	8,7	44,60	1,98	0,57	34,74
^{238}U	0,103	0,86	1,20	0,04	0,3	1,33
^{239}Pu$_u$	102	58,7	17,38	1,86	0,56	33,21
^{240}Pu	0,53	210,2	0,03	0,36	0,57	6,32
^{241}Pu	102,2	40,9	24,99	2,49	0,47	52,98
^{242}Pu	0,44	28,8	0,15	0,24	0,44	5,45
^{237}Np	0,52	33	**0,16**	0,32	1,7	**1,88**
^{241}Am	1,1	110	**0,10**	0,27	2	**1,35**
^{243}Am	0,44	49	**0,09**	0,21	1,8	**1,17**
^{244}Cm	1	16	**0,63**	0,42	0,6	**7,00**
^{245}Cm	116	17	**68,24**	5,1	0,9	**56,67**

10.1.3. Situation of the industry

All countries that have embarked on nuclear energy development have sought to master FR technology (except for Canada) and have rapidly integrated into their programs the development of this system (see Table 10.2).

The exclusion of water, moderator *par excellence*, has left as choice of coolant either gases or liquid metals. It is remarkable that all constructions have chosen this second option, undoubtedly owing to the perception of certain technological difficulties with the use of gas (for example, pressure). The FR system is now mature. Although initially some small size reactors used Mercury or NaK, all have since chosen sodium, for which the total characteristics balance was actually the most appealing.

With six reactors in operation in 2012, of which one was generating 560 MWe of electricity (see Figure 10.3), the sodium system only accounted for 0.2% of the world's production of nuclear electricity. New reactors are nevertheless under construction (see Figure 10.4).

10.2. Description of Superphenix

10.2.1. Principles

The reactor is of an integrated, or pool, type; In the integrated type, the entire primary system is contained inside the reactor vessel; this arrangement is fundamentally

Table 10.2. Liquid metal reactors in the world.

(O = in Operation, S = Stopped, C = under Construction)

	NAME	COUNTRY	POWER	CRITICALITY	STATUS
EXPERIMENTAL	CLEMENTINE (Hg)	USA	0,02 MW	1946	S: 1952
	EBR1	USA	1,4 MW	1951	S: 1963
	BR 1	Russia	0,03 MW	1955	S:
	BR 2	Russia	0,2 MW	1956	S: 1957
	BR 5 - BR 10	Russia	5/10 MW	1958/1973	S: 2002
	DFR (NaK)	GB	75 MW	1959	S: 1977
	LAMPRE	USA	1 MW	1961	S: 1965
	EBR	USA	60 MW	1963	S: 1993
	RAPSODIE	France	24/40 MW	1967/1970	S: 1983
	BOR 60	Russia	60 MW	1968	O
	SEFOR	USA	20 MW	1969	S: 1972
	KNK 1 - KNK 2	Germany	60 MW	1972/1977	S: 1991
	JEFR - JOYO	Japan	50 MW	1977	O
	FFTF	USA	400 MW	1980	S: 1992
	FBTR	India	40 MW	1985	O
	PEC	Italia	120 MW	NO	S: 1987
	CEFR	China	65 MW	2010	O
DEMO	EFFBR - FERMI	USA	100 MWe	1963	S: 1972
	BN350	Kazakhstan	150 MWe	1972	S: 1999
	PHENIX	France	250 MWe	1973	S: 2009
	PFR	GB	250 MWe	1974	S: 1994
	SNR 300	Germany	300 MWe	NO	S: 1992
	JPFR - MONJU	Japan	280 MWe	1992	O
	CRBR	USA	350 MWe	NO	S:
PROTO	BN-600	Russia	600 MWe	1980	O
	SUPERPHENIX	France	1200 MWe	1985	S: 1998
	PFBR	India	470 MWe	±2015	C
	BN-800	Russia	750 MWe	2014	O

concerned with providing the primary sodium with a containment that is geometrically simple, yet delimited, and having a mass of sodium with high thermal inertia around the core. In contrast, the primary vessel is large and its upper closure slab, which in particular holds the control rod mechanisms, circulation pumps, heat exchangers with the intermediate circuit and the assembly handling system, is a complex unit.

The debate between the proponents of this pool design and those who favour the loop solution, with outside circuits (see Figure 10.5), has been polarised, with Western Europe in favour of the pool solution, adopted for Superphenix and renewed for the 1500 MWe EFR project. It has also been chosen in Russia, Kazakhstan, India and China, whereas in Japan the loop system has been chosen principally for considerations related to the seismic strength of the reactor block.

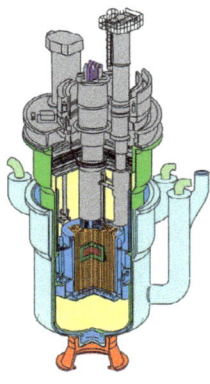 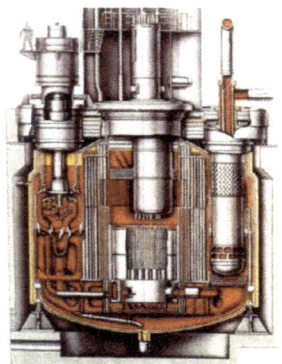 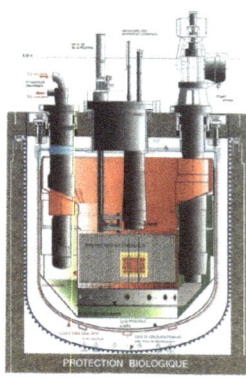

280 MWe Monju in Japan
[from Kazumoto ITO,
JAEA, June 2007]

600 MWe BN600 in Russia
[from OKBM, 2007]

250 MWe Phenix in France
[from CEA]

Figure 10.3. Schematic of several Sodium-cooled fast reactors under operation in 2009.

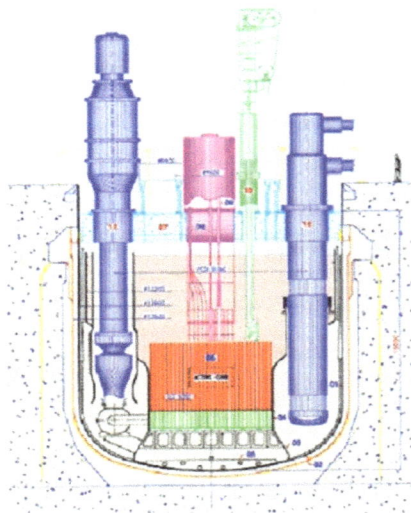 

The 500 MWe PFBR in India (criticality expected in 2015) [from S. Chetal, IGCAR, India, 2006]

The 800 MWe BN800 in Russia (criticality in 2014) [from V.I. Kostin et al., OKBM, October 2007]

Figure 10.4. Schematic of sodium-cooled fast reactors recently built (in 2014).

10.2.2. General design

Inside this primary vessel (Figures 10.6 and 10.8) there are:

- The core, a cylindrical pancake 1 m in height and 3.76 m in diameter, which rests on the base of the vessel via a supporting structure, and is surmounted by a structure bearing the instrumentation and the control bars - the core cover plug;

10 – Liquid metal cooled fast neutron reactors

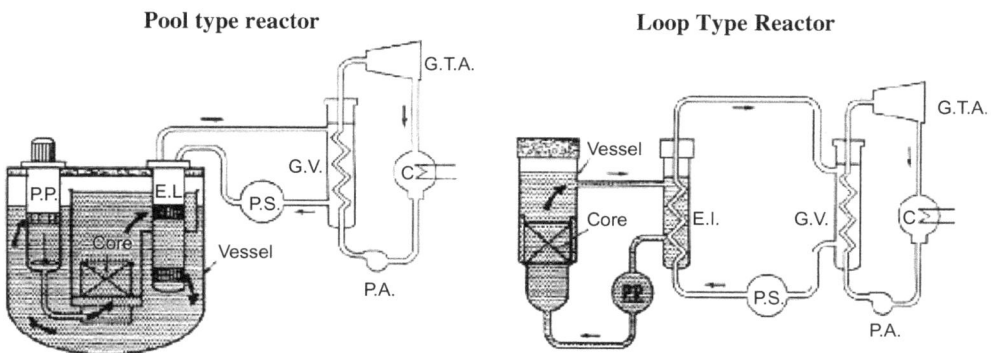

Figure 10.5. Pool and loop type reactor (P.P.: primary pump, E.I.: intermediate heat exchanger, P.S.: secondary pump, G.V.: steam generator, P.A.: feed pump, G.T.A.: turbo alternator group, C: Condensor).

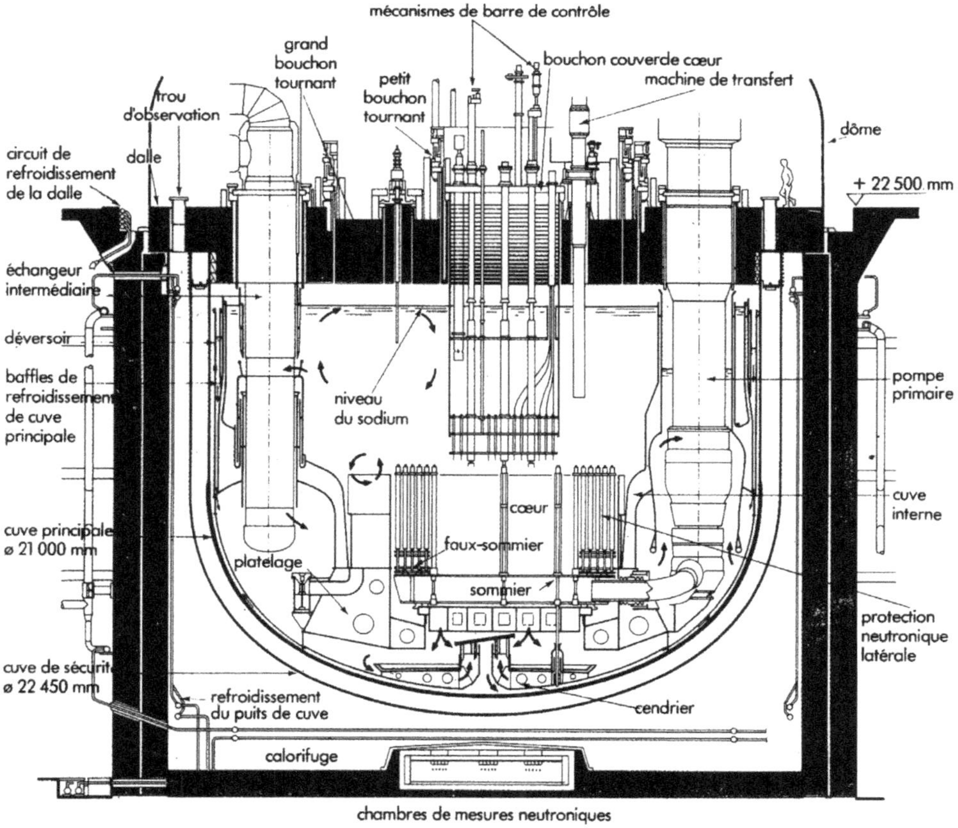

Figure 10.6. Section of the primary circuit of Superphenix.

- The primary circuit, delineated by an inner vessel, which separates the hot sodium from the cold sodium;

- 4 pumps, 8 intermediate exchangers;

- the sodium purification system.

The sodium is covered with an argon atmosphere. The intermediate sodium circuits are, with the exception of the intermediate heat exchangers, outside the primary tank; they transfer the heat to the steam generators (Figure 10.7). There are four loops, each incorporating a pump, two intermediate heat exchangers and a steam generator.

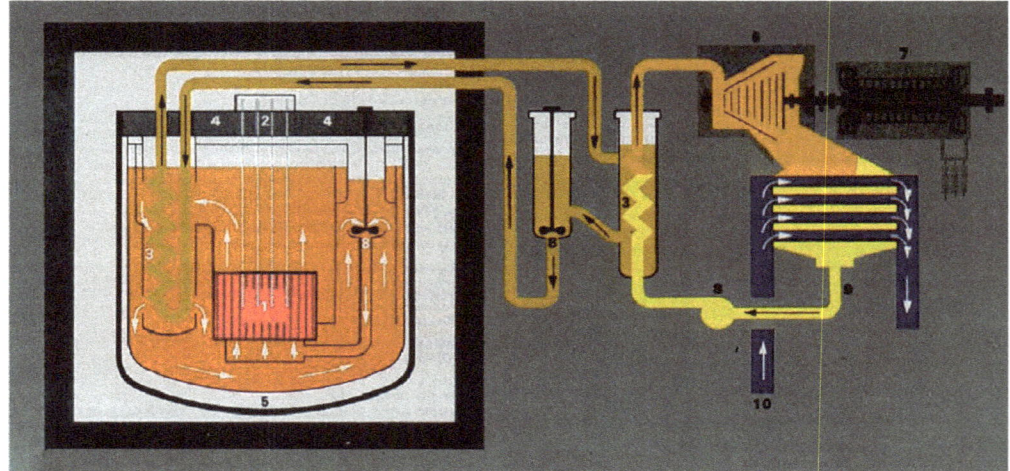

Figure 10.7. Block diagram of an integrated FR.

The steam generators are located in four distinct buildings separated from the reactor building for safety reasons. They feed two 600 MWe turbines; the condensers are cooled with water from the river Rhône.

10.2.3. Core and fuel

The core consists of a set of hexagonal subassemblies forming a compact grid; they are fixed vertically by their lower end or nozzle in a supporting structure named diagrid. Besides this geometric positioning function, the nozzle bears a flow control device.

Fuel subassemblies contain fissile matter; inside the hexagonal tube is a bundle of 271 cylindrical pins, arranged compactly as a triangle; these pins are spaced by helical wires (one wire per pin) delineating the cooling channels and mixing the sodium during its movement up the inside of the subassembly (Figure 10.9).

The pins are sealed by plugs welded to the ends; they contain the fuel in the form of a one metre high stack of ring-shaped cylindrical oxide pellets with mixed oxide of uranium and plutonium; this fissile column is framed by two axial fertile columns of depleted UO_2,

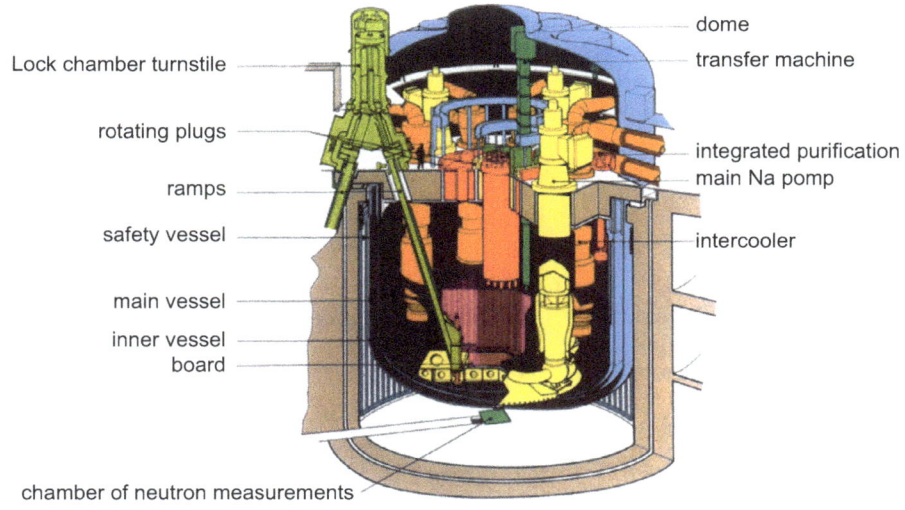

Figure 10.8. Exploded view of Superphenix.

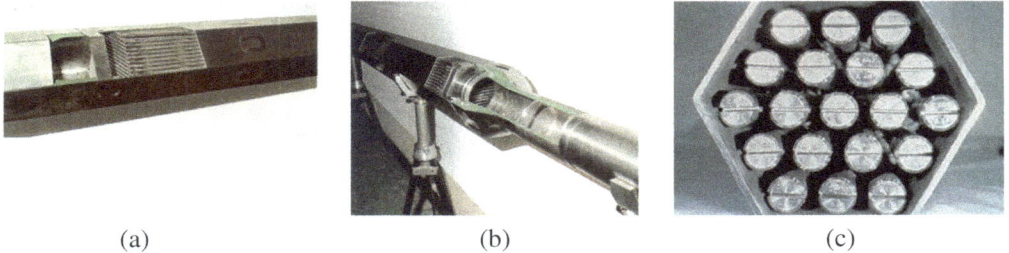

(a)   (b)   (c)

Figure 10.9. (a) Bundle of pins in its canister, (b) subassembly nozzle, (c) bundle of rods for a neutron collimator, seen from above.

each 30 cm, used for Pu production; the remainder of the volume accommodates the fission gases formed during irradiation. The design is the result of a compromise between the search for compactness and maintaining the stresses endured by the clad at a level ensuring the required lifetime.

The upper part, or sub-assembly head, is the gripping device; the presence of shielding materials reduces irradiation of the upper structures of the reactor block.

Around the fissile zone (two first rows, in black and dark grey on Figure 10.10) is arranged a fertile zone consisting of three subassembly rings (in grey) designed on the same principle as the fuels; the fertile material (depleted UO_2) is also in the form of cladded pellets, but the pins are larger and therefore fewer in number, given the low power which they produce.

24 absorber assemblies (in dashed grey) distributed within the fissile zone provide the control functions. They are divided into 21 assemblies constituting the main control system, and 3 emergency shutdown systems. They have in common a vertically moving part, which

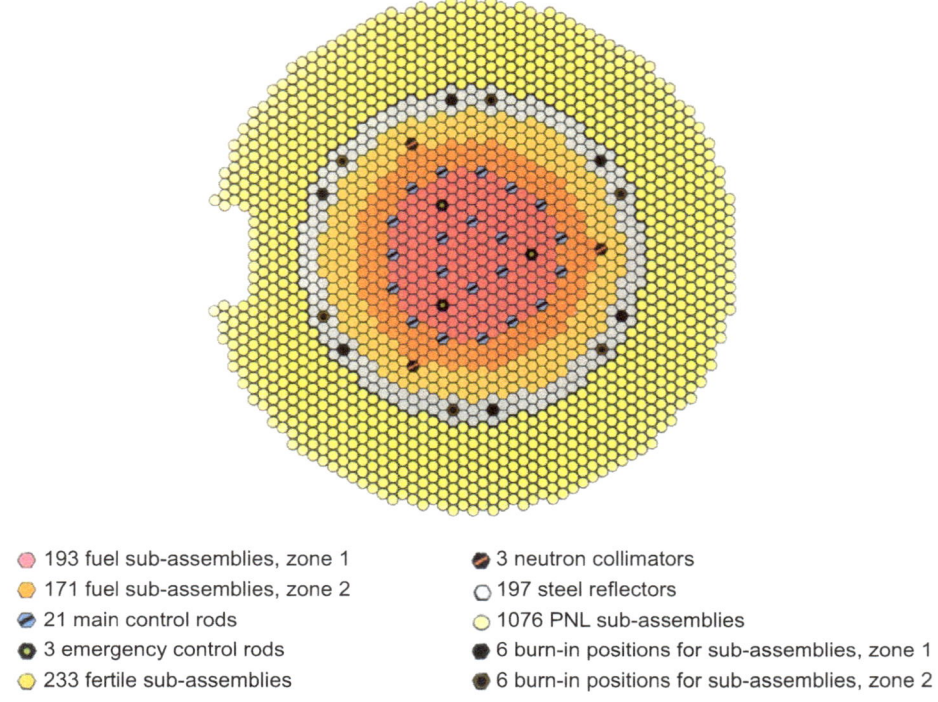

- 193 fuel sub-assemblies, zone 1
- 171 fuel sub-assemblies, zone 2
- 21 main control rods
- 3 emergency control rods
- 233 fertile sub-assemblies
- 3 neutron collimators
- 197 steel reflectors
- 1076 PNL sub-assemblies
- 6 burn-in positions for sub-assemblies, zone 1
- 6 burn-in positions for sub-assemblies, zone 2

Figure 10.10. Arrangement of the Superphenix core.

contains the absorber material which can be inserted in the core inside a hexagonal canister with the same outside dimensions as the fissile and fertile subassemblies. For the main system, this is a bundle of cylindrical rods; for the emergency one, three articulated stages facilitate shut down in a core which may be distorted.

Several steel neutron shielding subassembly rings (in light green) surround the core; they form a natural restraint system (maintaining the core in position without any other mechanical device), and considerably reduce the radioactivity of the internal structures and the vessels, as well as that of the sodium in the intermediate circuit. The radiation received by the vessel is not significant and is not taken into account in the evaluation of its lifetime.

Finally, if the power density of the fissile core is high, of the order of 300 MW/m^3, which makes it very compact, adding fertile elements and neutron shielding increases its volume considerably, typically by a factor of three.

10.2.4. Handling the assemblies

The introduction of fresh subassemblies and evacuation after irradiation of spent subassemblies must take into account the restrictions inherent with the use of sodium.

The closure slab covering the reactor has two off centre, rotating, intertwined circular plugs, (Figure 10.11); the combination of movements allows the positioning of two handling machines supported by the small rotating plug above any site in the core.

These machines move the assemblies vertically and they can therefore be positioned to suit. The transfer from and to the outside of the primary vessel is done via an inclined ramp which communicates by means of a rotating transfer lock with a symmetrical external ramp; this leads into a sodium storage drum (see the overall feature on Figure 10.12). The subassemblies leaving the reactor then pass via a cleaning facility which eliminates the sodium, and are then sent to storage in a water pool.

Figure 10.11. The closure slab of Superphenix.

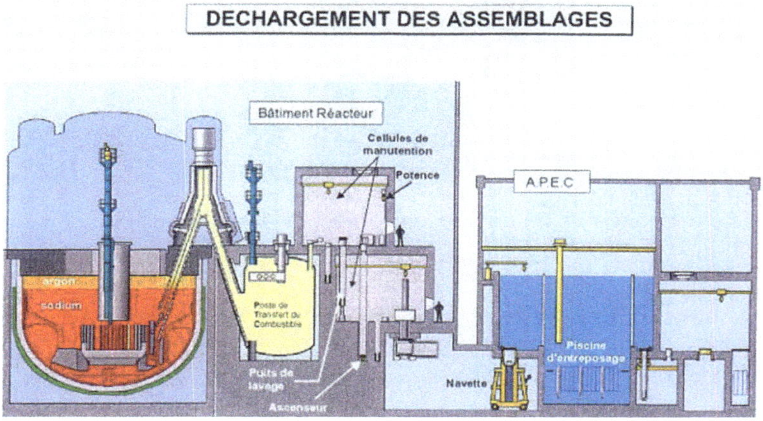

Figure 10.12. The fuel loading-unloading system.

Relative to this schema, inherited from Phenix, a notable modification has been contributed by the EFR project; the transfer to the outside is made through the slab via special casks. This simplification has been made possible by the reduction of the decay heat criteria for handling.

10.2.5. Reactor block

Inside the primary vessel, a complex metallic structure, the inner vessel, divides the sodium volume into two zones. The sodium, impelled by the primary pumps, passes from the cold zone to the hot zone through the core, from which it evacuates the energy, and from the hot zone to the cold zone via heat exchangers which transfer this energy to the intermediate sodium circuit. With the exception of the connection between the pumps and the diagrid, there is no piping. The importance of thermal-hydraulics for controlling flows and heat is apparent from this description of the whole, which characterises the integrated concept.

The inner vessel incorporates two walls. It is traversed by four chimneys isolating the primary pumps from the hot sodium and by the intermediate exchangers. A system of internal baffles provides a circulation of cold sodium along the wall of the primary vessel in order to maintain the latter at a low temperature, < 400 °C (Figure 10.13).

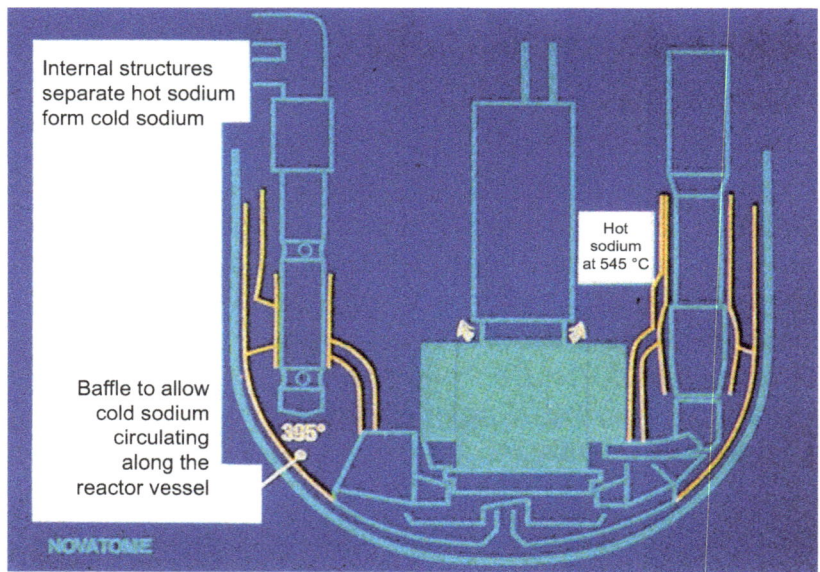

Figure 10.13. Internal structures of the reactor block.

A catcher plate is placed in the lower part to collect the molten core in the event of a severe accident.

The primary sodium purification system is used to control its purity. Two independent devices are immersed in the hot collector where they take the sodium which passes through metallic filters where, by cooling, the latter is cleansed of its impurities.

10.2.6. Sodium circuits

The pumps are mechanical with vertical shafts. The intermediate heat exchangers, also with a vertical shaft, are straight tubes.

The intermediate circuit piping is insulated and provided with sodium leakage detectors. These detectors operate on the principle of electrical contact with the sodium, which diffuses into the thermal insulation (see Figure 10.14).

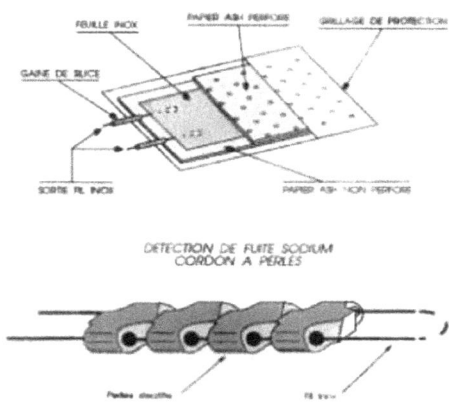

Figure 10.14.

10.2.7. Steam generators

These are the only components where water and sodium are present, separated by the wall of a tube. For this reason, they are considered the nerve centre of the power plant.

There are four steam generators (unit power 750 MWth) for direct passage, the steam in the upper parts flowing through a helical bundle of steel tubes with a high nickel content, the 800 Alloy, around which the sodium circulates on its way down (Figures 10.15 and 10.16).

If there is a leak, it must have sufficiently sensitive detection to minimise the consequences. Steam generators are provided with a system that is very sensitive for detecting the presence of hydrogen, a direct product of the reaction of sodium with water; they also have a passive discharge system (tare membranes) limiting the mechanical consequences of a chemical reaction on the secondary circuit.

The design of the EFR steam generators is different: they have straight tubes in ferritic steel.

10.2.8. Decay Heat Removal systems

There are two methods for removing the decay heat should the previous normal systems be unavailable. First, sodium/air heat exchangers are connected as derivation on the intermediate circuits. Second, four independent systems incorporating sodium heat exchangers are immersed in the primary sodium and evacuate the power via the sodium/air exchangers on their upper part.

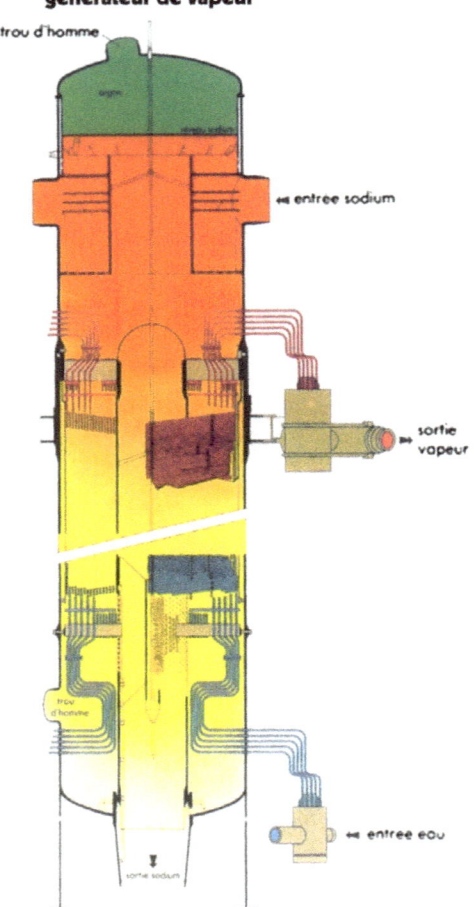

Figure 10.15. Schematic of a SPX steam generator.

Figure 10.16. Details of the steam generator tube in 800 Alloy.

10.2.9. Main Superphenix characteristics

Table 10.3.

Nominal unit power	3,000 MW
Net electric power (with two groups)	1,200 MWe
Net efficiency	0.40
Core and blankets	
Equivalent ^{239}Pu mass	5.56 tonnes
Mass of U in the core	30 tonnes
Mass of U in the blankets	≈ 77 tonnes
Breeding gain (total)	0.18
Maximum burn-up rate	70,000 MWd/t
Irradiation Time, in Equivalent Full Power Day	640 EFPD
Maximum neutron flux	6.1×10^{15} n/cm^2/s
Maximum/average volume power	360/290 MW/m^3 of core
Core height (fuel zone)	1 m
Fuel element	
Fuel material	Sintered mixed oxide
Mass enrichment (Pu equivalent)	16.2%
Total length of the pin	2.7 m
Total length of the subassembly	5.4 m
Clad material stainless steel	316 Ti, then 15-15 Tiε
Main shutdown system	
Number of subassemblies in the core	21
Number of absorber rods per assembly	31
Absorber material	B$_4$C
Main vessel	
Inner diameter	21 m
Height	19.5 m
Thickness of the closing slab	2.9 m
Primary circuit	
Total mass of sodium in the primary circuit	3,500 tonnes
Nominal flow rate	4 × 4.2 t/s
Core inlet temperature	395 °C
Core outlet temperature	545 °C
Maximum pumping power of a primary pump	3 MW
Secondary circuits	
Total mass of sodium in the 4 circuits	1,500 tonnes
Nominal flow rate	4 × 3.7 tonnes/second
Power of a secondary pump	1.2 MW
Steam generator exchange tubes material	800 Alloy
Maximum pressure of water/steam circuit	22 MPa
Steam outlet temperature	490 °C

10.3. Fast reactor fuel

10.3.1. Special characteristics

We have seen that the most interesting feature of the FR system is its ability to burn ^{238}U by its conversion into ^{239}Pu; this means that, in a FR fleet, closing the fuel cycle means that the fuel contains Pu and should be recycled. To these two fundamental characteristics, it must be added that sound economic performance means high burn-up rates, in order to reduce the economic burden related to the fabrication and recycling operations.

Operation in a fast neutron spectrum, for which the effective cross-sections are small, demands a high concentration of fissile material - and a high proportion of fertile nuclei if breeding is to be favoured - and certainly strict limitation of moderator or absorber materials:

- The power density in the fuel is actually: $P = \phi \Sigma_f J$
- ϕ neutron flow
- Σ_f effective macroscopic fission section
- J fission energy
- $\Sigma_f = \sigma_f N$ with N number of fissile nuclei per unit of volume

σ_f is small in a fast spectrum, so we are interested in having N maximum. To have a reasonable value of P and to avoid cores that are too voluminous, ϕ must be high.

10.3.2. Operating criteria

The fuel burn-up of a FR (i.e. the specific energy extracted from the fuel expressed in MWd/t or as a % of heavy atoms burned) must be very high to obtain satisfactory economic characteristics. This means that the fabrication and recycling processes, with the presence of Pu in a high proportion and very high specifications for the clad, are costly and weigh on the cost of the fuel cycle in proportion to irradiation time of the fuel in the reactor.

Another significant criterion is to avoid any melting of the fuel that may lead to a core transient overpower.

10.3.3. Stresses in service

The FR concept induces a series of stresses that are particularly demanding for the fuel element.

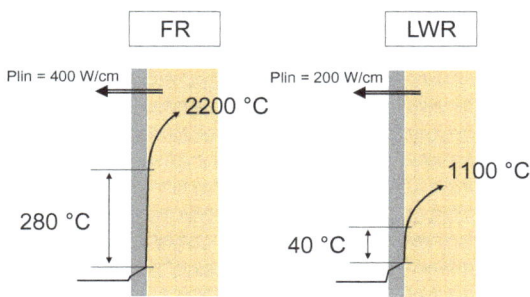

Operating temperature

The power density in the fuel is very high. This generates a level and significant gradients of temperature, which certainly depend on the chosen material. The temperatures of the structural materials are greater than those of a LWR, for example. The core temperature of a fuel pellet is greater than 2,000 °C.

Irradiation damage

The instantaneous flux of high energy neutrons, typically of some 10^{15} n/cm^2/s is greater by more than an order of magnitude than that of PWRs. This causes significant damage in metallic materials, in particular through swelling. The dose received by core materials is typically 40 dpa/year[2].

10.3.4. Fuel material

Successive constructions of FR nuclear system have called on several families of fuel materials.

Metallic alloys (UPuZr, UPuMo, etc.)

This is the solution which allows the density of heavy, fissile and fertile, atoms to be maximised. These are also the fuels with the best thermal conductivity, allowing high specific powers. Their drawbacks are a swelling rate three times higher under irradiation than oxide and a fairly low melting point, for example 1,160 °C for UPuZr.

The results of a few initial tests (on the DFR, EBR2 & FERMI experimental reactors) had led to this solution being discarded for power reactors. It is nonetheless a solution still preferred by some teams in the world, in association with recycling by pyrochemical methods.

---

[2] dpa: displacement per atom.

Mixed oxide (UPuO$_2$)

Being less dense, and having less capacity to conduct heat, this fuel is a priori inferior in performance., It has however proved capable of supporting a very high burn-up owing to moderate swelling and a high melting point at 2,760 °C offsetting its poor conductivity (see Figure 10.17).

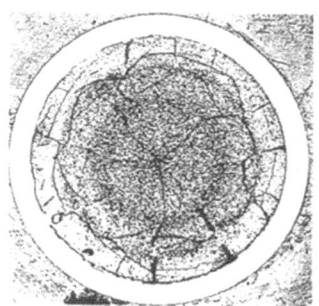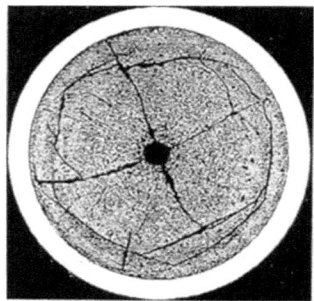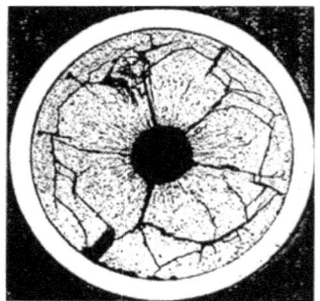

Figure 10.17. Restructuring of the oxide fuel under the effect of temperature produces a ductile central zone and a central hole which takes up the expansion and avoids too high a pressure on the clad (sectional view of an irradiated fuel pin with increasing temperature from left to right). [copyright CEA].

It reacts chemically with the sodium, which is a drawback if the clad breaks, but it reacts with a slow dynamic, and experience has shown that this situation is perfectly manageable in the reactor. The oxide fuel has finally emerged as the best solution and has been chosen for all demonstration reactors and industrial prototypes. The production and recycling of mixed oxide fuel has been mastered and is used effectively on an industrial scale. It is now the best proven industrial solution.

Other solutions – carbide, nitride, etc. (UPuC, UPuN)

These compounds have the potential advantage in that they swell less than metallic alloys, while being denser and better conductors than oxide. However, feedback is quite poor compared with feedback from oxide.

10.3.5. Clad materials and effects of irradiation

Fast neutrons have an energy which is high enough to move metal atoms from their crystalline site. Between the lacunae created in this process, and the interstitials, the recombination results induce a global swelling and a decrease in density, the amplitude of which depends on the material and the temperature. This effect increases with the dose incorporated.

This swelling generates structural deformations such as increases in the diameter of the clad, lengthening of the pins and the hexagonal tubes, bending of the subassemblies, etc., from which internal (swelling gradient) and external (interaction between components) stresses arise. The need to limit the resultant damage leads to restricting their time of use in the reactor. To this first effect, creep induced by irradiation and variation of the mechanical properties (ductility, elastic limit, resilience, etc.) is added.

The set-up of materials which demonstrate little swelling has represented one of the major challenges for the designers of fuel subassemblies. Progress in understanding the phenomena involved and exploitation of considerable experimental knowledge has allowed very substantial progress to be made in some fifteen years. Satisfactory solutions are now available for hexagonal tube materials in martensitic or ferritic-martensitic steels. For clad, that undergo greater stress, the best grades currently used are austenitic steels stabilised with titanium, mainly of the 316L type, but the objective for very high doses is to qualify ferritic steels with adequate mechanical properties, for example by reinforcing them with an oxide dispersion (oxide dispersed strengthened or ODS steel).

10.3.6. Characteristics of fuel elements and behaviour problems

We are referring here to oxide fuel. The fuel is formed of cylindrical pellets of small diameter, from 5 to 8 mm, stacked inside a metallic clad; a thin tube closed by welded plugs at both its ends. An empty expansion space is provided to collect the fission gases produced during operation. A "pin" is therefore produced that is sealed relative to the sodium. This is an imperative criterion in order not to contaminate the primary circuit with fission products, and avoid any contact between the fuel and the sodium.

The subassembly consists of a large number of pins (271 for Superphenix) combined in a compact bundle, separated by helical wires inside a hexagonal wrapper (Figure 10.18).

10.3.7. Fuel behaviour

The thermal, mechanical and physicochemical properties of fuel are varying continuously under the effect of increasing production of fission products. The chemical condition of the whole depends on the oxygen potential, which has also changed owing to the regular disappearance of fuel nuclei and the appearance of fission products.

In operation, fuel is the seat of multiple conversions, the principal causes of which are the high temperature gradient to which the material is subjected, and the production of fission products in quantities that increase with the burn-up. It is essential to note that the resultant phenomena are closely linked, even if they are presented separately.

Thermal behaviour of the fuel element

There is a limit of 650 °C on the maximum clad temperature which provides a margin to sodium boiling and ensures temperature stays in a zone in which the mechanical properties of the clad remain acceptable. The gap between the pellet and the clad is a significant obstacle to heat transfer, and the temperature of the pellets ranges at the beginning of their life between 800 and 1,000 °C on the surface and 2,000–2,200 °C in the core. A way of limiting this is to use annular pellets (initial central hole), which is the solution chosen for Superphenix.

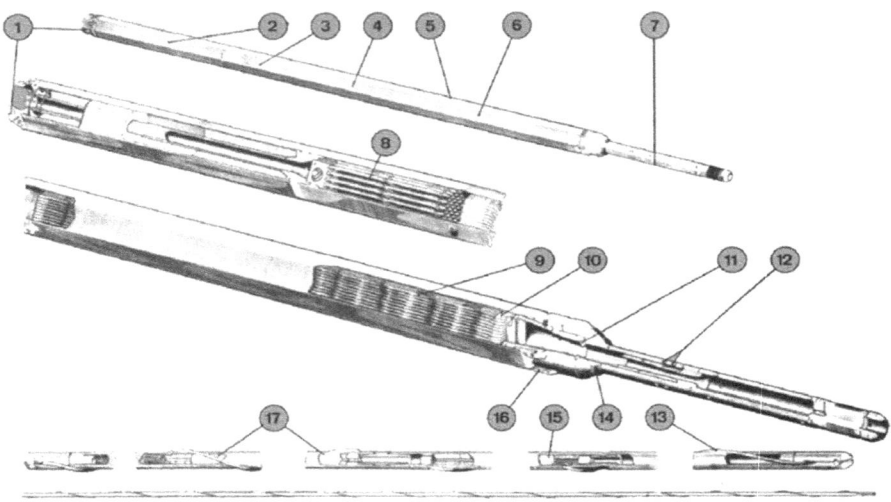

Figure 10.18. Fuel subassembly for the Phenix reactor. 1- prehension head, 2- upper neutron shielding, 3- upper fertile blanket, 4- fuel zone, 5- hexagonal wrapper, 6- lower fertile blanket, 7- nozzle, 8- fertile bundle, 9- fuel bundle, 10- pin hooking grid, 11- flow adjustment diaphragm, 12- sodium feed lights, 13- helical spacer wire, 14- range, 15- fertile pellet, 16- placement lock, 17- fuel pellet.

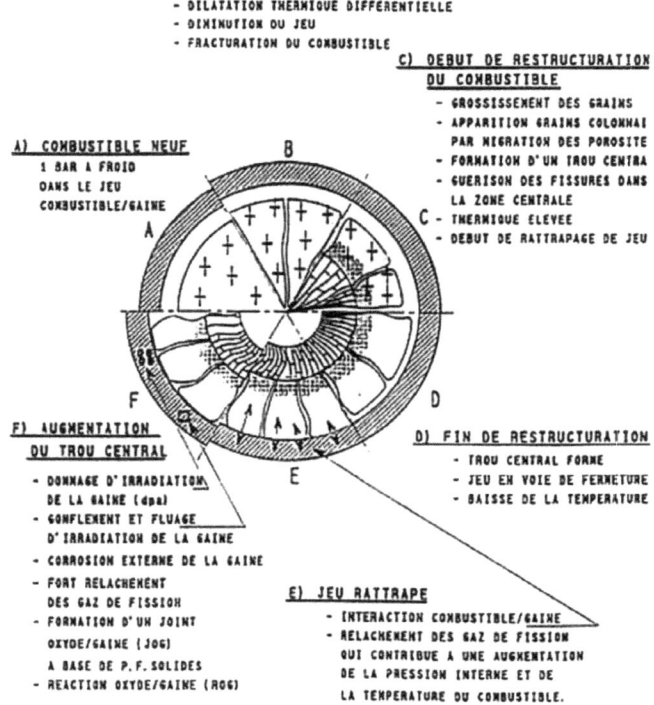

Oxide is a ceramic which does not resist the stresses caused by such gradients. The pellets break and jam some of the gap. In the hot zone, material is transferred from the hot surface of the pores to the cold one, and indeed the pores migrate to the centre. The result of these phenomena is that the heat regime falls rapidly by some hundreds of degrees. It will continue to change during irradiation as and when gap is lost by fuel swelling.

Interaction between clad and pellet

Fission products cause swelling of the fuel (approximately 0.7% per % of atoms burned) by accumulation of solids or gases in the matrix. The fuel will therefore absorb the gaps and tend to come into contact with the clad (and we have seen that everything was done to ensure that it deformed little or not at all). Over time, there is then a mechanical interaction of increasing amplitude, accelerated by operating transients. This phenomenon is another limitation of the irradiation time of the fuel.

Fission releases oxygen and varied fission product nuclei. There is accordingly a considerable change to the thermodynamic conditions of the whole which favours corrosion, by the Cs and the Te in particular. The desire to limit the thickness of the clad affected by this corrosion leads to a third criterion concerning the irradiation time.

10.3.8. Reprocessing

The oxide fuel can be reprocessed by the aqueous PUREX method: dissolution in nitric acid, separation and extraction by tributyl-phosphate. The core of the process is the same as that for Mox fuel for water reactors, which is not modified by specific characteristics such as the geometry of the fuel elements, the composition of the clad (steel instead of Zircalloy) or the high plutonium content. Dissolving in nitric acid is rather more difficult as the Pu content increases and this is why it is held up to this point at a maximum of 30% PuO_2, with an upper limit approaching 40%.

Depending on the degree of development of the reactor fleet, reprocessing can be done by dilution with light water reactor fuel or in dedicated facilities. A different route has been explored for metallic alloys: pyrochemistry. This is a method without water which uses molten salt solutions and separations by electrolysis. If it appears promising to a degree, although it has still not attracted any industrial development and still requires a significant R&D effort.

10.4. Fast reactor safety

This chapter refers to the measures taken for Superphenix or EFR.

10.4.1. Containment

Safety is based on setting up almost three barriers between the radioactive material and humans and the environment. These barriers need to be reliable, monitored in operation and protected by the corresponding safety systems, the principals of which are systems to shut down the reactor and evacuate the decay heat. For Superphenix, there are four of these barriers (see Figure 10.19).

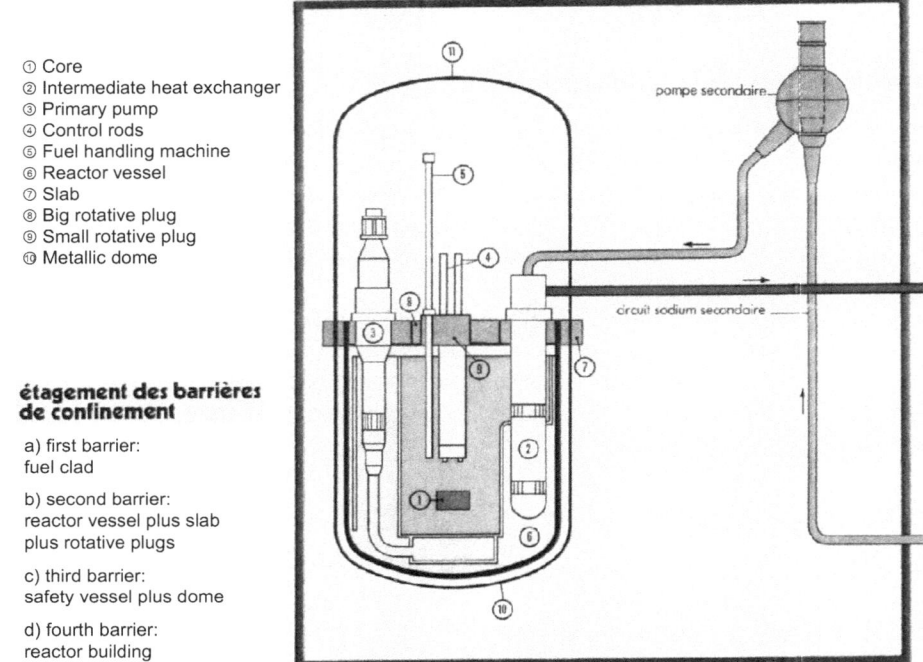

① Core
② Intermediate heat exchanger
③ Primary pump
④ Control rods
⑤ Fuel handling machine
⑥ Reactor vessel
⑦ Slab
⑧ Big rotative plug
⑨ Small rotative plug
⑩ Metallic dome

étagement des barrières de confinement

a) first barrier:
fuel clad

b) second barrier:
reactor vessel plus slab plus rotative plugs

c) third barrier:
safety vessel plus dome

d) fourth barrier:
reactor building

Figure 10.19. Staging of Superphenix containment barriers (source EDF).

As a rule, barriers are provided just as much for internal events as for external hazards such as a station black out. In particular, considering events which combine an earthquake with another fault generally defines the dimensioning events for the structures. In fact, those structures are subject to the relatively high temperature of the sodium through their thin thickness permitted by the low pressure of the coolant. The corresponding barriers and structures also have secondary safety functions, but these are nonetheless very important. Examples are:

- support of the core, in order to eliminate the risk of a rapid insertion of reactivity,
- retention of the primary coolant, to eliminate the risk of loss of flow in the primary circuit by lowering the free level or even removing the coolant from the core.

The required reliability of barriers and systems with safety functions needs an appropriate design and ease of inspection. Among the situations considered regarding the strength of the various barriers, we will note the following:

For the fuel clad

Loss of flow in the core, with rupture of the pump-to-diagrid pipe as the envelope scenario, with a risk of local boiling capable of leading to clad dry-out; an inadvertent control rod withdrawal, with the risk of melting fuel pellets in the core. In every case, the objective is to prevent propagation of any local fault to the whole core.

The consequences of accidents are made acceptable by the means of detection and by reactor shutdown. The main detection systems are the temperature measurements at the outlet of each subassembly and, above this, the delayed neutrons detection system with delayed neutrons being emitted by an open rupture in the clad or melting of the fuel element,. The design has a wrapper (hexagonal) tube for sodium FR subassemblies which allows the ruptured sub-assembly to be located.

For the intermediate containment

In addition to standard situations of Loss of coolant accidents, highly hypothetical cases of energetic accidents concerning sodium FR are studied, given their specific features (sodium voiding effect, recriticality) to guarantee confinement of radioactive materials. Thus, the Superphenix containment had been dimensioned to support an internal release of mechanical energy of 800 MJ. The reference scenario (a pump failure without control rod shutdown and without diesel restart) leads to a lower energy value.

For the secondary containment

Sodium fires are dimensioning situations for secondary containment rooms.

Given feedback from studies, measures are taken concerning the risks of dynamic loading (pressure) caused by droplet sodium fires (installing vents to relieve the pressure), thermal loading (pool fire on the slab, etc.) or hydrogen production (sodium concrete interaction).

10.4.2. Reactivity control

Transients affecting the core behaviour can be temperature variations at the inlet of the core, due for example to loss of a heat sink, of sodium flow, or of reactivity (control rod movement). For a core to be stable, its overall reactivity coefficients regarding these three types of transients should be negative, but of moderate amplitudes. However these negative reactivity coefficients are a problem in case of rapid cooling down. Any disturbance therefore materialises as a reactivity effect directly linked to the initiator or to core thermal reactivity effects, or to the rods shutdown. The objective is to achieve a new stable equilibrium ($\Delta \rho_{total} = 0$) with acceptable temperature levels. Moreover, the maximum reactivity achieved during the transient must remain well above the fraction β of delayed neutrons. The usual value of a FR core, $\beta \approx 360\,10^{-5}$, guarantees the core stability during operation.

As long as the sodium is liquid, the positive reactivity effect due to the sodium density remains low for a FR core during transient. During fast transients, effective reactivity effects are the Doppler effect and the fuel elements expansion effect. In the longer term, additional reactivity effects from differential expansion between rods and the core before scram and/or diagrid expansion act effectively. As the core of a FR is not its most reactive configuration, it is important to identify and prevent any initiator of local (melting by blockage, etc.) or global (seismic behaviour, etc.) melting of the core, of relative control rod withdrawal (slumping of the core support as a residual risk, etc.), and of sodium void effect (gas voiding or scenario with sodium boiling), this one being positive in the central part of the core.

We distinguish as initiators of reactivity insertion on a FR:

- slow insertions such as the inadvertent control rod withdrawal: detection is coming from temperature or power measurements;

- rapid insertions corresponding to displacements of subassemblies or control rod system or to a sodium voiding - this risk is important only if several central subassemblies are affected (gas bubble at the core inlet or whole core boiling).

Transients and accidents are classified using a probability/consequence curve, such that the worse the consequences, the lower is the probability of occurrence (see Figure 10.20). Below a probability deemed sufficiently low, events are no longer considered for sizing the reactor; this is the residual risk. FRs safety systems (from detectors to absorbers) are designed so as to site the scram failure in the residual risk category, i.e. with a probability $< 10^{-7}$/year/reactor. Systems are redundant for each function to be provided. Diversification of equipment is stipulated at all levels to cover the risk of failure due to a common cause or common mode (articulated rods, etc.) Control rod systems are the only means of controlling reactivity. In the EFR project, a particular effort was made concerning supplementary systems with the addition of a 3rd shutdown level so as to force the reactor into a safe configuration despite the failure of the normal shutdown systems, the objective being to eliminate from the design the core meltdown consideration.

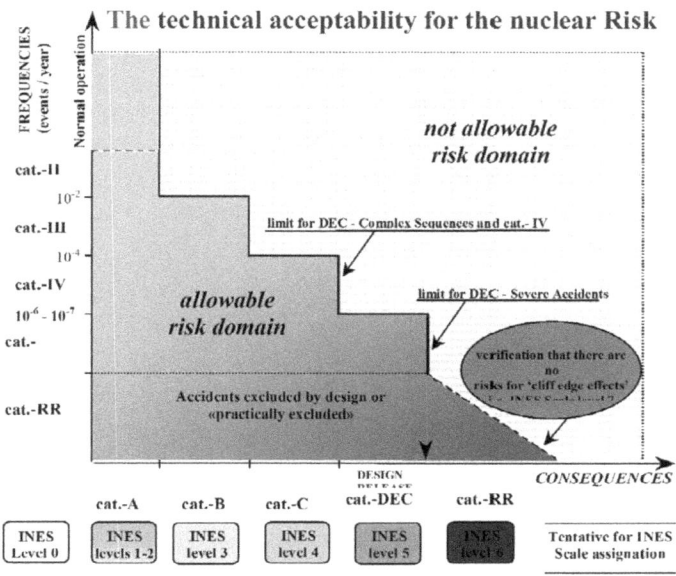

Figure 10.20. Probability versus risk diagram used to classify undesired events.

10.4.3. Decay Heat Removal

Under normal or accident configurations, decay heat is evacuated by heat exchanges. There are three types of shutdown cooling system:

- Normal system: by feed water to the steam generators and then up to the outside cold source (sea or river). Below the technical minimum for the operation of the steam generators, a switch is made to sodium/air heat exchangers.

- Backup systems to face with the loss of normal systems of intermediate loops; these must comply with temperature limit criteria to guarantee the integrity of main structures. These maximum temperatures depend on concepts. The order of magnitude is 530 °C in an incident situation, 620 °C in an accident situation and 650 °C in a design extension condition.

- Dedicated means: some components with safety functions close to the primary circuit need cooling to guarantee their integrity. These are principally the concrete structures of the reactor cavity (cooled by a water circuit) and the closure slab.

10.4.4. Considering accidents involving fuel melting

Whether or not to take into account core melt down accidents is one of the questions which characterises the safety approach of each type of reactor designs.

Compared with PWRs, for which safety procedures marked a change after the accidents at Three Mile island and Chernobyl, the safety procedures associated with the FR took into account very early the core meltdown accidents, owing to the risk of rapid transient overpower related to even a small compaction of the core. The figure opposite illustrates the different phenomena studied during a core meltdown. The first objective of this type of procedure is to design the containment for a mechanical energy release, hence the design basis accident of the Superphenix containment set 800 MJ by decree. To confirm the correct design relative to this specification, a scenario of unprotected loss of flow was chosen for its character deemed envelope concerning the consequences, thought it did not represent the most probable sequences.

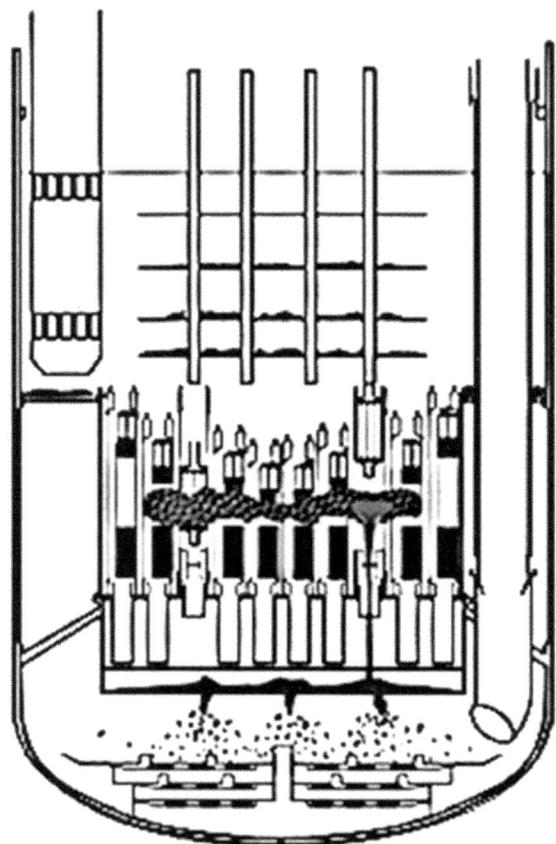

Large-scale experimental programs were produced in support, for example on the CEA/Cadarache reactors. The Cabri program simulated fuel melting in a pin during a power transient and its ejection into the sodium channel after a clad rupture and the Scarabée program simulated the development of a Total Instantaneous Blockage of a subassembly that induces its melt down.

A safety approach based essentially on the prevention of whole core meltdown was undertaken in the EFR project: the reference accident was a Total Instantaneous Blockage of a fissile subassembly and the demonstration of non-propagation of the melting to the whole core. Nonetheless, European regulatory organisations have recommended dimensioning the containment at 500 MJ and today current projects are examining in depth the various core degradation scenarios with the objective of avoiding any accidental radioactivity release outside the plant site.

10.5. Sodium technology

10.5.1. Sodium

Sodium is produced from sea salt by electrolysis (see Figure 10.21). This chemical element is extremely widespread throughout the world and is necessary for the functioning of all living organisms.

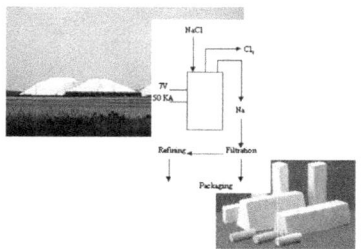

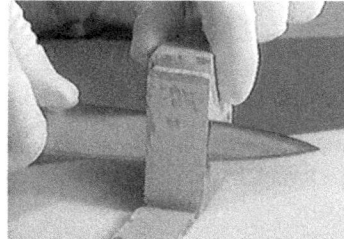

Figure 10.21.

It reacts with water to form caustic soda, which allows the use of techniques for washing and treating sodium waste that, once diluted, can be released into the environment.

10.5.2. The choice of sodium

The criteria for the choice of coolant are numerous. It must neither moderate nor absorb neutrons, and must have the capacity to extract heat at high transfer rates. Liquid metals are essentially effective coolants with a high thermal conductivity and specific heat. The principal candidates are sodium, lead and their alloys NaK and Pb-Bi. Mercury, which was used initially by small facilities, can no longer be considered for an industrial reactor due to its very high toxicity.

If sodium has received universal agreement, it is because of its exceptional thermal qualities. Its high conductivity generates good thermal exchanges which allow high heat generation to be extracted from the fuel pins. Its low density and high dilatation favour natural convection regimes and hence accommodate accident situations. It has other interesting characteristics:

- it is liquid over a wide temperature range (from 98 °C to 883 °C at atmospheric pressure) and its vapour tension is low. This allows it to operate without pressure in high performance thermal regimes, while at the same time ensuring a margin before boiling point is reached;

- low energy systems are sufficient in practice to keep it liquid, and the ability to easily solidify is of value for some maintenance operations;

- its viscosity at 450 °C is comparable to that of water; this minimises the pumping power needed to cool the system.

Sodium is also completely acceptable for the core physics, compatible with steels and, as we shall see, its potential drawbacks can be overcome. Table 10.4 above assembles some data relating to sodium, water and lead.

Table 10.4. Properties of sodium; Comparison with water and lead (the physical properties are given for water at 40 °C, for sodium and lead at 500 °C).

	Water	Sodium	Lead
Melting point (°) – at atmospheric pressure -	0	98	327
Boiling point (°C) – at atmospheric pressure -	100	882	1737
Density (kg/m^3)	992	832	10 390
Coefficient of expansion (/K)	$3.9\ 10^{-4}$	$2.8\ 10^{-4}$	$1.15\ 10^{-4}$
Dynamic viscosity (Pa.s)	$5.9\ 10^{-4}$	$2.3\ 10^{-4}$	$18.9\ 10^{-4}$
Surface tension (N/m)	0.07	0.16	0.43
Specific heat (J/kg.K)	4180	1262	149
Thermal conductivity (W/m.K)	0.63	67	15
Heat exchange coefficient (W/m^2.K)*	17,000	36,000	23,000
Electrical resistivity (ohm.m)	8,105	$2.7\ 10^{-7}$	10^{-6}
Capture cross section (barn)**	22	$2.8\ 10^{-3}$	$4.7\ 10^{-3}$
Moderator effect ($\xi.\Sigma d$ -/cm)	1.28	0.30	0.06
Radioactivity induced by reaction to neutrons (half-life)	H$_3$ 12.3 years	Na$_{22}$ 2.6 years Na$_{24}$ 15 hours	Pb$_{207}$ 52 hours

*for 3 m/s in a 25 mm tube; **for a mean energy of 2 keV.

10.5.3. Sodium chemistry and purification

Sodium has considerable chemical affinity for most chemical elements. It burns in air and reacts explosively with water. A first precaution is to purify it permanently. In particular, its oxygen, hydrogen and carbon contents must remain extremely low, of the order of only a few ppm[3]. Original methods of purification have been developed, based on variations of solubility of oxygen and hydrogen with temperature. This is the technique known as cold trap, which lowers the temperature of the sodium sufficiently to crystallise impurities on a metallic trellis (see Figure 10.22).

Sodium pollution has a number of significant consequences, including an increase in its corrosive power on metals M in structures following principally the following reaction:

$$Na_2O + 1/2\ M \rightarrow 1/2\ NaMO_2 + {}^3/_2\ Na$$

---

[3] ppm = part per million.

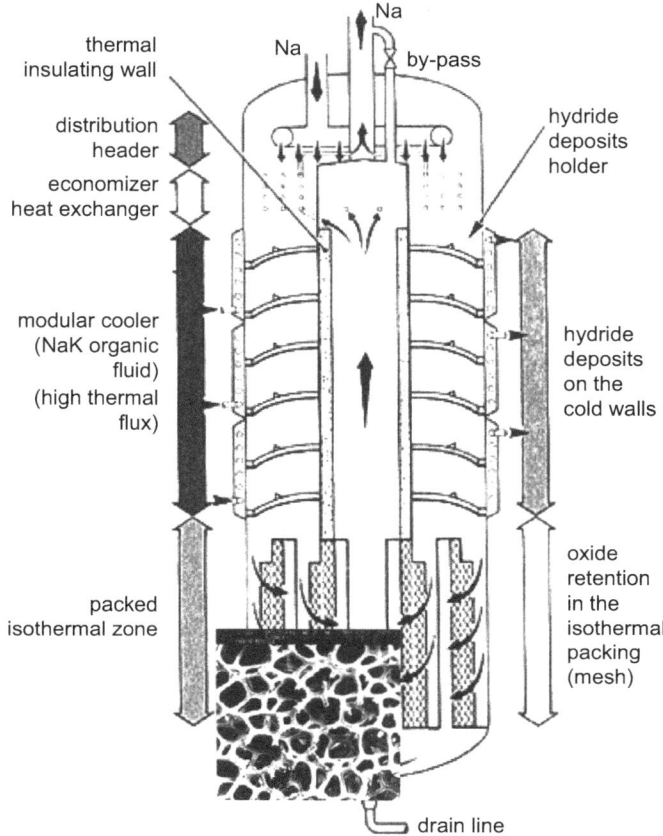

Figure 10.22. Sodium cold trap equipment.

By maintaining a low oxygen concentration, any corrosion is avoided. In French sodium cooled FR, the temperature of the cold traps is set at 120 °C, i.e. 22 °C above the melting point of pure sodium.

10.5.4. Compatibility of sodium with materials

It is essential to ensure compatibility of sodium with the materials constituting components and circuits, instrumentation systems, and with products such as oils, greases, sealing materials and insulation. Today, we have appropriate materials and products that are duly qualified.

10.5.5. Circuits and instrumentation

Sodium must be moved by conventional mechanical pumps, with which it behaves like water. This is the solution chosen for high pumping powers. It can also be pumped by electromagnetic pumps as it is an electrical conductor with good magnetic permeability. The interest lies in having pumps without moving parts and without sealed passages. Nonetheless, their power is presently limited. Flowmeters and level sensors have been installed on the same principle.

As sodium is in liquid form only above 98 °C, it must be maintained above this temperature at all times in the circuits and components of the reactor to prevent it solidifying. Systems are therefore insulated. In general, the heat released by pumping is sufficient to keep it liquid in the circuits.

Reliable, high-performance leak detection systems can provide a high level of safety for the equipment.

10.5.6. Interventions, inspection, repair

To check or repair a component, the difficulties to be overcome arise from sodium's opacity, its high temperature, its incompatibility with air, water and many other elements, along with the associated risks of corrosion (formation of soda or sodium oxide) and finally also from its radioactivity in the primary circuit. Specific technologies have been developed, which now allow the absolute control of the use of sodium in FRs (see the principles chosen on the EFR Figure 10.23).

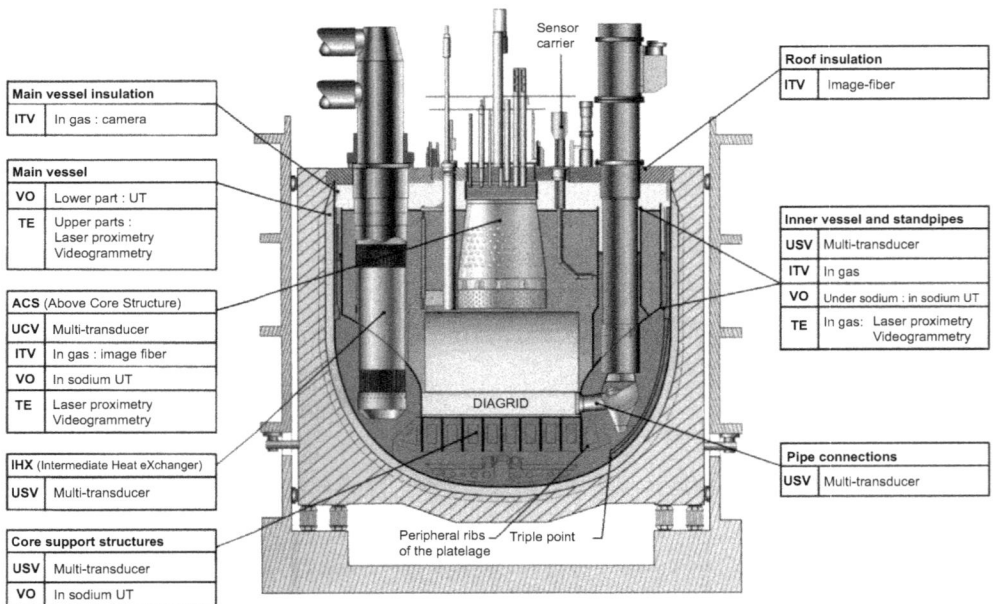

Figure 10.23. Principles for inspection and repair proposed in EFR.

Ultrasound methods are used in order to "see" through sodium. The sensors tolerate sodium and a high temperature level, and software interprets the ultrasound measurement signals. An example of industrial application was the inspection of the welds supporting the

Phenix core, in 2000, made by ultrasound that propagated in the core supporting structure from the outside of the primary vessel, and which allowed the operator to inspect several junctions immersed in sodium at a distance of several metres from the sensors.

Washing and decontamination procedures were set up to treat a component extracted from the sodium and to authorise its handling, i.e. cleansed of the sodium and the surface radioactivity, and finally able to be re-used in the reactor. Sodium-cooled FRs have been operated and, as a consequence, worked on for a number of years, the information from this work is valuable. A noteworthy example is that of extraction, repair then re-introduction and re-use of the intermediate heat exchangers from Phenix, victims of leaks due to the impossibility of accommodating differential expansions; comparable experience has been acquired on primary pumps.

Today, interventions in a sodium environment are well under control. However, the inspectability of reactors will improve when measurement methods and ultrasound sensors become capable of carrying out controls (distance measurements and non-destructive testing) even within liquid sodium.

10.5.7. Safety

The main risks are those of a sodium water reaction and of sodium fires.

Sodium/water reaction

It is not difficult to design installations that virtually eliminate the risk of sodium coming into contact with water and only steam generators have posed a real problem. Indeed, the reaction is strongly exothermic and releases hydrogen:

$$Na(liq) + H_2O(liq) \rightarrow NaOH(liq) + 1/2\ H_2\ (gas) \quad -141\ kJ/mole\ Na\ (NTP[4])$$

We can now describe the consequences of a leak from a tube which is considered tolerable so long as detection is sufficiently advanced in precision and rapidity. We can detect the hydrogen produced or the acoustic signals emitted by the reaction. The system designed for EFR combined the two techniques.

Sodium fires

Depending on the type of leak from a sodium pipe into the surrounding air, several potential consequences are expected from the reaction:

$$2\ Na + 1/2\ O_2 \rightarrow Na_2O$$

For a small leak, the sodium is oxidized without any notable effect. For a significant flow, the sodium flows to the ground and can cause a pool fire and, depending on the physical support, an attack on the concrete, generating a large quantity of aerosols. From numerous analytical tests which have been carried out, one can model the thermal and physicochemical effects on the containment (ventilation included) and the environment.

---

[4] NTP = Normal temperature and pressure.

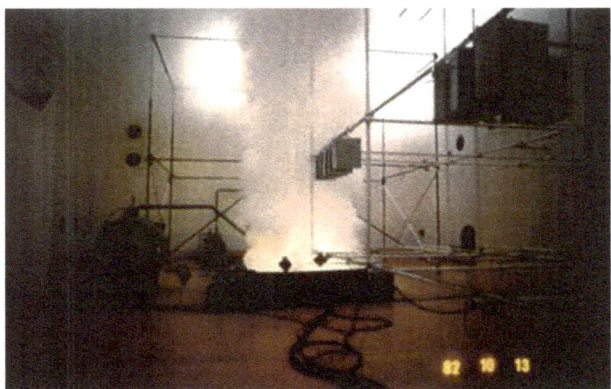

More technology-based tests (such as on the photo opposite of a pool fire in the CEA/ Cadarache Esmeralda facility) have allowed the setting up of preventive and counter measures, as well as aerosol filtration and concrete protection systems. Concretes that are inert with regard to sodium have even been developed. Rules for intervention and repair of facilities after fire have also been issued, including instituted techniques for treating combustion residues. To fight a fire, a preventive arsenal is currently in place: bin systems isolating the sodium from the air; availability of a special extinguishing powder and aerosol filters, as well as effective protective clothing and guards.

10.5.8. Overall assessment of the use of sodium

Sodium technology has been mastered allowing the industry to reap the benefits of its excellent characteristics and to confirm its use on an industrial scale.

Nonetheless, the use of this technology brings with it specific burdens in controlling reactivity with air and water and to bypass the opacity. This leads to important design restrictions, for example the installation of an intermediate sodium circuit between the primary circuit of the reactor and the steam generator, and instrumentation of steam generators. These limitations are reflected in the high investment cost of sodium systems.

10.6. Alternatives to sodium

10.6.1. Liquid metals

The combination of requirements actually offers, among the liquid metals, a single alternative to sodium: lead or the lead-bismuth eutectic alloy. Table 10.5 below gives the main properties of these three metal coolants.

Pb, or Pb-Bi, has the advantage that it reacts neither with water nor with air, nor with the oxide fuel. It is not such an effective coolant as sodium, but it has acceptable properties to be used in the design of a reactor. Its neutron characteristics are good and its boiling point is clearly higher than that of sodium, which gives it another advantage from a safety point of view. On the other hand, it has the drawback of having a higher melting point,

Table 10.5. Lead, lead-bismuth eutectic and sodium coolant properties.

Properties (at atmospheric pressure)	Coolant		
	Pb	44.5% Pb 55.5% Bi	Na
Melting point, °C	327	123	98
Boiling point, °C	1,745	1,670	881
Density at 500 °C, kg/m^3	10,470	10,050	833
Thermal conductivity at 500 °C, W/m.K	15	14	66
Specific heat at 500 °C, kJ/kg.K	0.15	0.15	1.25

a handicap particularly sensitive for lead (327 °C); given the necessary margins, it must be considered that a lead-cooled reactor will have to be permanently kept at over 400 °C.

The Lead-Bismuth Eutectic, also called LBE, lowers the melting point from 327 °C to 127 °C and reduces the operating limitations; however, it uses Bi, which is expensive and natural resources of this metal are limited. It also is activated in the core to form radioactive ^{210}Po, a hard alpha emitter.

The high density of Pb and LBE may have an impact on the seismic resistance of pool-type FRs and the possibility of introducing instruments within the liquid metal for measurements or maintenance.

Finally, all liquid metals are opaque, so in service inspection is difficult. Ultra sonic equipment must be adapted from sodium to lead.

10.6.2. Corrosion by heavy liquid metals

The main difficulty to overcome when using a heavy liquid metal is corrosion of solid structure (see Figure 10.24). Figure 10.25 below shows the different zones of interaction between lead (or LBE) and a metallic structure. If the oxygen concentration in the liquid is less than a very small given value, steel is dissolved. For a higher concentration of oxygen, a protective oxide layer is formed on the structure, but its stability is not guaranteed. For a concentration greater than an upper limit, oxygen is no more soluble in lead, and solid particles of lead oxide are formed.

Figure 10.24. Corrosion of a steel tube by liquid lead-bismuth.

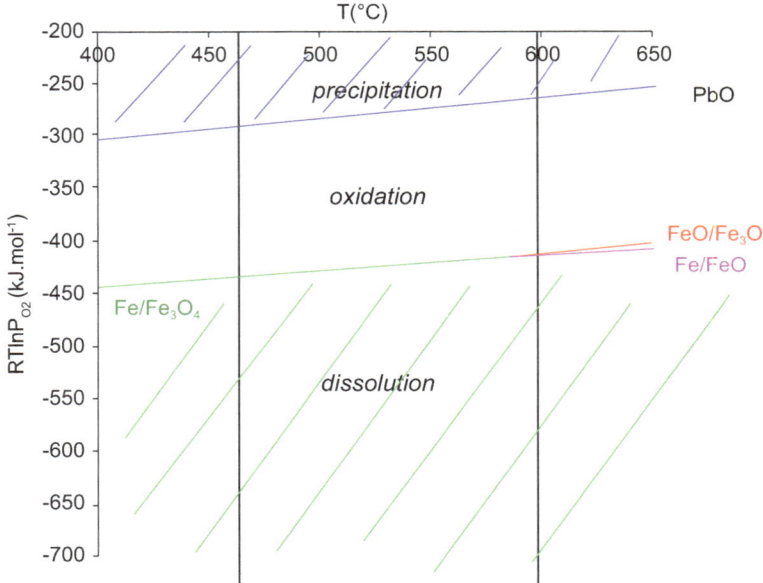

Figure 10.25. Liquid Lead interaction with a solid metal, at different temperatures, depending on the oxygen concentration in lead (logarithmic measure).

The correct zone in a reactor is the one called "oxidation" in Figure 10.25, which is temperature dependent. Note that a typical value of oxygen concentration is in the order of 10^9 (10^3 ppm), which is very small, so difficult to control. Techniques of protective layer deposition using alumina are also developed that can protect steel.

10.6.3. Lead-bismuth reactor feedback experience

They have been several LBE-cooled reactors under operation; they were Soviet submarines in the 1980's (see Figure 10.26). Called the Alpha class according to a NATO classification, eight submarines and two prototypes on land were used from 1971. Thanks to the very high power achievable in a small volume with a fast neutron core (155 MWth), their performance was very good.

Unfortunately they have had several problems regarding the LBE with easy solidification at 123 °C requiring constant heating of the reactor all times even during cold shutdown. Presence of active ^{210}Po was also a problem during maintenance and finally the strong solicitation of the core from low to high power and the restricted in-core fuel cycle led to their limited use. They were stopped in 1996 and were being dismantled in 2012.

10.6.4. Lead-cooled reactors

Design constraints

To successfully operate a lead-cooled reactor, care needs to be taken to prevent corrosion of structures by liquid heavy metal. In the core, the coolant velocity is limited to less than

Figure 10.26. Soviet submarine powered by a lead-bismuth-cooled nuclear reactor (source: Shipbucket.com/Guns).

2 m/s which is quite a low value, and to get a sufficient flow rate, coolant volume ratio is raised, which decreases fuel fraction. Finally, designers typically present designs in which the average volume power is around 70 MW/m^3. To enhance the neutron efficiency, a dense fuel is also proposed, carbide or nitride (see Table 10.6), which is denser than oxide but with a good temperature stability, and several radial fuel zones are proposed to flatten the radial power.

Table 10.6. Comparison of different fuels.

	Carbide (U,Pu)C	Nitride (U,Pu)N	Oxide (U,Pu)O$_2$	Metal (U,Pu,Zr)
Theoretical density (g/cm^3)	13.58	14.32	11.5	14
Melting point (°C)	2420	2780	2750	1080
Thermal conductivity (W/m/K)	16.5	14.3	2.9	14
Swelling	1.6 to 2%		0.8%	
Thermal stability	Stable	Stable up to 1600-1800 °C	Very stable	–

In the end, the necessity to limit corrosion necessitates a large core volume for a given power that limits the unit power. On the other hand, the absence of the reaction of lead with water allows simplification of the design by putting the steam generators in the main vessel, for instance.

Reactor projects

Two different types of design are primarily proposed; either a battery reactor or a medium size one.

The battery reactor is proposed by Russia and the US as a low power transportable reactor of 2 to 200 MWe. A good example is the SSTAR project (see Figure 10.27) with 20 MWe (45 MWth) and a very long-life ^{235}U loaded core of 15 to 30 years, with the idea being not to unload the fuel on site. It is a pool type concept, cooled by natural circulation of lead. Fuel material is nitride and the power conversion system uses supercritical CO_2.

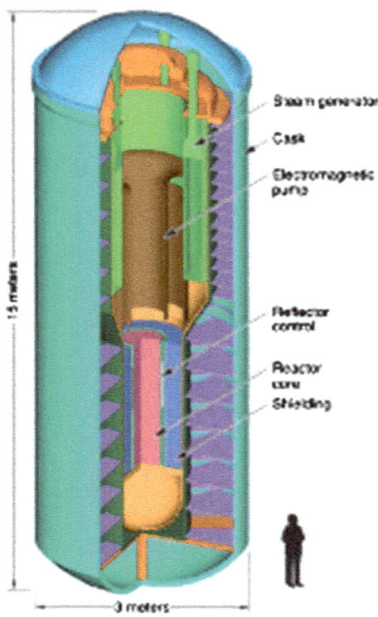

Figure 10.27. The SSTAR -Small Sealed Transportable Autonomous Reactor- project from Los Alamos National Laboratory in the US.

The medium size reactor is illustrated by the BREST-300 Russian project (see Figure 10.28 and Table 10.7) initiated in 1990. It uses a mixed U and Pu Nitride fuel with three radial zones of different fuel pin diameter. The reactor vessel is supported by the ground, due to the high weight of lead.

In 2010, this project was resumed as part of a federal Russian program on closed fuel cycle and fast reactors.

1. Pump
2. Vessel
3. Heat insulation
4. CPS
5. Core
6. Plating
7. Inner vessel
8. Internal storage
9. Steam generator
10. Concrete reactor cavity
11. Rotating plug
12. Steam safety valve
13. Handling machine
14. Inner vessel support

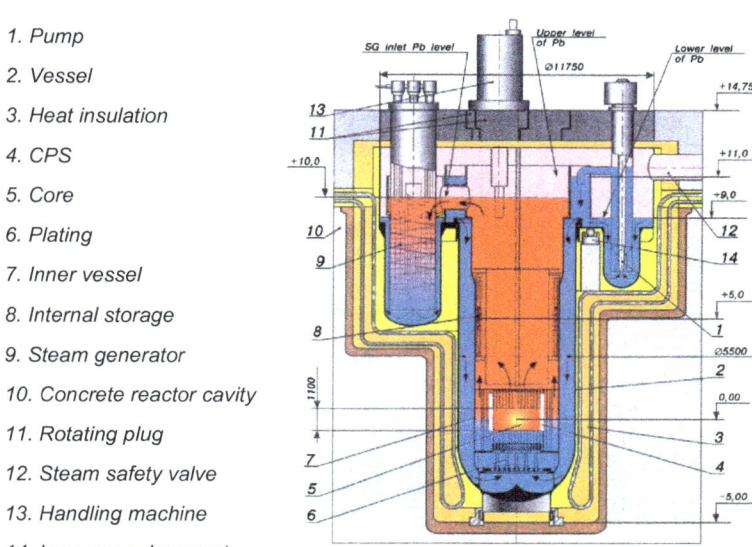

Figure 10.28. The Russian lead reactor project BREST-300 (source: A. Filin et al., RDIPE State Scientific Centre of Russia).

Table 10.7. BREST-300 main characteristics.

Thermal power, MW	700
Net electric power, MW	300
Fuel Assemblies in the core	185
Core diameter, mm	2300
Core height, mm	1100
Fuel element spacing, mm	13.6
Fuel element diameter (three zones), mm	9.1/9.6/10.4
Core fuel	UN+PuN
Average Pu content	13%
Fuel lifetime, yrs	5
Refuelling interval, yrs	1
In/Out lead temperature, °C	420/540
Maximum lead velocity, m/s	1.8
Efficiency, %	43
Core breeding ratio	~1
Lifetime, yr	30

10.6.5. Conclusion

We may think that the lead-cooled reactor is a variant that differs little from the sodium-cooled reactor (same opacity, same cooling and neutronics characteristics) while bringing some advantages such as undergoing no reactivity with air or water. But given the new drawbacks that it presents (corrosion and toxicity), it is not clear that it will represent breakthroughs. A demonstration reactor is needed to qualify effective operation over a long enough time period.

10.7. Development prospects

10.7.1. Current context

After an increase in the seventies, new construction has plateaued. Today in the 2010s, India, China and Russia have undertaken new construction and several projects are planned in countries with nuclear power; in France with a prototype in 2020 and in Japan with the JSFR, an industrial prototype succeeding Monju.

The main factor in this trend is the predictable crisis in world uranium resources should nuclear energy in the future occupy a place at least equal to that which it has today. One can estimate that accessible resources will be exhausted between 2060 and 2100 depending on the growth of the world nuclear fleet as shown on Figure 10.29, with other estimates at 2120 and later. Use of FRs is therefore a necessity in the future. The question is to know whether to start the development of a new system today or in twenty years time.

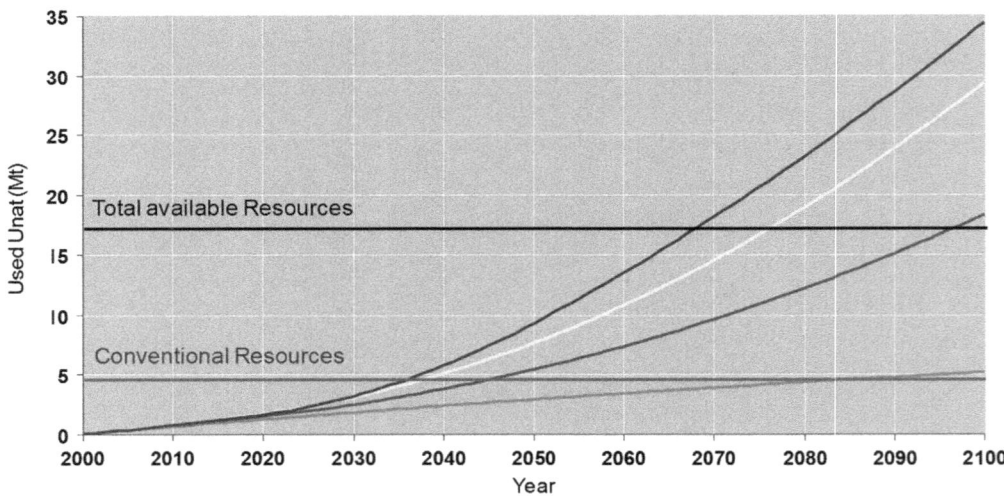

Figure 10.29. Possible future uranium consumption (source OECD/NEA).

The Generation IV International Forum was instigated in 2001 to organise an international R&D program (13 countries in 2010, including France) covering reactor systems deployable from 2030. New requirements have been formulated concerning economy, safety, resource and waste management and non-proliferation. The sodium-cooled fast reactor (SFR) is one of six concepts chosen by this forum, and now enjoys much broader support from the various organisations participating in the forum.

10.7.2. Economy of sodium-cooled FRs

The objective of the EFR project in the 1980s was to design, in the Superphenix line and its European equivalents, a reactor that is economically competitive with a PWR. Finally, it is proven that this reactor has a cost for the electricity produced which is almost 15% greater than that of EPR (see Figure 10.30) specifically due to a much higher investment cost.

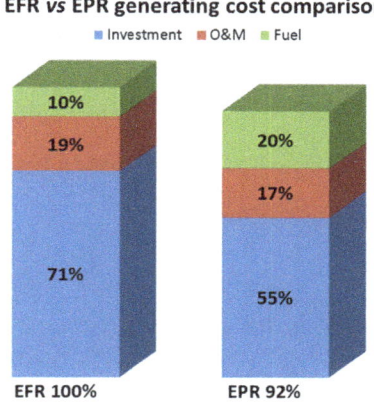

Figure 10.30.

This is why efforts today are in part focusing on reducing the investment cost of this system. In parallel, the increase in safety requirements are another part of the development studies, specifically with the development of cores less sensitive to sodium boiling and severe accidents. Finally, the prospects for growth of the uranium market will dictate the conditions for industrial use of this nuclear system in users' countries.

10.7.3. FR plutonium burner and radioactive waste transmuter

The excess neutrons available after fission can be used for plutonium regeneration. One can also use this excess to consume plutonium or convert radioactive waste, minor actinides or long lived fission products.

For plutonium to be consumed, uranium must first be eliminated as far as possible, being the source of fertile to fissile conversions. Finally, Pu consumption increases with the Pu content of the fuel and is at a maximum with pure Pu. The burner FR path therefore uses specific fuels. For the most familiar one, mixed oxide, there is a limit located at about 40% Pu beyond which dissolution in nitric acid during the recycling after irradiation is no longer possible. Studies have resulted in a reference core with annular pellets with 40%

PuO$_2$, inert pins in the sub-assemblies and dilution sub-assemblies in the core. This core could be installed in an EFR reactor that demonstrates the flexibility of FRs.

For transmutation of nuclear waste, there are two major categories of solutions:

- the homogeneous solutions that consist of dispersing elements to be destroyed in the fuel. Minor actinides then contribute little to fissions in the core but at the same time degrade the safety characteristics. This solution is limited on the one hand by the capacities of the fuel reprocessing plant to accommodate these actinides, which generate high gamma radiation and neutrons, and on the other hand by the degradation of the safety characteristics of the core, mainly in the decrease of the delayed neutrons fraction. A typical content of minor actinides in a homogeneous core is less than 5%;

- the heterogeneous solutions that consist of the assembly of the elements to be destroyed in specific sub-assemblies placed in the core or in the radial blanket. The concentration of elements to be destroyed can then be much higher, up to 40%. It is, however, limited by the manufacturing process and needs a particular processing and production facility, different from that of the driver fuel.

10.8. Conclusion

In a strategy for the sustainable development of nuclear energy, it appears that FRs cannot be ignored. They allow the use of all the uranium raw material, or even the thorium, by breeding, thus extending the use of nuclear energy by several thousand years. Core flexibility, which can be designed for breeding or burning, can contribute both to control of the inventory of plutonium produced in a nuclear fleet and to elimination of radioactive waste, minor actinides and long lived fission products. The experience acquired with sodium FRs has shown that the technology of these reactors has been mastered and that breeding is possible industrially.

If constructions are less numerous today, it is because the nuclear development scenarios will play out some quite time into the future, after 2040, a time at which the use of FRs will be forced upon the industry. This situation does however differ from country to country along with their access to uranium resources.

The benefits outlined above, however, justify extra R&D effort expended on this system. The international program remains important, in particular the Generation IV initiative. The final objective is to have the best industrial product at the time when FRs will be introduced in nuclear power fleets.

References

[1] A.E. Waltar, A.B. Reynolds, *Fast breeder reactors.* Pergamon press, 1981.
[2] G. Vendryes, *Les surgénérateurs.* Que sais-je ? PUF, 1987.
[3] G. Vendryes, *Superphénix pourquoi ?* Nucleon édition, 1997.
[4] H. Bailly, D. Menessier, P. Prunier, *Nuclear fuels for pressurized water reactors and fast neutron reactors.* CEA synthesis series, Eyrolles, 1999.

[5] *Fast Reactor Database*, 2006 Update, IAEA-TECDOC-1531 (ISBN: 92-0-114206-4), 2006.
[6] *Status of Fast Reactor Research and Technology Development*, IAEA-TECDOC-1691 (ISBN: 978-92-0-130610-4), 2013.
[7] J.-F. Sauvage, *Phénix. 35 years of history: the heart of a reactor*. CEA/Valrhô Edition, July 2009.
[8] J. Guidez, *The Phenix experience feedback*. EDP Sciences, 2013.
[9] M. Schneider, *Fast Breeder Reactors in France*. Science and Global Security **17**: 36–53, 2009.
[10] T.B. Cochran et al., *Fast Breeder Reactor Programs: History and Status*. Research Report 8, International Panel on Fissile Materials, 2010.

11 The gas-cooled fast reactor

P. Anzieu

11.1. Introduction

It is possible to design a gas-cooled reactor with fast neutrons (Figure 11.1). Helium or CO_2, or even N_2O_4, can be chosen as a coolant. With no moderator in the reactor core, the neutron spectrum becomes hard and favours the breeding capabilities.

This system benefits from some similar advantages to HTR: the coolant has no phase change and is optically transparent, thus simplifying inspection and maintenance activities. The HTR technology for most of the components outside of the core can be used with few modifications.

As with SFR, it benefits from the ability to breed plutonium and to transmute radioactive waste. It also eliminates the chemical risk associated with sodium.

The main discrepancy is the low coolant quality of gas. It must be pressurised (at around 5 to 10 MPa) and any loss of pressure accident must be correctly managed.

11.2. History

The gas-cooled fast reactor was first studied in the 1960's and 1970's, when it was considered to be a gas variant of the sodium-cooled fast reactor. The idea was to eliminate the sodium risk and to raise the temperature and so the power efficiency, while maintaining a high breeding gain. The fuel element was that of a SFR: a pin bundle, with oxide fuel and metallic clad. Some innovations were introduced such as a roughened surface of the clad to increase the heat transfer to the gas (see Figure 11.2), which unfortunately raises in parallel the pressure drop in the core. Additionally, a complex pressure equalisation system was needed to correctly manage the tendency of the gas to flow through the less resistant channel with little flow in the other channels.

The main projects that have been studied at that time are listed in the Table 11.1 below.

The feasibility of those gas-cooled systems has been established, but they did not show any real advantage over sodium-cooled systems. While the sodium technology was under development at that time, those projects were stopped at the beginning of the 1980's due to the poor prospects for large industrial deployment of the fast neutron reactors.

The GBR studies have evaluated both a pin type fuel element and a particle bed fuel assembly. Figure 11.3 shows this innovative fuel element, where an amount of small coated fuel particles are radially cooled by the coolant flow from the outside of the assembly.

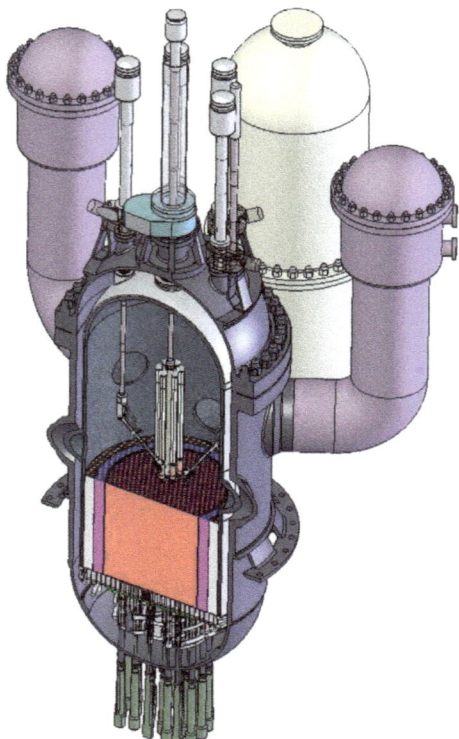

Figure 11.1. Primary circuit of a gas-cooled fast reactor. The core in red is located in a pressure vessel. An intermediate heat exchanger of the primary circuit is in white, while two decay heat removal systems, in purple, are plugged into the main vessel. [source: CEA].

Today the gas-cooled fast reactor takes the name of GFR and the system features a high temperature helium-cooled fast spectrum reactor. It is associated with a closed fuel cycle. The GFR benefits from a less mature technology than does the SFR, but with greater potential performance for a longer term industrial deployment. It uses the same fuel recycling processes. The GFR can also be seen as a sustainable version of thermal spectrum helium-cooled reactors, which also benefits from a more mature technology, with fuel recycling and optimal use of mining resources.

11.3. The GFR, a Generation-IV system

The main GFR core design specifications are the use of helium gas as a coolant, a fast neutron spectrum with a self- or positive-breeding gain and a homogeneous recycling of all actinides. With no separation of plutonium from other actinides (for proliferation

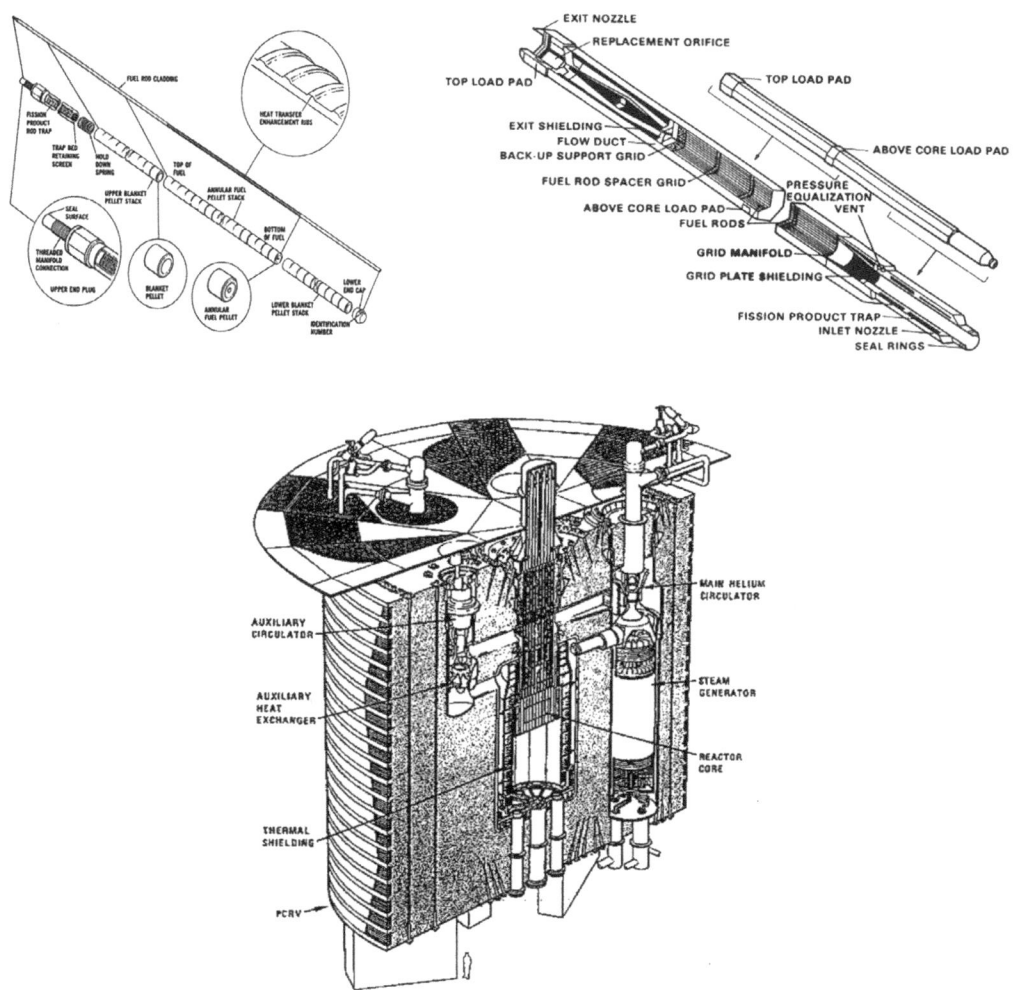

Figure 11.2. The GCFR design of General Atomics in the 1970's, with a detail of the fuel pin element (top left) and the fuel sub-assembly (top right). (Figure reproduced from Simon et al. [1]).

resistance), and a core plutonium inventory not exceeding 10 tons/GWe, a reactor fleet with high fuel burn-up could realistically be deployed in a few decades.

It appears feasible to design a core with such performance, and the research shall focus on safety management. In HTRs, the use of a graphite moderator increases the thermal inertia of the core, thereby limiting the maximum temperature during transients (see Figure 11.4). On the other hand, GFR cores have relatively low thermal inertia; design characteristics aimed at overcoming this apparent unfavourable feature include a fuel element based on refractory materials and high thermal conductivity, with the ability to ensure

Table 11.1. Early projects of gas-cooled fast reactor.

Project acronym	GBR	ETGBR	GCFR
Full name	Gas breeder reactor	Existing technology gas-cooled breeder reactor	Gas-cooled fast reactor
Company	European GBR Association	UK national program	General Atomics Atomics
Time period	1968–1978	1970's	1962–1980
Power, MWe	1000–1200	1320	1240
Coolant	He or CO_2	CO_2	He
Breeding gain		1.2–1.4	
Vessel		Pre-stressed concrete	
Power conversion		Steam Generator	

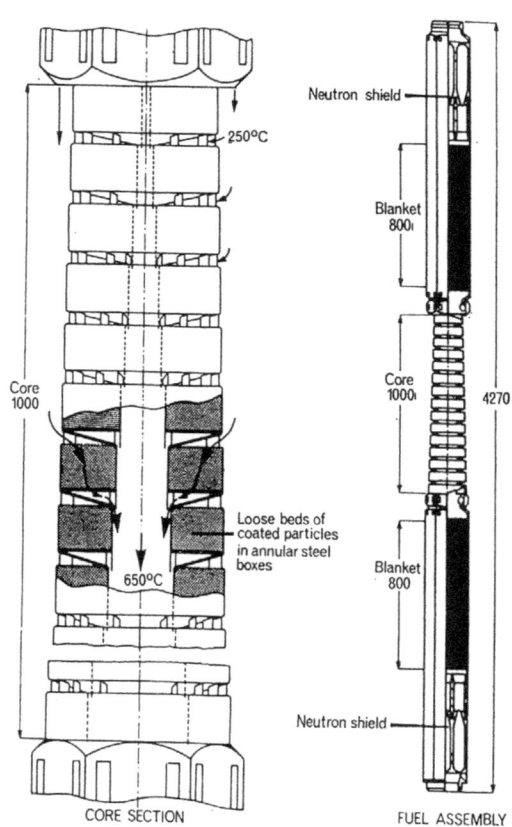

Figure 11.3. The GBR2 fuel assembly with fuel particle bed.

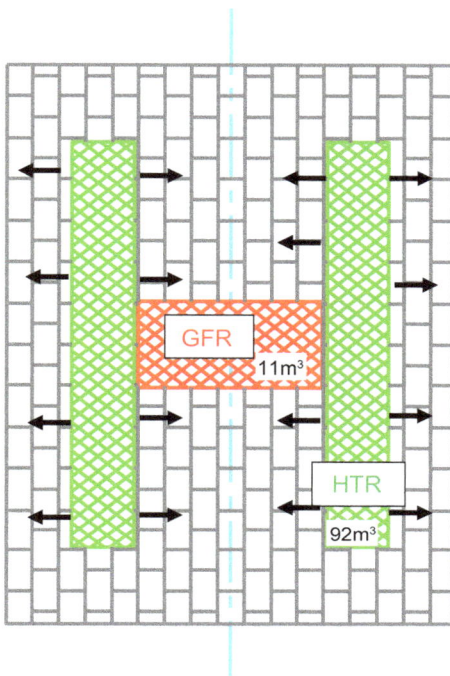

Figure 11.4. Relative geometry of a 600 MWth GFR core (in red) compared to a HTR core (in green) surrounded by its graphite moderator (in grey). Dense GFR volume power requires active gas cooling every time. [source: CEA].

radioactive material confinement up to very high temperatures. It also requires a primary circuit design based on upward core cooling and a moderate pressure drop for all the primary components and circuits involved in accident scenarios. One essential parameter for safety system performance is the gas pressure. The primary helium is pressurised to around 7 MPa under nominal conditions. So a gas tight envelope enclosing the primary circuit has been added in order to limit the loss of pressure in case of loss of primary coolant. Maintaining high helium density allows the decay heat removal system to rely on moderate pumping power and even on passive natural convection in some situations.

The fuel element must be able to withstand high operating temperatures and transients associated with the poor heat capacity of the gas coolant. Design of the reactor and the fuel element is based around some reference temperature criteria:

- an operating temperature of around 1 000 °C, that provides a sufficiently ample margin for failure;

- a boundary temperature of 1 600 °C below which fission product release is prevented;

- an upper temperature of 2 000 °C below which the core geometry can be safely cooled down.

11.4. GFR design options

As of 2015, no GFR has ever been built. Basically, all the GFR high level characteristics are typically the ones shown in Table 11.2, with the objective being to develop a 1100 MWe industrial reactor for electricity production.

Table 11.2. Main standard GFR characteristics.

Unit power	2400 MWth
Efficiency	45–48%
Coolant	Helium
Fuel material	Mixed Carbide or Nitride
Clad material	Ceramic or refractory metal
Average fuel burn-up	5% FIMA
Minor Actinides loading in the fuel	1.1%
Core power density	100 MW/m^3
Clad material	Ceramic or refractory metal

11.4.1. Fuel element

At least two fuel concepts have the potential to fulfil the above temperature requirements, that is: a ceramic plate-type fuel element and a ceramic pin-type fuel element (Figure 11.5). On a laboratory scale, the present reference material for the structure is reinforced ceramic, a silicon carbide ceramic matrix composite. The fuel compound is made of pellets of mixed uranium–plutonium–minor actinide carbide. A leak-tight barrier made of a refractory metal or a Si-based multi-layer ceramics is added to prevent diffusion of fission products through the clad. It can therefore be seen that the fabrication of the fuel element is complex.

 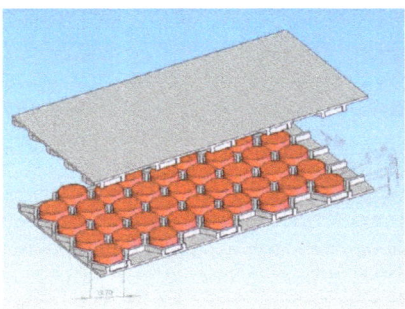 

Figure 11.5. Possible GFR fuel elements, ceramic matrix composite cladded. Pin (left) or plate (center). Example of composite ceramic SiC structure (left). [source: CEA].

11.4.2. Core design and performance

The core layout (246 fissile subassemblies, 24 control rods) is proposed so as to be consistent with the maximum power derived from thermo-mechanical and thermal-hydraulic analysis, the requirements of the reactivity control system and the optimised power distribution. The main characteristics of a reference core are summarised in Table 11.3.

Table 11.3. GFR 2400 MWth typical core characteristics.

GFR 2400 MWth core, with fuel pins	
CORE – SUB-ASSEMBLY	
H/D fissile core	0.387
Inter-assembly gap (mm)	3
Fissile height (mm)	1650
Sub-assembly width (mm)	175.3
FUEL ELEMENT	
Clad thickness (mm)	1.08
Internal Liner (μm)	40 + 10 = 50
Pellet Diameter (mm)	6.71
Pin pitch (mm)	11,56
OPERATING CONDITIONS	
Core Pressure drop (MPa)	0.14
Tmax fuel (°C)	1280
Tmax clad (°C)	990
CERAMIC PIN CORE – MAIN FEATURES	
He/clad/gap/fuel volume fraction	42.9/26.8/2.4/27.9
Trans-Uranium element enrichment (%)	17.5
Pu loading weight (t/GWe)	10.2
Core management (eq. full power days)	3 × 480 = 1440
Average discharge burn-up (at% FIMA)	5.0
Breeding Gain	0.0
Delayed neutron fraction (%)	0.360
Core voiding reactivity (%)	0.322

One can verify that the gas voiding reactivity effect in the core is naturally not significant, i.e. lower than the delayed neutron fraction.

11.4.3. Primary system

The reactor vessel is similar to the HTR vessel in terms of engineering studies. It is a large and thick metallic structure with an internal diameter of 7.3 m, an overall height of

20 m and weight of about 1000 tons with a thickness of 20 cm in the belt line region. The material selected, a martensitic 9Cr1Mo steel (industrial grade T91, containing 9% by weight of chromium, and 1% by weight of molybdenum) undergoes negligible creep at the operating temperature (400 °C). The reference material for the internals is either 9Cr1Mo or stainless steel, typically SS316LN. The global primary arrangement is based on three main loops of 800 MWth each, each fitted with one heat exchanger–blower unit, enclosed in a single vessel (Figure 11.6).

1. Primary cross-duct
2. Secondary pipes with isolating valves
3. Control rod drive mechanisms
4. Primary blower and associated motor
5. Compact Heat Exchanger modules
6. Pipe connections for Decay Heat Removal systems
7. Primary isolation valve

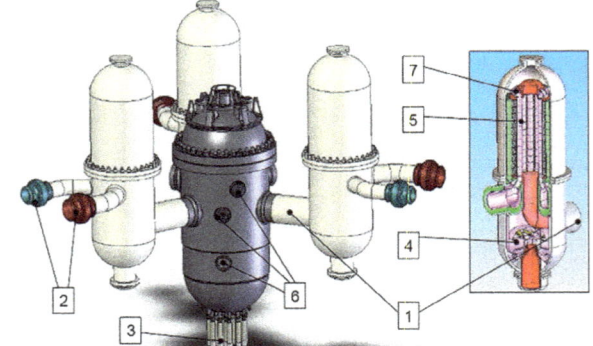

Figure 11.6. GFR primary system [source: CEA].

The shutdown system is derived from sodium-cooled reactor projects, with two redundant and passive shutdown systems (no power supply, gravity drop of absorber elements). Each main control rod and shutdown device and diversified shutdown device is individually driven, forming two independent groups each connected to a dedicated instrumentation and control system.

A gas tight envelope, acting as additional close containment, has been designed to provide and maintain a backup pressure in case of a large gas leak from the primary system. It is a metallic structure, initially filled with nitrogen at slightly over atmospheric pressure that reduces the possibility of air ingress (see Figure 11.7).

This component limits the consequence of a concomitant first and second safety barrier rupture (the fuel clad and the primary system respectively).

Specific loops for decay heat removal in case of emergency are directly connected to the primary circuit using a cross duct piping, in extension of the pressure vessel, and are equipped with heat exchangers and forced convection devices.

This system arrangement allows the residual power to be extracted in any accident situation. In addition, thanks to the low pressure drop of the core design, a passive gas natural circulation can be used in most of the situations, including small primary breaks.

The fuel handling system is based on a jointed arm system, with fuel element loading and unloading using a fuel storage drum *via* lock chambers, with the vessel closed, as shown in Figure 11.8. A dedicated forced convection device, located outside the reactor vessel, is designed to cool the spent fuel sub-assembly during its handling.

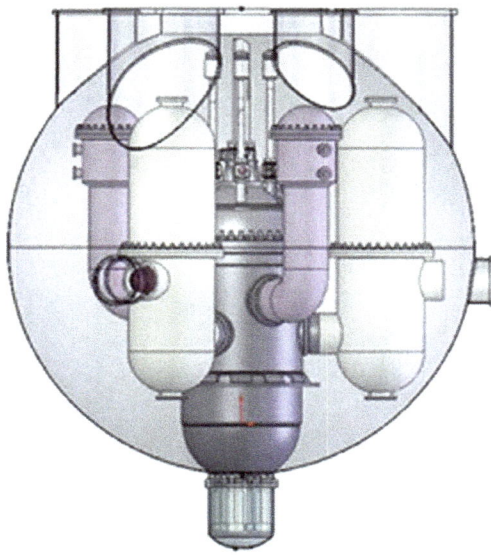

Figure 11.7. GFR primary circuit enclosed in a containment vessel [source: CEA].

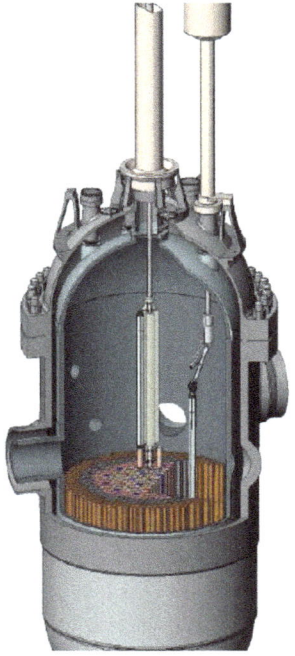

Figure 11.8. The GFR fuel handling system inside the primary vessel. [source: CEA].

11.4.4. Power conversion system

One choice can be the indirect combined cycle with He-N₂ mixture for the intermediate gas cycle. The cycle efficiency is approximately 45%, based on assumed component efficiencies and pressure drops. A schematic view of this power conversion system is shown in Figure 11.9 below.

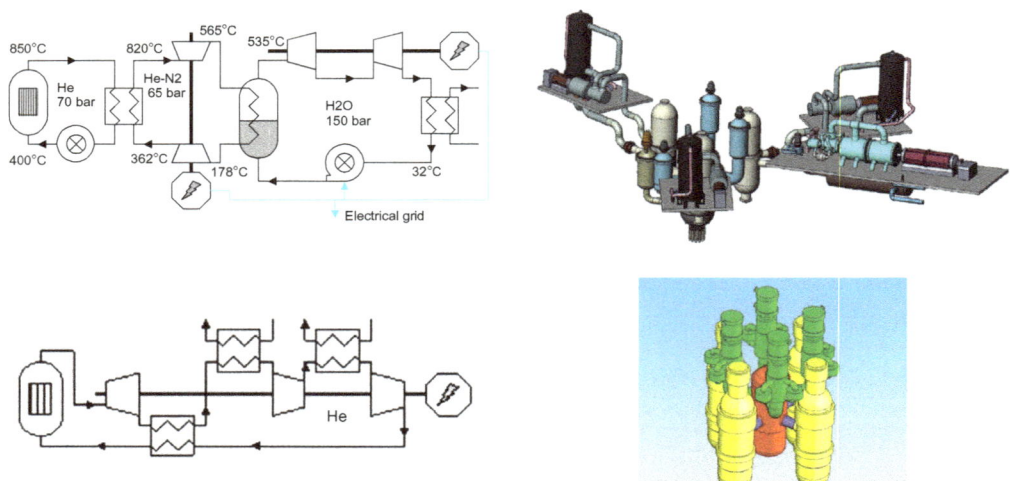

Figure 11.9. GFR power conversion system: indirect combined cycle (top) or direct gas cycle (down). [source: CEA].

Another choice can be the direct cycle, with the helium exiting the core going directly to a turbine. This comes from the GT-MHR, a HTR compact design, and calls for the technology of a vertical large power turbo-machinery, not yet available in the 2010's. Such a gas cycle is very compact and can raise the global efficiency up to 50%. But a safe design is also more difficult to achieve due to the important coupling between the core and the power conversion system.

11.4.5. Towards a demonstration reactor

A first GFR will of necessity be a small demonstration reactor to demonstrate the viability of the GFR system line, no reactor of this type having ever been built before. In the 2010's a project named ALLEGRO was proposed.

With a thermal power of around 80 MWth, it will not produce any electricity. At first, it is foreseen that ALLEGRO will incorporate, at a reduced scale, all the architecture and the main materials and components foreseen for the GFR, excluding the power conversion system (see Figure 11.10). Its safety principles are those proposed for the GFRs: core cooling through a gas circulation in all situations, ensuring a minimal pressure level in

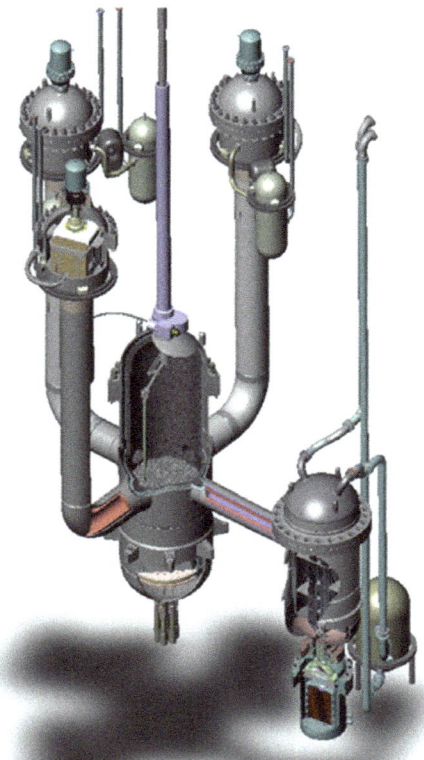

Figure 11.10. ALLEGRO, a first possible GFR demonstration. [source: CEA].

case of a leak thanks to a specific close containment surrounding the primary system. It will also contribute to the development and qualification of an innovative refractory fuel element that will withstand high temperature levels, which is one of the key points by which the GFR system will be assessed.

References

[1] R.H. Simon, J.B. Dee, W.I. Morita, *Gas-cooled fast breeder reactor demonstration plant,* Gas-cooled reactors: HTGR and GCFBR, pages 336–354. Gatlinburg, USA, May 7-10, 1974. ANS.
[2] P. Richard et al., *Status of the pre-design studies of the GFR core.* PHYSOR 2008 3. proceedings, Switzerland, 2008.
[3] C. Poette et al., *GFR demonstrator ALLEGRO design status.* ICAPP'09 proceedings, Japan, 2009.

[4] J.Y. Malo et al., *GFR 2400MW. Status of the conceptual design studies and preliminary analysis.* ICAPP'09 proceedings, Japan, 2009.
[5] N. Chauvin et al., *GFR fuel and core pre-conceptual design.* GLOBAL 2007 proceedings, USA, 2007.
[6] F. Bertrand et al., *Preliminary safety analysis of the 2400MW GFR.* ICAPP'08, USA, 2008.
[7] J. Somers et al., *The Gen IV GFR Fuel and Other Core Material Project.* Proceedings of the GIF Symposium 2009, Paris, France, September 9-10, 2009.

12 BWR: specific features, trends

J.B. Thomas

This chapter will focus on specific features and trends related to the future of LWR and to their potential for improvement in the areas of safety and fuel cycle, while keeping the main advantages of the reactor "type".

The chapter is closely connected with the one dedicated to "The Advanced Gen III LWRs".

12.1. History, principles and architecture

The Boiling Water Reactor NSSS is directly coupled to the turbine, obviating the need for an intermediate system consisting of heavy equipment (the steam generators), which, on the other hand, constitutes a static, permanent second barrier and – sometimes – provides useful decoupling between the nuclear heat source and the two-phase thermal-hydraulic section. However, the coupling can have some advantages (see the "operation" section).

Historically, in light of the experience acquired with naval propulsion pressurised water systems, the BWR appeared to represent a simpler, lighter approach appropriate for less stringent operating specifications. It was General Electric that developed and promoted them as a competitor to pressurised water reactors, once industrial-scale units became commercially available.

However, in the fifties and sixties (and even later, particularly in France, when the time came to select a single technology), the principle of the **Boiling** Water Reactor itself raised doubts that resulted in the deferral of its adoption, and it actually being ruled out. The main doubts related to:

- stability, the means of controlling it, and the upper bound consequence in the event of an accidental excursion (BORAX),

- operation with direct coupling, meaning the absence of a barrier protecting against the results of corrosion-activation, fuel leakage due to cladding failure etc.

- the complexity of the fuel and its effect on the costs;

- the absence of soluble boron for controlling reactivity.

The specific difficulties identified with BWRs were finally overcome, sometimes due to a measure of luck (success achieved empirically in a rather unpredictable manner), to daring innovations and greater willingness to adapt than was the case with pressurised water technology. Long lasting success was achieved by adapting to successive requirements relating to nuclear safety and operations, while maintaining competitiveness and increasing the scope for further evolution.

At the present day, the BWR installed power represents about 20% of the nuclear power generated worldwide with levels of performance comparable with those of the PWR (the USA leading the race where the two types are in direct competition, and Japan – before the Fukushima accident – close behind). However the first advanced or "two-plus/third generation" LWR were ABWR (in Japan and Taiwan), and the ABWR has received NRC certification. General Electric Hitachi has announced total construction times not exceeding 48 months. In the context of BWR technology, the ESBWR and KERENA have considerable safety and economic potential resulting from some architecture simplifications and from the extensive use of passive safety features. BWRs were therefore in a strong position. The Fukushima accident should not affect this positive trend in the mean term. It actually strengthens the need to respect the defence in depth principles and their consequences (in plant design and operation), challenging both BWR and PWR technology. Most of the modern designs fulfil the related requirements and an optimisation will lead to the best available mix of active and "passive" systems, both in normal operation (pumping power vs. natural circulation, for instance) and in the case of accidents.

Actually, the real world perspective is currently not so bright for BWR: only a handful of BWR are under construction and the majority of new build projects are PWR. One of the reasons, apart from the fact that successful Russian VVER are PWR and that ROSATOM is not preparing a shift to diversification, is that the (major) BWR improvements came stepwise. The reactors currently under scrutiny (starting with Fukushima) are older design implementations from the 60's, which were already criticised in the USA in the 70's, and some of them did not have a comprehensive safety update. While PWR changed marginally (and incrementally – without architecture rearrangement or component redesign) over a period of decades, several major innovations were implemented in BWRs: integration of the pumps in the vessel, inducing the suppression of long recirculation loops, reducing at the same time the radiation dose to the workers during maintenance as well as the potential frequency of occurrence of LOCA with a large break area and thus the seriousness of the consequences; modification of the shape and of the materials of the containment; important improvements in fuel design, etc.

One of the main criteria for the future is cycle flexibility: the ability to burn plutonium and possibly certain minor actinides, as well as the ability to increase the conversion ratio to prepare for the future increase in the cost of natural uranium and for the switch to breeder technology. In this respect BWR are well placed for mono and multi-recycling of plutonium, with advanced fuel concepts (which need to be assessed more fully in terms of technological feasibility and for competitiveness).

The main differences (in comparison to PWR) in architecture (see Figure 12.1) and design are the following:

- reduced pumping power and a trend towards natural convection (ESBWR),
- no steam generators or pressuriser,
- reduced containment vessel volume (the pressure being reduced by condensation in the dedicated pressure suppression pool in case of a LOCA), and a strong scrubbing

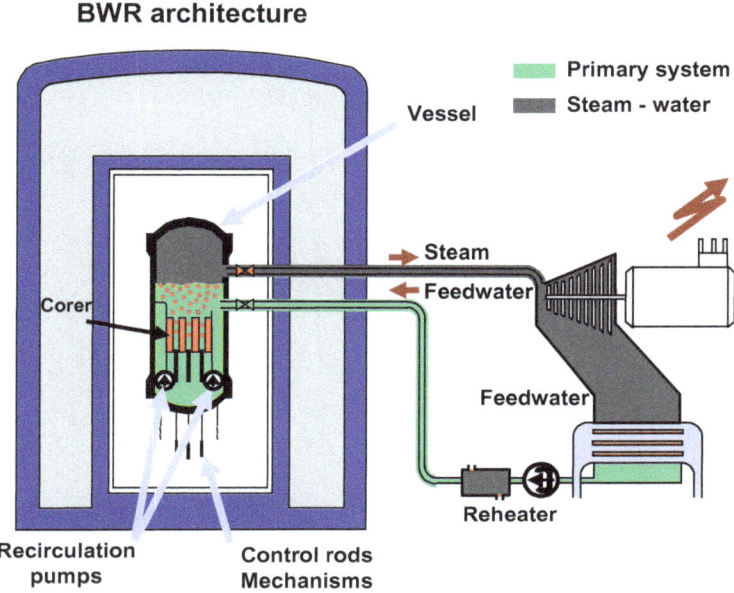

Figure 12.1. Boiling Water Reactor Architecture.

influence on the fission product (FP) retention performance (excepting the noble gases like xenon) in case of a severe accident leading to core fusion and a to a massive release of FP (and more specifically of cesium and of iodine), as in the case of Fukushima. This works well so long as there is no break in the "wet-well" and so long as the water inside the pool is in sub-cooled situation. On the other hand, just as in any under-optimised design, a larger containment (in PWR) gives a little more resilience (the same holds in PWR, for instance, with the large water amount in the secondary side of the steam generators).

- larger vessels (power per unit volume of 50 kW/L, half that of PWR at 100 kW/L), but with thinner walls (pressure reduced by half),
- reactivity control by cross-shaped control rods and burnable poisons, with no soluble boron,
- radial flattening of the flux driven by two-phase thermal-hydraulics in closed channels and by the void coefficient,
- axial control of flux **and spectrum** (through void fraction variation), hence more flexibility on the breeding and burning of plutonium by moving the cross-shaped control rods and by varying the pumping power (spectral shift),
- power regulation by the flow of the recirculation pumps as well as by moving the cross-shaped control rods,
- more finely divided fuel (smaller bundles), more complex fabrication (enrichment and poison zoning) offering greater operating margins in certain cases and leading to

higher fabrication cost than in PWR (typically 300 $/kg heavy metal with an over-cost between 1/4 and 1/3),

- direct coupling to the turbine with related constraints on operation and cleanliness,

- cross-shaped control rods inserted from underneath with electrical and hydraulic drive mechanisms.

The following Figures (12.1, 12.2 and 12.3) schematically show the design of a boiling water reactor.

- Plant output, Mwe 1350 nominal can be uprated to 1500
- Single, stand-alone unit
- Licensed in three countries
- Meets EPRI URD
- Has been reviewed against EUR

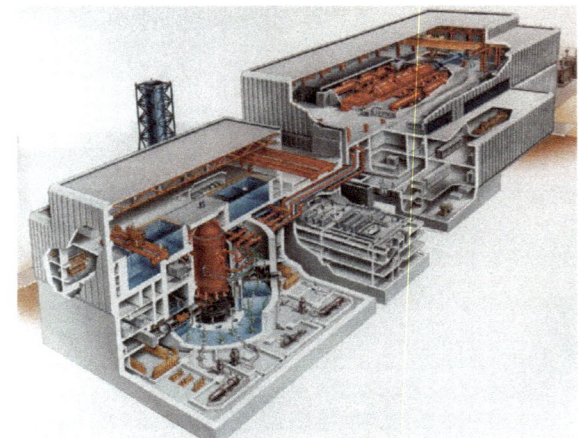

Figure 12.2. The ABWR type (from General Electric).

It is difficult to achieve a global balance of the advantages and drawbacks due to the differences between PWR and BWR. Nonetheless, two examples will be given:

- the reactivity control without soluble boron means about three times more rods than for PWR. On the other hand, the individual reactivity worth is lower and the core periphery coverage is denser and fuller, leaving less room for an exceptionally high reactivity cluster of fuel assemblies to create (locally) a critical "core". Moreover, there is no perturbation of the reactivity feedback coefficients by the variations in bore concentration, from BOC (Beginning of Cycle) to EOC (End of Cycle). This is also fortunate, taking into account the important role of the void coefficient in safety, as well as in operation flexibility. Last but not least, there is no risk of a reactivity insertion by incidental injection into the core of a few dozen of liters of clear water, deprived of soluble boron, which is a risk in PWR.

- The assembly casing is needed in order to control the two phase flow through the core (and a casing can even be implemented in a potential chimney above the core, as in the ESBWR design), as well as for assuring a smooth channel and a guaranteed gap width between the external rods of neighboring assemblies. The casing induces a higher fabrication cost for a higher number of assemblies, some sophisticated fuel

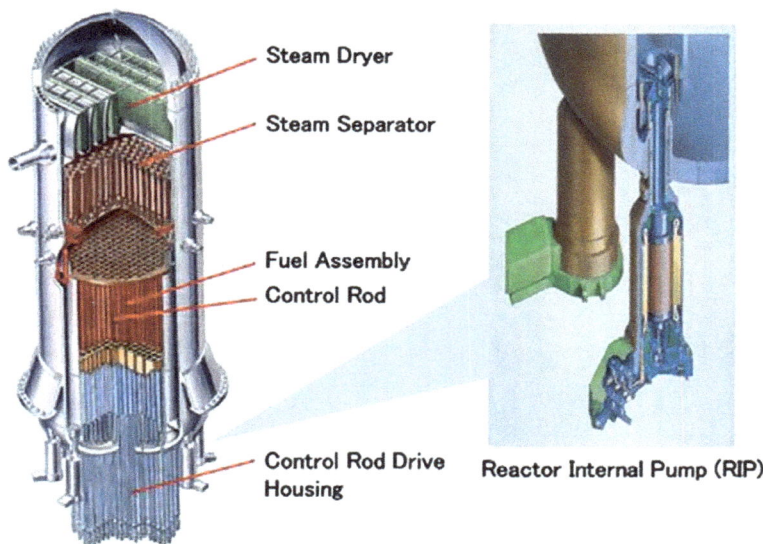

Figure 12.3. ABWR internals and Internal Pumps (General Electric).

enrichment and poisoning zoning inside the assembly to reduce the local power peaking due to the water gap. It leads also to about 70% greater release of hydrogen in the case of massive zircaloy burning during the early phases of a severe accident. On the other hand, the casing creates a powerful thermal-hydraulic – neutronic decoupling of the assemblies in the core, making it easier to introduce innovative fuel assemblies into a core loaded with various types of older fuel elements.

Finally, some fortunate convergences between mandatory properties of different nature occur. For instance, the large room above the core, mainly filled with steam in normal operation, offers a larger expansion room inside the vessel. It contributes to damping, slowing down incidental reactivity ramps due to the rapid closure of the steam valves, for instance.

Figure 12.4 shows the main flow balance and some thermodynamic values in the ABWR NSSS.

12.2. Neutronics, absorbers, fuel

12.2.1. BWR vs. PWR: moderation ratio

The neutronics of the BWR core are not so far removed from those of a PWR.

As far as moderation ratio is concerned, due to the high void fraction in the upper part of a BWR core (typically 50%), one has to introduce two indicators:

- The "geometrical" moderation ratio, defined as the V_{H2O} / V_{UO2} volume ratio of the volume occupied by the water or steam/water over the volume occupied by the fuel oxide inside an assembly, in the core as a whole. This geometrical indicator is higher

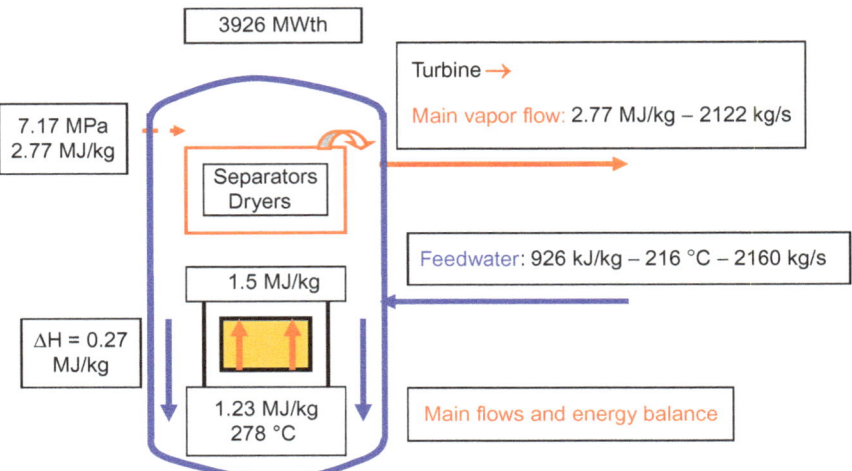

Figure 12.4. Thermodynamic properties in the NSSS of the ABWR (General Electric).

in the BWR than in the PWR, for reasons partly due to thermal-hydraulics and partly due to neutronics (see below);

- The "nuclear" moderation ratio, related to the numbers of hydrogen nuclei vs. the number of heavy nuclei in a given plane transverse section of an assembly, of the core. This ratio is strongly dependent on the height of the considered section as well as on the fuel assembly design (for instance with part-length fuel rods) and on the operation regime connected to the recirculation pump rotational speed controlling the axial void fraction distribution. From this viewpoint, even if the whole core geometrical indicator is higher in BWRs, in the core upper part and more specifically inside the fuel casing, the nuclear moderation ratio can be low and can be varied at will in a limited range (see below the spectral shift operation mode usage of this flexibility). The nuclear moderation ratio in a given region of the core ($N_H/N_{HeavyMetal}$) has to assure the necessary sub-moderation for the temperature and void fraction feedback coefficients so as to always be negative and thus ensure stability and safety.

On the other hand, the V_{H_2O} / V_{UO_2} volume ratio in a BWR is higher than in a PWR (2.5 to 2.7 compared to 2 to 2.1). For this reason and due to the critical flux limits (around two thirds of those of a PWR), the core of a BWR is larger than that of a PWR of equal power (typically 50 kW/l instead of 100 kW/l).

12.2.2. Core structures and fuel assemblies, Reactor Pressure Vessel (RPV)

When compared with Nuclear Propulsion fuel, channels were eliminated from PWR (even for very large cores, far larger than the propulsion cores) but retained in BWR to maintain the compartmentalised axial circulation of the two-phase flow. The casings are made of

Zircaloy-2, like the fuel rod cladding. The number of fuel rods has been increased from 7 × 7, to 8 × 8 and finally to 10 × 10 (Figures 12.5 and 12.6) or even 11 × 11 (Figure 12.7), making it possible to increase the specific power and the burn-up while providing greater safety margins (see the "Operation, fuel and plutonium" section below).

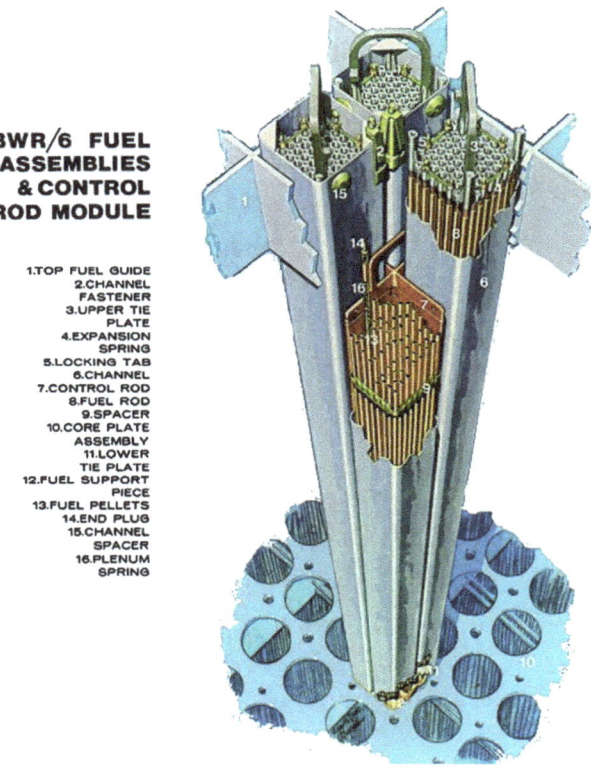

Figure 12.5. Cutaway view of a fuel assembly (GE).

Furthermore:

- the number of control (cross-shaped) absorbers is typically 205 in a BWR compared to 73 rod bundles in a 1450 MWe PWR;

- core loading is performed by assembly in a casing, using greater subdivision than in a pressurised water reactor (four to five times more for the same power).

- the cost of the casing corresponds to 20% of the fabrication cost. Severe accident studies indicate the casings contribute to the formation of 70% more hydrogen than in a PWR;

- the diameter of the pressure vessel (RPV) of the ABWR is 7.1 m compared to 4.5 m in a 1450 MWe pressurised water reactor. The diameter contributes to obtaining a high steam retention volume in the dome of the vessel, which notably slows down

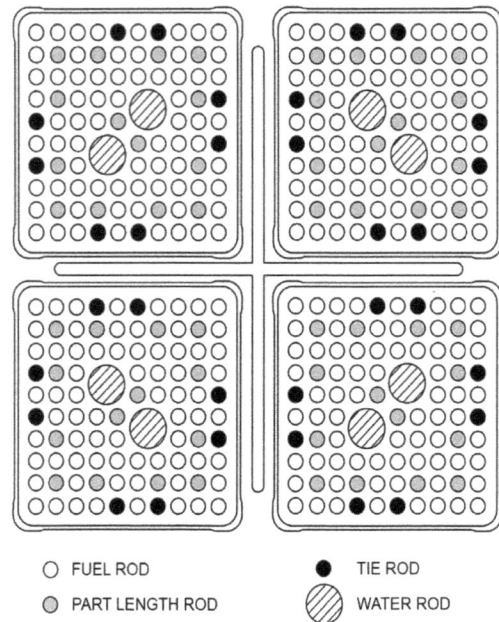

Figure 12.6. Cell consisting of four 10 × 10 fuel assemblies and a cross-shaped control rod (General Electric).

the pressure rise in the event of closing of the isolation valves. The downcomer width of approximately 0.9 m allows the recirculation pump impeller extraction and confers on the ABWR reduced damage of the pressure vessel wall by fast neutron irradiation.

Total negative reactivity potential is 20,000 to 25,000 pcm, twice that of a PWR control rod bundle. In the absence of soluble boron, the rods contribute to compensation for the "reactivity swing" devoted to irradiation, completing the addition of gadolinium to the fuel.

The ABWR has 205 control rods, which rise up from the bottom of the core. Removal of the most effective rod, adds less than one dollar. This is the rod releasing the highest amount of reactivity in case of removal, with a given set of absorbers inserted in the core (and specifically all the other ones in some hypothetical design accident situations).

For the rods, preference has been given to hafnium and boron carbide as absorber material. Pressure rise transients require rapid movement of the rods. Another way of controlling the reactivity is through void fraction, driven by the pump rotation speed, at a given power level (see below: operation).

12.2.3. Distribution of enrichment and of poisons

In the vicinity of the water gaps outside the channels, i.e. at the periphery of the fuel assemblies, more of the neutrons are thermal. Power is flattened by using a higher enrichment at

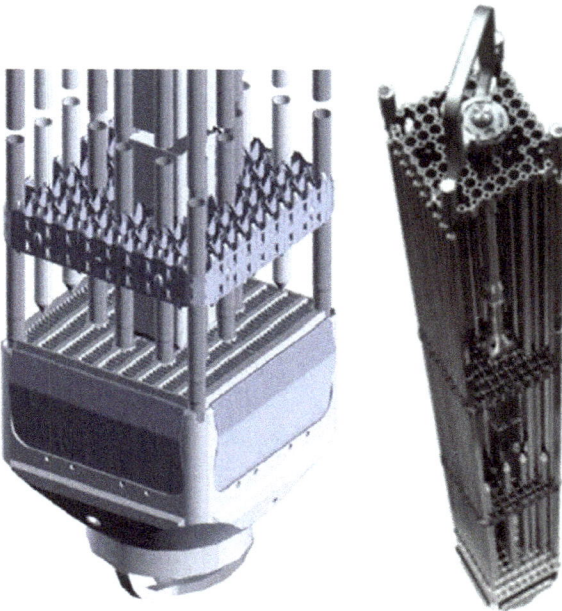

Figure 12.7. AREVA ATRIUM fuel assembly. ATRIUM™ (11 is a BWR fuel assembly with a 11 × 11 square lattice, the internal quadratic water channel occupying nine (3 × 3) lattice positions. The fuel assembly comprises a total of 112 fuel rod positions, among them 12 short part-length fuel rods and eight long part-length fuel rods.

the center, a lower enrichment at the periphery and a very low enrichment in the corner rods. Fuel assemblies contain 3 to 9 different enrichment levels, including that of natural uranium.

The Gd_2O_3 mixed with the sintered UO_2 reduces the thermal conductivity and cannot be added to the most-enriched rods. To get a well-balanced axial flux, steps of concentration and zoning can be used. Currently, higher enriched pellets are being placed in the upper half of the core. There are also water rods, cross-shaped internal water gaps and "atrium" designs.

Thus, to constitute an 8 × 8 or 10 × 10 fuel assembly, it is necessary to correctly place rods of seven to ten different sorts, making pellet composition a sophisticated matter and increasing the complexity of quality control and raising the cost of fabrication. However, the extra cost resulting from this complexity (channel, division, zoning of enrichment and poisons) does not appear to significantly reduce competitiveness because:

- the increase (about 1/4 of the fabrication cost), affects a small part of the busbar cost;

- this increase is compensated for by a finer management of the fuel (see the section below on operation), reducing significantly the uranium and the SWU consumption for the same power production.

12.3. Thermal-hydraulics and its tight coupling with neutronics

12.3.1. Recirculation ratio

Steam quality at the core outlet is limited by the risk of cladding dryout. The recirculation ratio is about 7.

The steam outlet quality is itself linked to the outlet temperature and pressure (T_{sat} (p)), as well as to the power level and the flow rate.

At rated power, there can be an incentive to depart from 100% nominal flow in the core, either at the start of the cycle, by diminishing it to reduce the reactivity by undermoderation or at the end of the cycle, raising it to as much as 115%, which is possible with the 10 pumps operating at full output, in order to adjust the cycle duration by stretch out. This results in variations in the recirculation factor.

12.3.2. Coupling between neutronics and thermal-hydraulics

The column of water rising up the channels in the core remains liquid over the first 40 to 70 centimeters before boiling. At rated power and flow, steam already occupies an average of 50% of the available space in the channels, half way up the core. At the channel outlet, the proportion of steam by volume is on average 65% (75% in the most heavily loaded channel) as compared to an average of 40% in the core as a whole.

The negative void reactivity coefficient is about −100 pcm (−1‰) per percent void fraction around the end of a cycle (EOC) at equilibrium. Towards the end of a cycle, the global void effect is limited to −3000 pcm.

The strong coupling between the thermal-hydraulics and the neutronics has a number of consequences:

- A boiling water reactor has a high radial natural/intrinsic flattening feedback mechanism through the tight coupling between neutronics and two-phase flow thermal-hydraulics; even when exacerbated by the xenon effect, slow radial flux oscillations are countered by the high negative void effect;

- The orifice plates preclude static instability (Ledinegg-type instability, see below), and no radial control is necessary, even with cores 5 m and greater in diameter. This radial stability eliminates one of the main constraints on the loading pattern,

- Conversely, the "natural" axial power offset is higher than in a pressurised water reactor, due to the high mean void fraction in the upper part of the core, with a high negative void reactivity feedback coefficient. In the first BWR, the power peak in the lower part of the core was mainly substantial at the beginning of a cycle (BOC) with fresh fuel. The effect was limited by inserting the control rods at the bottom and the (nevertheless) faster burning of the fuel at the bottom of the core gradually reduced the axial offset. Axial steps in the proportion of burnable gadolinium in the fuel were then introduced, with a later development being higher enrichment in the upper half of the core.

- The effects of feedback associated with the negative void coefficient are sufficient to offset the axial effects of xenon.

In normal operation, pressure is kept constant by controlling the steam turbine regulation valve.

The turbine is protected against over-speed by a valve in the steam inlet, which closes if the generator trips. Its closing causes the bypass to the condenser to automatically open.

The release of steam can also be interrupted by closing the isolation valves. With the ABWR, in the first seconds after a scram, the core continues to produce steam at a rate of 58.5 m^3/s. This causes the pressure to rise at a rate of 8.5 bar/s. The pressure coefficient is estimated to be 40 pcm/bar.

12.3.3. Thermal-hydraulic instability

Static (Ledinegg-type) instability: the core of a power generating BWR, unlike that of an electronuclear PWR but similar to that of a propulsion core, contains closed, parallel channels. All the channels have the same inlet and outlet pressures. Pressure drop Δp is therefore the same in all the channels and the flows adapt accordingly.

This type of nuclear steam supply system presents a risk of static instability due to the fact that the two-phase pressure drop Δp does not increase monotonically as a function of flow Q in the channel.

Two different solutions exist for the same Δp:

- A low flow one, with a severe overheating (and thus cladding failure) potential for a whole fuel assembly enclosed in a casing;

- A high flow one, over-cooled, which wastes flow and pumping power.

The Ledinegg instability can be countered by installing inlet orifice plates at the channel inlets.

The recirculation pumps impose/dissipate an increased pressure loss upstream from the core, causing function Δp (Q) to increase more monotonically.

There is also another potential source of dynamic instability of a different nature. A disturbance at one point can have deferred consequences upstream and downstream. The energy surplus supplied in the mixture can give rise to self-sustaining oscillations. The geometry of the channel, which is square in cross-section and 13.4 cm of a side with a fissile length between 366 and 381 cm and pressure within it set at around 72 bar leads to an oscillation period of 2 s.

A phenomenon of this type occurred in a General Electric boiling water reactor in La Salle 2 in 1988. After loss of two recirculation pumps, the switch to natural convection put the reactor in an unstable condition in which in-pile flux readings showed power oscillations of 18 to 118% relative to the rated value. This triggered a scram.

Figure 12.8 shows a transient that occurred at Forsmark 1 in 1989; Figure 12.9 illustrates the principle of the negative feedback loops in a boiling water reactor.

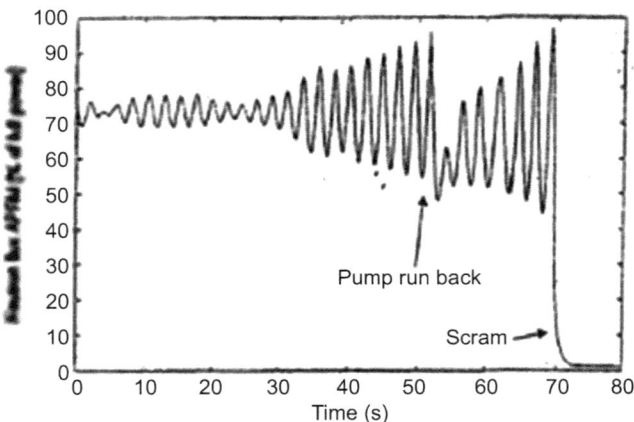

Figure 12.8. Oscillations at F1 in 1989 – Forsmark Kraftgrupp.

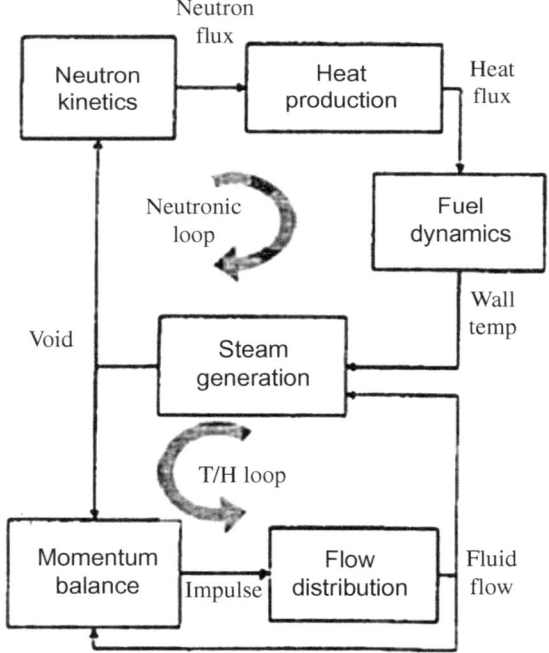

Figure 12.9. Feedback loops in BWR.

Figure 12.10 shows the ABWR stability effects on operation. The stability limit is close to the natural convection capability zone, which is located from 0 to 20% of the nominal flow and from 0 to about 50% of the nominal power, at the left side of the chart (see also the operation diagram in Figure 12.11). Tripping a pump while operating just above this zone leads into the instability zone. The fuel loading strategy has to take into account this type of behaviour.

Figure 12.10 shows the operating constraints associated with instability.

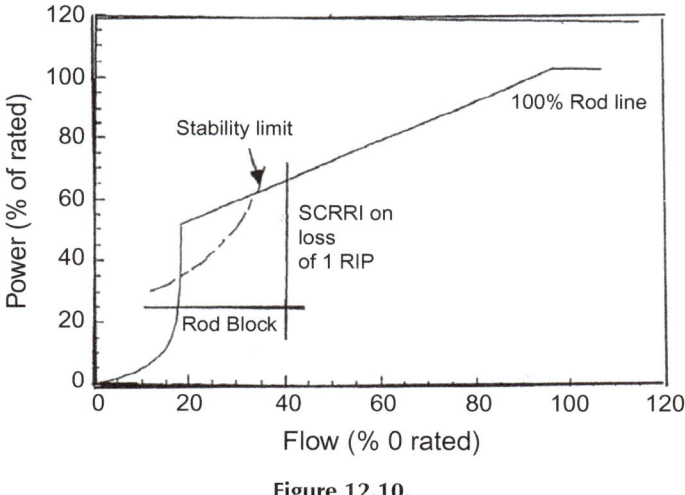

Figure 12.10.

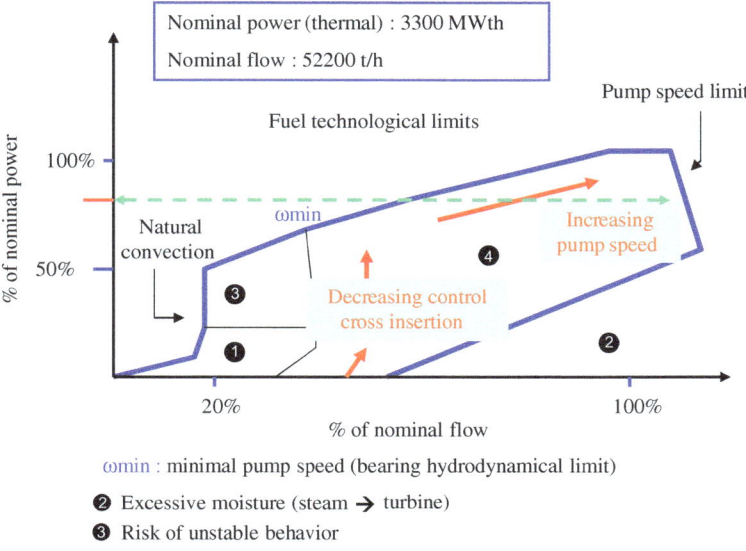

ωmin : minimal pump speed (bearing hydrodynamical limit)

❷ Excessive moisture (steam → turbine)
❸ Risk of unstable behavior
❹ normal operation

Figure 12.11. Operation diagram showing the operation envelope and the control principles (from GE & Techniques de l'Ingénieur).

12.3.4. Stability loops; conceptual scheme of a sequence of feedback effects

The Doppler coefficient is around −2.3 pcm/K at the beginning of cycle, with uranium oxide fuel.

During a typical sequence of events, for instance when reactivity increases by $\Delta\rho$, some of the feedback mechanisms illustrated in Figure 12.9 are triggered, acting sequentially due to different time constants:

1. the energy release in the pellets increases, so does the temperature of the fuel: the immediate stabilising Doppler feedback is intrinsic;

2. the wall heat flux increases with a time constant of a few seconds, and so does the steam bubble generation at the wall. At an initially constant pressure, the void ratio increases. With a strong void coefficient (typically −1‰ per 1% void increase), this stabilising feedback is powerful but slower than the Doppler effect.

3. If the valve regulating the supply of steam to the turbine doesn't open, the increase in the steam flow causes the pressure to increase in the restricted space available.

Saturation temperature T_{sat} increases with the pressure. Sub-cooling (T_{sat} − Te) increases at the core inlet, the height of the liquid-phase-only water region at the bottom of the core increases and, overall, the average void fraction in the core diminishes, which causes the reactivity to increase. This has a strong destabilising effect but, its time constant being longer, the reactor remains controllable by forward regulation using the control rods and the recirculation rate.

12.4. Operation

12.4.1. Principles

Grid power increase causes the steam valve to open, causing the saturation temperature and pressure in the system to drop, the volume of the bubbles to increase and the reactivity to fall. Instead of meeting the power demand like in the PWR case, the reactor chokes.

Increased demand is thus satisfied by accelerating the circulation pumps and/or withdrawing the control rods.

The turbine moderating valve position is slaved in short loop to maintain the upstream pressure constant. When, on withdrawing the control rods for instance, the pressure in the reactor rises, the moderating valve opens to allow more steam to escape. This mode of operation is called forward regulation.

To control the power of the reactor, regulation modifies:

- the position of the control rods,
- the flow Qe of the water entering the core, where:

    $\partial\rho/\partial Qe = +14$ pcm/% at nominal flow.

Furthermore:

$\partial\rho/\partial Te = -70$ pcm/K;

$\partial\rho/\partial Ps = +40$ pcm/bar.

Varying water flow Qe has the advantage of avoiding local disturbance of the neutronic power. Control rod action, which must be rapid, is generally second to recirculation flow action, at least in modern equipment where the pumps are integrated into the bottom of the pressure vessel and are of reduced inertia.

Eliminating the pumps and relying only and separately on natural circulation for cooling and on mobile absorbers for reactivity and flux shape control, entails a loss of flexibility.

12.4.2. Operating envelope

The two control parameters are shown in Figure 12.11:

- the flow through the core is on the x-axis and the power on the y-axis;
- the position of the control rods is shown by arrows representing the decreasing insertion of the control rods.

The temperature of the feedwater and the pressure are kept constant, and are not shown in the diagram. The position of the rods and the flow rate determine the power of the reactor. The same power level can be obtained either at rated flow with a few control rods inserted or at reduced flow with all the rods out. This second mode of operation can only be used if it is possible to accept high axial offset at the beginning of cycle (which assumes that there is a linear power density margin, or that this can be avoided by axial zoning of enrichment), but it is advantageous to replace the sterile captures in the rods with fertile captures that produce plutonium (which is burnt *in-situ* at the end of cycle (EOC)).

The diagram shows the natural circulation curve. The flow is then greatly limited by the fact that the stopped recirculation pumps result in significant flow resistance. This would not be the case in a design relying on natural convection, without pumps and with the help of a chimney.

The diagram shows the thermal-hydraulic instability Region III.

Region II is also avoided, as here the quality of the steam leaving the core is too low for the steam separator/dryers.

Full rated flow is maintained with one of the ten pumps stopped. In the ABWR, this is also achieved with the ten pumps running at 90% of their rated speed, i.e. 590 kW on the shaft (777 kW at the full rated speed of 1,500 rpm). This means that it is possible to achieve 111% rated flow at reactor rated power, and even 116% at reduced load.

This flexibility provided by the broad power range (100 to 64%) controllable by pump speed alone without control rod movement is one of the key features of the ABWR, facilitating load regulation and load following on the power grid.

12.4.3. Operation, fuel and plutonium

The sub-division of BWR fuel (which for a long period of time had 7 × 7 rods bundles per channel) now goes beyond what is strictly necessary to address a LOCA situation.

The transition to 10 × 10 rods has been accompanied by a great reduction in linear power, allowing high axial offset which can be made use of to:

- increase operating flexibility,
- vary the spectrum of the neutrons,
- increase the enrichment and the average discharge burn-up.

Spectral shift: an application of the operational flexibility to improve the use of uranium is methodical burning of the fuel, level-by-level, beginning at the bottom.

- At BOC, the fresh fuel can withstand high power peaks at the very bottom due to its sub-division and, with a core flow limited to 75%, the median and upper part of the core have a very high void fraction and a hard spectrum. There is therefore very little burning of uranium-235 but much breeding of plutonium-239 by resonant fertile capture by uranium-238.
- The lowest level is first to be depleted of fissile uranium-235 and plutonium-239. This results in a power shift towards the top of the core where the plutonium produced takes over.
- Towards the end of the cycle, when there is a general depletion of fissile isotopes, the core flow is progressively increased to 115%, which reduces the void fraction. At the top of the channels, the spectrum, which becomes softer, allows further consumption of the plutonium formed at the BOC.

There is thus considerable variation in the spectrum (spectral shift) at the top of the core which is a key factor in the improved use of uranium.

- BWR can be quicker to refuel than PWR. This feature encourages greater sub-division on refueling (five-batch pattern). The reduced specific power (150 t of uranium in-pile in an ABWR compared to 104 t in the 1300 MWe PWR of Paluel Power Station, which is of comparable unit power), makes it possible to also resort to longer cycle durations of 18 and 24 months. There is a common trend with both BWR and PWR to reduce the maximum linear power to increase core management flexibility and facilitate the recycling of the plutonium with, in both cases, enrichment levels raised to 4.5–5% and burn-ups tending towards 60/65 GWdays/t.
- Uranium consumption and cycle costs are similar in both types. According to AREVA NP GmbH which supplies fuel for both PWR and BWR: in 1973, BWRs consumed 32 t of U_3O_8 per TWh (247tUnat per Gwyr), that is 10% less than the PWRs, and 13 SWUs (of isotope separation), which is 20% less than the PWRs.

While the fabrication of BWR fuel is more complex, open cycle (once-through) costs are about the same. The lower enrichment of BWR fuel mainly reduces the isotope separation work but the lower levels of discharge burn-up, coupled to higher fabrication cost, had a negative effect on cost.

In 1993, Siemens set a goal of 45 GWdays/t burnup for BWRs containing 10 × 10 fuel, i.e. 28 t of U_3O_8 per TWh (216 tUnat per GWyr, 10% less than for PWRs with fuel removed at 52.5 GWdays/t and 13 SWUs (18% less than for the PWRs). Although it had dropped by 30% over twenty years, the cycle cost remained equivalent in both types. Since this time, improvements are still being made in both types, with increasing burn-ups, adaptation of the fuel, and increased burn-up for MOX fuel. Increasing cost of uranium will spur further improvements, probably down to 150 or 120 when taking into account a strong tail enrichment decrease as well as recycling.

This is a very important trend and could well be followed by a modification of the size and of the shape of the assemblies as well as of the lattice pitch, in order to increase the conversion ratio and to enter into a new regime of efficient and complete closing of the plutonium fuel cycle (as for plutonium alone at first) by multi-recycling (without plutonium isotopic composition degradation penalty) in under-moderated cores. This could lead to levels of well under 100 tUnat per GWyr, as soon as a strong and sustainable economic incentive arises from natural uranium cost increase.

12.5. Chemistry of water and materials

It is an essential issue for design and operation in a directly-coupled reactor whose system must remain clean (no leakage of fuel, no corrosion-activation deposits). Progress has been mainly empirical, with some lucky breaks.

The main issues are as follows.

12.5.1. Radiolysis

Radiolysis breaks water into nascent oxygen and atomic and molecular hydrogen and provokes other reactions involving atmospheric nitrogen entering via the condenser, resulting in the creation of NO^-_2 and NO^-_3 oxidizing ions, accompanied by other species such as OH^- and H_2O_2 in the recirculating liquid phase, which affects the corrosion of the metals present.

12.5.2. Cladding

Nucleate boiling in the presence of active oxidizers is a severe condition for the cladding. The discovery (by chance) that an alloy of zirconium containing 1.5% tin slightly contaminated with stainless steel could give rise to a resistant material (Zircaloy-2) was a decisive factor in the success of the boiling water reactor.

In addition to corrosion on the water side, there was internal corrosion caused by fission products, clad-pellet interaction and embrittlement of the cladding by hydriding. The (numerous) measures included for instance the use of "barrier fuel" by General Electric, i.e. cladding with nine tenths of its thickness being Zircaloy-2 and one tenth pure soft zirconium on the inside to accommodate the clad interaction with hard, rough pellets.

12.5.3. Intergranular stress corrosion

The system purity requirements dictate some of the construction options and make it necessary to use certain materials. In particular: the reactor vessel is clad with austenitic stainless steel, the pressure vessel internals are made of austenitic stainless steel.

Many methods have been used to avoid intergranular stress corrosion cracking, such as the introduction (and accurate adjustment) of hydrogen into the feedwater.

12.5.4. Activation and gamma-emitting deposits, radiation protection in the turbine hall

A number of isotopes originating from Co, Ni, Fe, Cr, Zn emit gamma radiation in the 1 MeV range.

Any deposits of radioactive isotopes in the systems are difficult to remove by chemical means, as may be desirable prior to field work relating to refueling, maintenance and safety inspections being carried out. The thickness of the steel of the loops is not so sufficiently great as to adequately attenuate gamma radiation from internal deposits. Therefore, with BWR, the recirculation loops were responsible for greater staff irradiation than with PWR. When these loops were deleted, the situation was improved.

The reduction of the activity of the systems necessitates reduction of the cobalt content (which forms cobalt-60), but the nickel content (which forms cobalt-58) cannot be significantly reduced.

Water chemistry is also carefully controlled. Schematically, beginning with oxidizing, as pure as possible water, oxygen, hydrogen, iron and zinc have been injected.

As its half life is very short (7 s), ^{16}N (produced from ^{16}O, through a (n, p) reaction) is not problematic in maintenance outages.

Deposits of cobalt-60 etc. are more awkward. Cladding failure can also complicate maintenance work in the turbine hall.

Finally, the progress made with both the fuel and the water chemistry reduced significantly the doses to staff in the turbine hall.

12.6. Safety

12.6.1. Containment barriers

A BWR has three containment barriers:

- the fuel cladding,

- the pressure boundary of the NSSS, consisting of the pressure vessel, its extensions (the casings of the control rod drives etc.) and the pipes connecting the pressure vessel to the feedwater isolation valves and non-return valves and the double steam isolation valves at the containment vessel penetrations,

- the containment, which forms a leak-tight enclosure completely surrounding the pressure vessel and the associated piping of the second barrier up to the isolation valves.

In a PWR, the steam generator tubes form part of the reactor coolant system pressure boundary (a static barrier). In a BWR, the second barrier is only effective once the double steam isolation valves are closed. The second barrier is thus referred to as dynamic. The valves are provided with protection to ensure that a missile cannot damage both valves of the same line.

12.6.2. Containment pressure reduction

General Electric has developed a pressure reduction system using bubbling and condensing in a pool, which is referred to as passive as it does not require human intervention or any external source of power. From the dry part of the containment where the feedwater and live steam systems are located, pipes lead to dip tubes in a wet chamber (wet well) half full of water. Figures 12.12 and 12.13 compare the PWR and BWR design principles.

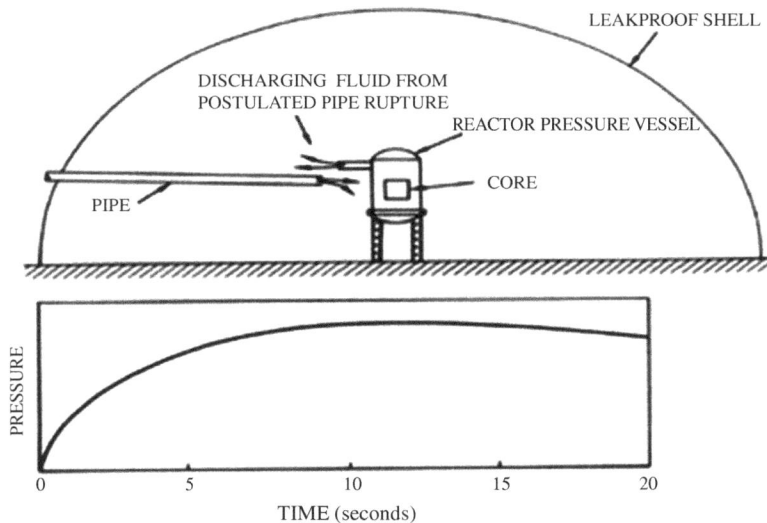

Figure 12.12. The PWR containment building principle: a large building and a slow depressurisation by aspersion (the Figure does not show the bulky components leading to a very large containment, anyway. The situation is different in the case of the BWR compact NSSS, paving the way to the search for a "smart" condensing solution).

If a water or steam line breaks, around half the steam released is condensed in the 4,000 m^3 of cold water. The product of the volume of the containment and the design pressure is thus reduced by a factor of two compared to a pressurised water reactor which reduces the cost of the containment, yet at the cost of greater complexity of the structures.

Use is made of this internal water reserve, which is available in the event of an accident, to also condense any steam that may escape via the system safety valves in certain transients.

The reserve is also a source for safety injections, replacing the reserve of borated water which, in current PWR designs, is located outside the containment (but not in the EPR).

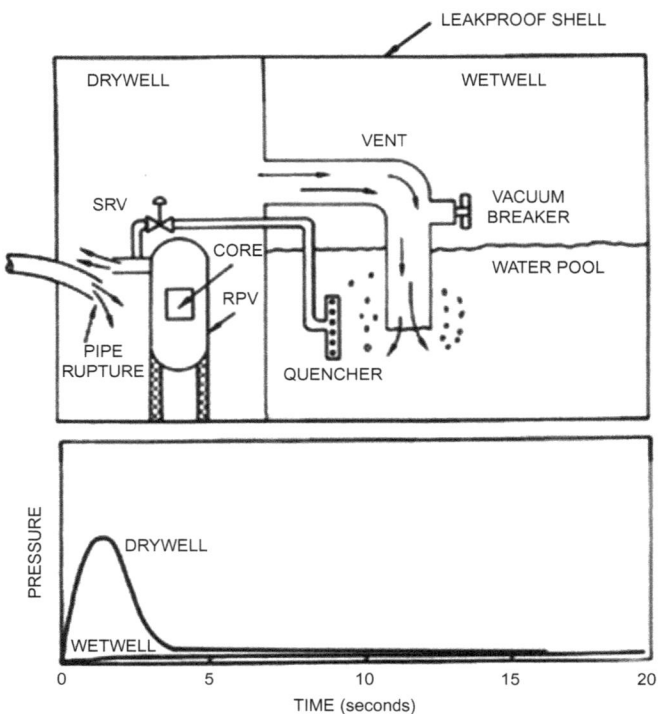

Figure 12.13. The BWR compact way with two parts: a small drywell "tightly wrapped" around the NSSS and a fast response pressure suppression pool (wetwell – WW).

On the other hand, with no cold heat sink being located over the core (in the older design), as in the case of the SGs in Westinghouse PWR design, no natural convection can be established with the pressure suppression pool alone in closed loop. In case the recirculation pumps fail to feed the core, the following loss of water must be compensated for in another way. These drawbacks don't affect the KERENA or the ESBWR – type design (see below).

Last but not least, a powerful scrubbing effect is obtained in the case of a severe accident by the steam exiting from the damaged core, highly loaded with gaseous (condensable or not) fission products and condensing in the pool. This scrubbing achieves a significant retention of condensable FP. The benefit was visible in Fukushima for iodine and cesium.

None of the containments built have yet had to handle a pipe break accident. Large-scale experiments were carried out when the BWR were under development.

However, there were instances when releases from the steam lines caused damage to the structure of the early steel containment vessels. Steel was finally abandoned in favor of concrete.

Figure 12.14 shows the main features of the ABWR containment.

For their ABWR implementation, the Japanese constructors chose a thick single-wall cylindrical containment, with high-density steel reinforcement, constructed monolithically on an enormous foundation slab carrying the entire reactor building.

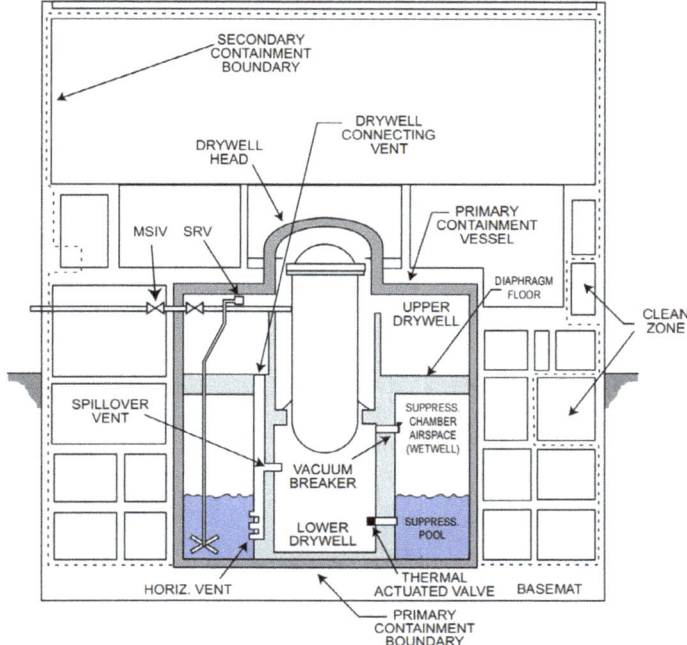

Figure 12.14. ABWR containment design (from GE).

Some key features of this implementation of the general principles are:

- Reinforced concrete containment vessel (RCCV)

- Steel liner for leak tightness

- Design pressure of around 3.1 bar; failure point around 3x design

- Designed for 0.3 g earthquake; more stringent specifications for specific implementations (in Japan, for instance)

- Reactor building serves as second containment boundary

- Right circular cylinder geometry making it easier to construct

During the Fukushima accident, the wet-well (a torus-shaped one made of steel (see Figure 12.15), in the case of these old GE Mark-1 reactors – then GE shifted to concrete building) played an important role and showed some limitations in the severe context of a complete and long blackout where the reactor was deprived of power, of fresh water make up, with some blocked valves closing paths necessary to passive system operation, and with hydrogen explosions inside the reactor building.

Nevertheless, apart from the speculative case of a (possibly) structurally damaged and thus leaking torus, the wetwell was very efficient as long as enough subcooled water was available. A powerful scrubbing effect was established from measurements in the environment outside the plant, reducing the external release of many fission products (including caesium and iodine but not the noble gases) leaking from a massive core melting.

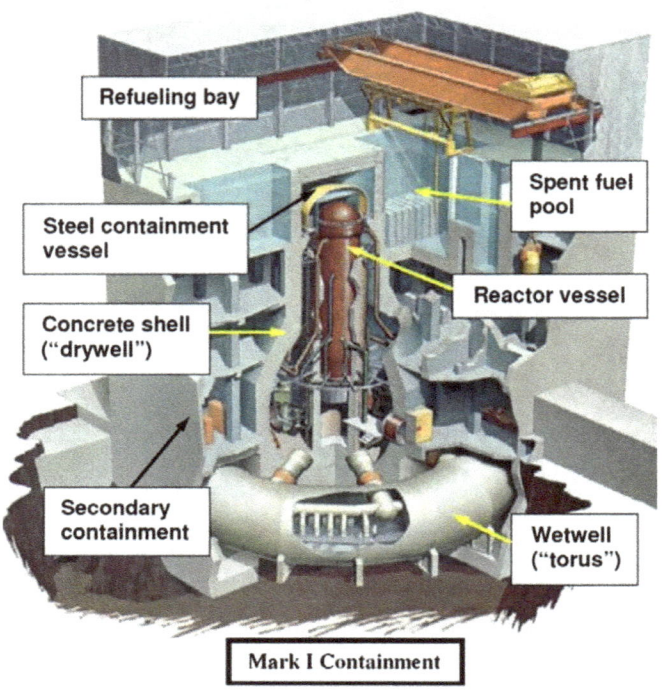

Figure 12.15. Overview of Mark-1 type BWR (Unit 1, 2, 3 and 4). ©DR.

12.6.3. *Safety injection, core meltdown and long-term containment*

In BWRs with recirculation loops (up to and including the General Electric BWR-6), the design basis accident is a break in a loop 250 to 300 mm in diameter. This results in dryout of the core, which is then cooled from above by a core spray system. In BWRs without recirculation loops, the reference accident is the breaking of a flux measurement or control rod drive mechanism penetration at the bottom of the pressure vessel. No such break can cause dryout of the core. Core spraying from above has nevertheless been retained in the ABWR.

The ABWR possesses a number of effective high-pressure flooding systems, and the overall risk of core meltdown is extremely low, with a slight residual risk with the reactor depressurised in the event of long-term cooling failure. This is a case which has been under scrutiny since Fukushima and which has to be solved one way or another in the perspective of the "hardened core" approach with augmented resilience complemented with the "Rapid Action Nuclear Force" equipped with mobile safety devices (see below).

Reaction of the zirconium with steam could, in a BWR, produce 1.7 times more hydrogen than in a PWR. Apart from the deflagration risk, there is a risk of detonation, particularly in the drywell. The explosion in Reactor 3 in Fukushima showed the difference between the 3D, isotropic, "normal" explosion that occurred in Reactor 1 for instance and which blasted the walls and the roof of the last upper stage of the building, and the axially vectored, accelerated, devastating explosion in Reactor 3, throwing multi-ton metallic parts of heavy equipment at a height of 80 metres. Total or partial inerting with nitrogen increases the level of safety, even if inerting would not be imposed a priori for modern BWRs with pre-stressed concrete containments, but discussed on a case by case basis. Obviously, hydrogen recombiners, filtered venting, double containment walls with extraction and filtering of the intermediate atmosphere and inerting, can be adequately combined to reduce the risk.

During the Fukushima accident, an extreme "Beyond Design Basis" context challenged the mean-term – passive – cooling of the cores and the injection of fresh water.

After the early loss of power from the tsunami effects, about one hour after the earthquake, passive safety systems were available for further cooling, giving a valuable opportunity (more than one day long, in principle) for recovery.

In the Unit 1, a passive Reactor Core Isolation Condenser system started but stopped operating after a few minutes.

In the other three units (but in Unit-4, the fuel was in the spent fuel pool, not in the reactor core), the RCIC (Figure 12.16), which was manually operated, uses a turbo-pump: the reactor steam drives a generator which in turn drives a pump injecting water from the

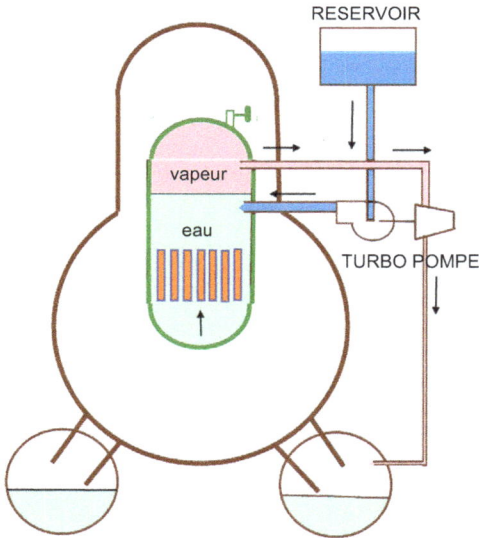

Figure 12.16. The potentially long-lasting passive cooling of the core by a turbo-pump driven water circulation.

condensate storage tank and then from the wetwell, into the vessel. Its operation lasted several days in Reactor 2 but stopped after about one day in Reactor 3, leading to core uncovering, clad oxidation then combustion, with release of hydrogen which ultimately caused the severe explosion that occurred in Reactor 3. This explosion in turn indirectly caused the loss of control of the Reactor 2 cooling by destruction of the trucks delivering fresh water.

Over two days, the Decay Heat Removal can be achieved by the passive operation of the pumps driven (through a generator) by the steam produced by the decay heat itself. The operation stops when:

- either the steam pressure is too low, but at the same time this means that the decay heat has significantly decreased;

- Or the make-up water, taken from the condensate storage tank and then from the wetwell, is no longer available.

There is however no sufficient condition but a set of necessary conditions (for a problem to be resolved) and in the case of Fukushima they were not fulfilled for several reasons, some of them still unknown in their detail, but involving batteries (soaked or empty), circuit lining and related valve status and speculative earlier damages to the core and to the circuits (breaks). Therefore the passive systems could not be used to their full potential. They stopped operation respectively: quite immediately after starting, after one day and after two days, the latter performance being quite good and at the same time, accidentally interrupted. Even the full potential achievement might have been insufficient in avoiding a long core uncovery leading to the melting of significant core portions, depending on the detailed scenario of the accident.

The further cooling of the reactor cores has been performed in a "feed and bleed" mode. An external pump, to be moved ASAP onto the site and connected to a power source, to an "unlimited" available water source and ultimately to the vessel at a proper location injected cold water (sea water in the case of Fukushima over a period of days, with a growing risk of flow blockage inside the vessel and the core due to salt crystallization). The water is turned into vapour.

The water flow consumption is simply governed by $Q = P_{decay}/(L + c_p * (T_{sat} \# 100\ °C - T_{cold}))$, the denominator being clearly dominated by L, the latent heat of vaporisation.

The core pressure is controlled by periodic opening of the safety valve and the pressure is limited in the containment by periodic venting into the upper part of the reactor building. This venting brings incondensable gases escaped from the wet torus into the upper part of the BR (made of "light" structures of concrete and steel, and containing the spent fuel pool). These gases include mainly hydrogen (and therefore have a high explosive potential), noble gases (Xe for instance), and a small fraction of iodine, caesium, etc. released from the damaged core.

The further objective is obviously to come back, as soon as possible, to cooling in a closed loop. By the way, the very physical state inside the vessel will be observed later, like in the TMI case).

Meanwhile, TEPCO is providing updates on unresolved technical matters ([5]). In a report, issued August 6, 2014, TEPCO said there had been no explanation for the premature failure of the unit 3 RCIC (Reactor Core Isolation Cooling), which pumps water into the core using a turbine driven by steam (Figure 12.16). The Unit 2 RCIC operated for several days following the earthquake and the tsunami. "However, the unit 3 RCIC failed after 20 hours, even though battery power needed for control of the system remained". The cause of the system's turbine trip may have been an RCIC design feature aimed at protecting the RCIC system from damage in case of high pressure in the steam line to the turbine. "A turbine steam stop valve is equipped with an electrical mechanism to protect the RCIC". The new information will help operators and regulators to "take a look at these operational characteristics and see if they should modify any procedures", for instance by overriding the high turbine exhaust pressure trip setting during an accident.

This example shows how a brand new explanation could foster a tight interplay between design and control "hardware" on one hand, and with operation procedures on the other hand, for instance for Kashiwasaki-Kariwa ABWRs, as a direct and immediate application. It shows also that it is not sufficient to have a passive Decay Heat Removal system implemented on the plant, even a smart and simple one, and that "the devil is in the detail" of implementation and/or through a systemic overflow generated by unexpected events.

New information has also been given regarding the unit 3 melted fuel believed to have left the RPV and get collected in a sump pit. The fuel debris are supposed to have eroded about 0.7 meter of concrete, stopping about one meter above the steel liner plate of the containment structure.

Nevertheless, some analysts think that the new information "does not strengthen the case for core catchers". On the other hand, many people think that short to medium term design should consider EVR (Ex-vessel retention, thus generally core-catchers) and that more R&D and qualification is needed before potentially shifting to IVR.

In France, a recent interview ([5]) of the chairman of French regulator ASN helps to delineate the principles of what is known as "hardened core". "The hardened core is to protect against extreme events that are unforeseen", he said..

There are three key elements of the hardened core: access to water for cooling, pumps to move the water where it is needed, and electricity sources to operate the pumps. "Those three need to be really protected and available even under very extreme conditions".

A feedback on the design seems to confirm that it is more difficult to implement a "hardened core" on Gen-2 reactors than on new design reactors and that a significant number of the required actions, as well for the plant hardware as for the safety "software" (procedures and crisis management) have been made at the early Gen-3 design stage, with learnings taken from the TMI and Chernobyl accidents. Adding a "Rapid Action Nuclear Force", known as FARN in French, securing and hardening the quick plug-and-play capability of the related mobile safety devices, completes the system.

Coming back to BWR, the Gen-3 design looks as adequate as the design of the same generation PWR. Despite the Fukushima event, BWR and PWR are still leading the race. Moreover, their short term adaptation and improvements, even if apparently costly at first sight, will generate more operation feedback in the advanced configuration, widening the pre-existing gap with potential competitors, for instance newcomers of the Gen-4 selection. As for the next step, do the BWR have some higher potential for useful fuel cycle adaptation (efficient plutonium multi-recycling and the induced increased thriftiness)? The point will be addressed in the last part of the next section.

12.7. Trends

The recent trends are summarised in the chapter devoted to the advanced LWR, and also partly in the present chapter devoted to assessing the potential of advanced, adapting LWR, to bridge the gap and assure a smooth fuel cycle transition to the Gen IV systems (see next §).

The main topics are safety and fuel cycle improvement:

12.7.1. Safety, in the aftermath of Fukushima

- The combined improvements in safety (leading to very low core fusion frequencies in ESBWR and KERENA PSAs) and competitiveness (by reducing the steel and concrete mass balance and further reducing the number of components and systems, hence the investment cost and the construction time) make the Gen-III BWR very attractive. The architecture and design of ESBWR (Figure 12.17) and of KERENA (Figure 12.18) build upon (instead of dealing with a posteriori) the outstanding thermodynamic (latent heat of vaporisation) and thermal-hydraulic (natural convection, "heat pipe" capabilities) properties of two phase water flow. An annular space with water available at every axial level surrounds and overhangs the central nuclear core of the plant (NSSS). It is enclosed in a tight drywell protected by the reactor building. The whole system is topped by what has been dubbed as the "containment lake" (after C. Fribourg). The shapes are as simple and as smooth as possible. The ultimate simplification (but possibly a step too far) seems to be the removal of the recirculation pumps. Last but not least, one has to be careful with the use of IVR (In Vessel Retention) at very high unit power.

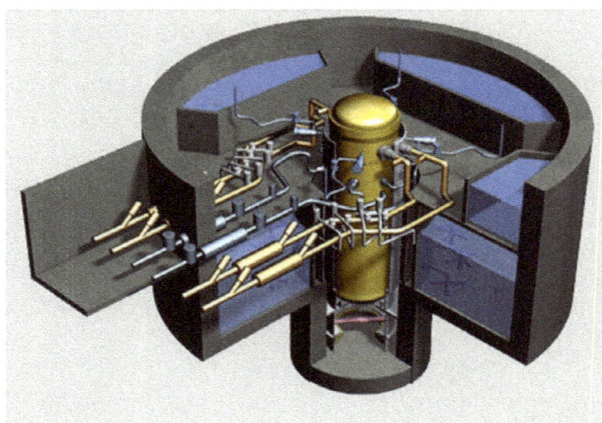

Figure 12.17. Overview of the safety related topological features of the ESBWR.

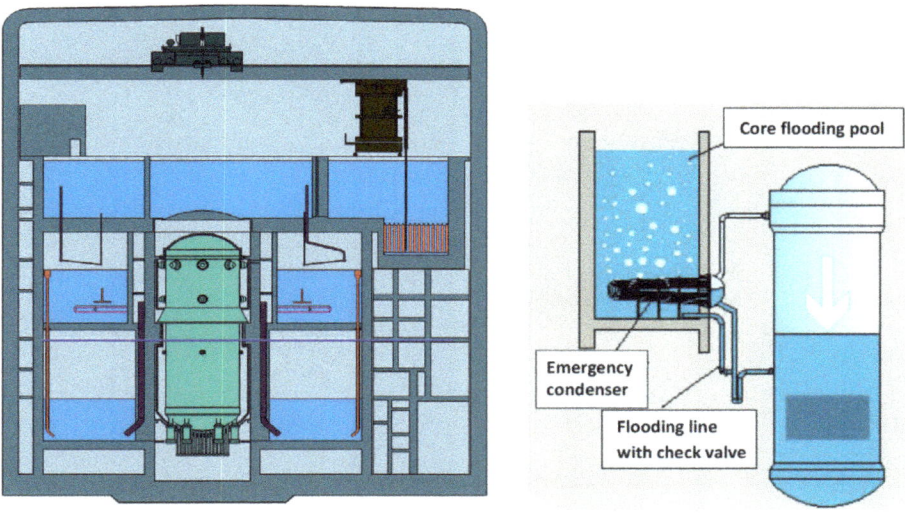

Figure 12.18. Design principles illustrated; passive and active safety systems integration in the containment; the example of the emergency condenser system (from AREVA).

- In any case, the extreme choices illustrating a principle are not necessarily the ultimate ones. For instance, keeping some pumps and combining passive and active safety features and devices could finally lead to the optimum, flexible, safe and robust BWR design (KERENA and potentially some GE-Hitachi New Generation BWR design).

One of the lessons learned from Fukushima and completing the previous ones from operation feedback and from TMI and Tchernobyl is to go further in how the defense in depth principle is applied, as much in terms of design as for operation. This includes the crisis management "software", vision and leadership preparation and implementation, the availability of well trained fast intervention teams equipped with dedicated, "pooled" hardware adapted to achieving the main cooling and injection functions in remote and troubled areas.

Within this set of tools, the design is still the frontline and must be strengthened, taking into account natural as well as human intentional threats which can combine several causes (the last example being a total and long blackout, an earthquake, the flooding of critical devices (including batteries) and the loss of fresh makeup water). These can last for a long time and affect several plants simultaneously.

The KERENA design (Figure 12.18) includes very attractive features. It shows also how passive and active safety system integration in the KERENA containment can be achieved, as well as a focus on the emergency condenser system. Figure 12.19 shows the integral facility (INKA in Karlstein) dedicated to the experimental qualification of KERENA passive safety systems. Figure 12.20 shows the ESBWR Passive Containment Cooling System (condensation) qualification facility.

Technical data relative to KERENA:
Scaling factors:
- 1:24 in volume
- 1:1 in height
- 1:1 in component sizes

Safety components:
- Emergency condenser
- Containment cooling condenser
- Passive core flooding valve
- Passive pressure pulse transmitter
- Vent pipe, DN700
- Fuel pool cooler

Vessels:
- Flooding pool vessel, 210 m³
- Drywell vessel, 190 m³
- Flow rate: 200 kg/s (sat. steam at 85 bar)
- Pressure suppression pool vessel, 350 m³
- RPV simulator, 125 m³

Figure 12.19. Karlstein Integral Test Stand (INKA) – AREVA GmbH – Qualification of components of safety passive systems.

Figure 12.20. ESBWR Passive Containment Cooling System (condensation) qualification facility – GE – HITACHI.

12.7.2. Fuel cycle improvements

The potential for further improvements in fuel cycle is largely related to the presence of channels and orifice plates enabling the coexistence of very different fuel bundles into the same core, as well as to the two phase flow regime. Combining these specific features makes it possible to design tailored 3D heterogeneous (as for the spectrum) cores. Perhaps advanced BWRs could become convertible reactors enabling efficient multi-recycling of plutonium, at due time over the course of the next decades, paving the way to a robust transition to Gen IV fuel cycles and reactors.

For some years, Hitachi has been working on the development of so-called "Resource-renewable Boiling Water Reactors" or RBWRs. They could potentially belong to a New Generation of BWRs (dubbed NGBWRs in the next section), after the ABWR and the ESBWR. These NGNBWs could differ from their predecessors in several aspects: unit power (from 1350 GW for the ABWR to 1550 for the ESBWR and up to a potentially higher unit power for the NGBWR); recirculation method (forced circulation for the ABWR, natural circulation for the ESBWR, back to a forced circulation for the NGBWR); safety systems (active, passive, then shifting to a hybrid safety system).

On the other hand, the very purpose of the RBWR is to burn efficiently the TRU (plutonium and potentially some minor actinides). The related advantages would be to reduce the long term radiotoxicity level of the residual waste to the one of the long term FP, as well as to reduce the net consumption of natural uranium through multi-recycling with an increased conversion ratio. The use of TRUs from used fuel as fuel along with uranium, would be a win-win option, assuming as the same non-core components found in current BWRs, including safety systems and turbines, could be used. This would be the case even if new core fuel and structure concepts (such as Y shaped control crosses, triangular tight lattice, "short" and axially heterogeneous, assemblies, see Figure 12.21) are required. A limited unit power NSSS could fit within the ABWR pressure vessel. The RBWRs (and similar reactors) are unique in that "experience accumulated through the application of BWRs can be leveraged to achieve efficient nuclear fission in TRU" ([4]). Simply changing the core portion of the plant would be an attractive option but it would be necessary to check that the flexible operation and the design rating could be kept while fulfilling the safety criteria. Moreover, the whole fuel cycle would be altered and a much higher mass of fissile nuclei be loaded in the core than in a thermal spectrum reactor (getting closer to the inventory of a fast reactor core) – this in-core inventory having to be complemented with about the same mass for out of pile operations before loading the recycled TRUs after cooling down, (transport), processing, fabrication.

It is an advantage to work at moderate power density and moderate linear power, and to have a moderate decay power density in used fuel, simplifying the unloading, which is easier under water than under sodium, lead or gas, and the cooling down (adequate location for cooling down, transport specifications, residence time before moving to used fuel processing). On the other hand, similar performances can be achieved in PWRs (Russian studies regarding a high conversion ratio PWR, for instance, as well as French studies on BWR and PWR), but BWR seem to have a structural advantage and offer simpler routes to adaptation. A heavy constraint will be the need to maintain a significantly negative void reactivity feedback coefficient. Specifically in BWRs, if their flexible operation mode is to be kept, this coefficient must be made use of along with safety considerations and flexible operation (see above). On the other hand, from a thermal-hydraulics viewpoint (critical

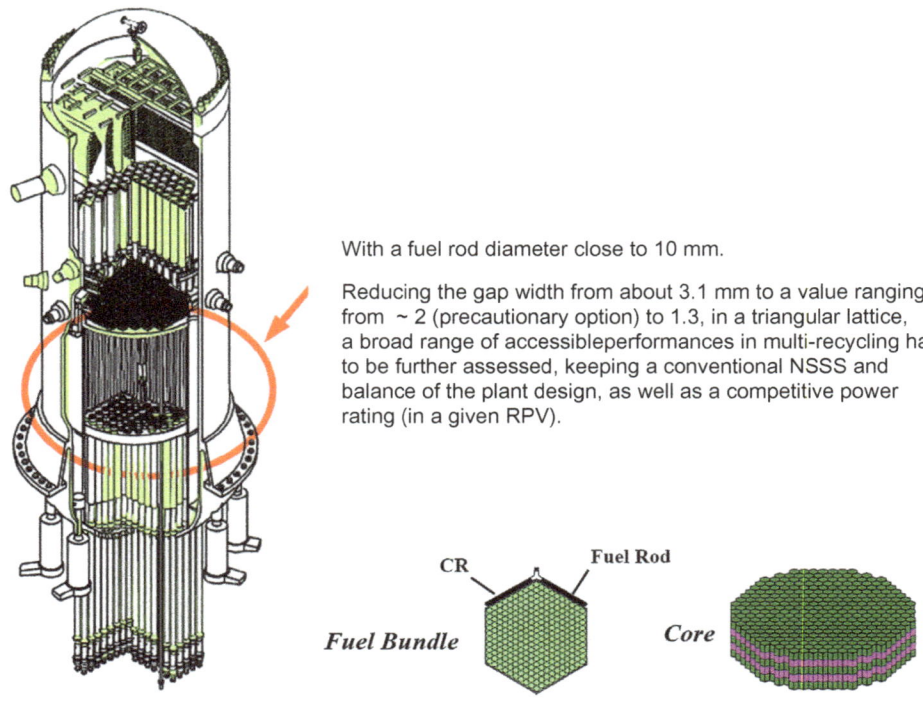

Figure 12.21. Overview of Innovative RBWR ("Resource-Renewable BWR") by GE-HITACHI. The message is: a High Conversion Ratio Reactor can be realised with BWR technology. "Only change the Core Portion".

heat flux), a conservative approach would be to start with a fuel gap width close to or no less than 1.8 to 2 mm. The reduction of the conversion ratio when comparing with more daring options would not be de-motivating from a fuel cycle viewpoint and a second step of innovation/improvement may be a mean term R&D topic.

References

[1] Les Techniques de l'Ingénieur/Energies/Génie nucléaire/Typologie des réacteurs nucléaires/ « Réacteurs à eau bouillante ordinaire » – BN3130 et « KERENA: réacteur à eau bouillante avancé de puissance moyenne à sûreté passive – B3135.
[2] The Thermal-hydraulics of a Boiling Water Nuclear Reactor (second edition) by R.T. Lahey and F.J. Moody, American Nuclear Society, 1993.
[3] Bertrand Barré: private communication.
[4] WNN – Nuclear Fuel – vol. 39/Number 19//September 15, 2014: Hitachi-university research program to develop TRU-burning BWR.
[5] WNN – Nucleonics Week – Vol. 55/Number 34/August 21, 2014: Tepco report on Fukushima explains loss of cooling during accident.

13 The place and the potential of Light Water Reactors in the transition from Gen-III to Gen-IV

J.B. Thomas

13.1. Introduction

The current chapter is not of the same nature as the previous ones, apart from the Introduction. It forms a link between the overview of existing systems with their history and SWOTs on the one hand, and the presentation of fusion and of "futuristic systems" on the other hand. As such, it has two objectives:

- delineating an "ARIANE" thread, or the navigation of a path through the maze leading from the known phylum of reactor systems and cycle options to a credible, efficient and competitive, sustainable nuclear future. All such paths cannot be explored or simulated. The path followed in this chapter is not a new dogma. It shows, among others, a (hopefully) possible road to success. Its **quantified** exploration aims at making the constraints controlling all possible ways more visible. It introduces a method to identify and use the necessary data and information needed to set up a new investigation on any type of strategic option;

- this can only be achieved through the second objective which is to calculate the main parameters controlling these dynamics – dynamics which are still unknown as is the case in every prospective strategy exercise. These calculations have a second incentive: putting to work the knowledge and the tools gathered during the conferences in the various physics and "system plus cycle" disciplines.

Summarizing the thread

In order to follow the thread, we have to start with the active rewriting and with the (updated) "weighing" of the Generation IV key statements. It works in a similar vein to the process of "active reenactment" used by the historians, because, if we miss the red line, the unifying thread of the intrigues, we can be trapped in the bottomless pit of the infinitesimal level (after Paul Veyne, a French historian).

Then we have to question the assumptions related to the specific limits of LWR on the way to Generation 4. These limits are primarily intrinsic but their expression during the current century (like the expression of the genes in biology) is controlled by exogenous factors: actual scarcity of the uranium resources and speed of development of nuclear power throughout the world.

Then, if we focus our exploration on the natural uranium–plutonium closed cycle, we have to assess the design drawbacks related to the fast spectrum.

Next, one has to make it possible to set up a synergy between advanced LWR and the emerging fast breeders. The fast breeder fleet growth rate will be limited by their (long) intrinsic breeding doubling time and by the scarce inventory of plutonium that will be available around the year 2040 worldwide; not to mention the long qualification period required to build up an operation feedback giving to the utilities the assurance of a reasonable availability factor. 2040 has been defined as the starting date of Gen-4 already shifted by a decade since the year 2000. So we need to question the capability of the "Gen-III$^+$" LWR to adapt to "smart recycling" (multi-recycling without a quick degradation of the plutonium quality in terms of isotopic composition). Satisfying this "mere" condition leads to a significant increase in conversion ratio making the multi-recycling profitable and efficient enough in terms of thriftiness – as for natural uranium consumption – to shift by a century or more the imperative need of a majority of breeders or quasi-breeders (with an optimal mix of uranium – plutonium and thorium fuel) in the nuclear fleet, worldwide. Meanwhile, the majority fleet of power-burner LWR will generate plenty of plutonium (and possibly ^{233}U), and act as the driver of the rapid growth of the breeder fleet. Until then, an optimized Gen-4 fleet could integrate stepwise breeders that have to be qualified ASAP.

Other paths can be imagined and studied. The current scheme gives the opportunity to develop the capability of quantitatively assessing the advantages and drawbacks of such strategic options.

Some basic assessments/discussions of assumptions (as well as more or less well founded assertions), are intertwined with calculations which are actually performed during tutorial classes between the lectures.

As a consequence for this specific chapter, the plan (following the thread) and the content (intertwining simple calculations with information gathering and with argument presentation and assessment) are unusual, and probably burdensome.

The structure and the presentation could be optimized in order to reduce the burden for the non-student reader, but the price to be paid would be to separate the quantitative/active arguments from the global intrigue. Two exceptions have been made: for the calculation of a typical once-through cycle in a PWR as well as the calculation of some variations related to this "standard" option on the one hand; for the definition and the calculation of several concepts of doubling time for breeders, on the other hand. These paragraphs form two Annexes at the end of the chapter.

Nevertheless, I must apologize for the effort that will be required of the reader.

13.2. The stable and plentiful ground of physics and a changing world

The last decade has seen many smooth evolutions and a few disruptive events.

The so called "nuclear renaissance" is still a relevant concept, but there has been a drift from the pure Gen-IV perspective to a somewhat more balanced vision of the future.

Over the last five years, four major orientations or events have given a new colour to this vision:

- the financial, then economic systemic crisis, influencing the financing and the criteria for choosing new investment targets, with opposing consequences depending on the regional and national context, just like in the wake of the major oil crises;

- the re-assessment of the capital cost of LWR "newcomers". Once more, the level of the gap between optimistic projections and real world construction depends on several parameters, affecting constructions in Asia and projects in the USA and Europe to different extent;

- the US nuclear strategy, both at home and abroad;

- the shale gas emergence;

- the Fukushima major accident.

Nevertheless, with a requirement for multi-criteria adaptation that is more stringent than ever, the LWR reactor types (PWR and BWR) still seem to hold the key to the future, and will do for many decades to come, possibly up to the end of the century and beyond (in operation, if not in further construction).

Their main supposed weakness is related to core and fuel cycle physics, namely:

- the conflicting relationship between the role of ordinary water as a moderator and as a coolant seems to preclude the access to a "hard" if not genuinely fast spectrum enabling breeding with the fertile/fissile pair associating U-8 and plutonium, in acceptable conditions of safety, competitiveness, as well as of comfortable and robust operation;

- for the same reason, it is suspected that useful transmutation of selected minor actinides (still under the uranium/plutonium fuel cycle assumption) would challenge their tight neutron balance.

Considering more thoroughly these limits and challenges on the one hand, and the potential of LWR to address these issues without losing their superiority amid the set of nuclear reactor systems, as well as their ability to compete with "King Coal" for base-load power production) on the other hand, the next sections will describe the main relevant parameters and their interplay in the game of problem solving. Two conclusions emerge:

- a right mix of systemic adaptation and of (limited) technological and design innovation will make it possible to achieve the useful improvements;

- a synergy emerges from the analysis of the LWR evolution; a synergy between their stepwise evolution (market driven) and the development of a worldwide fleet of breeders.

Finally, even if for some reason things do not work out that way, another way is left open, in which LWR have a substantial advantage, with a thorium-U-3 perspective, being less sensitive to the (fast) spectrum issue.

13.3. The Gen-IV vs Gen-III specification gap: the specifications for sustainable nuclear power

13.3.1. Introduction

The specifications for sustainable nuclear power underlie the definition of Gen-IV. However, the nuclear facilities of Gen-III reactors and fuel cycle facilities are required to comply with them insofar as technology and competitiveness constraints allow, making good use of the advantages of the LWR that are currently producing nearly 90% of nuclear power.

The Gen-IV systems must come ever closer to the ideal defined by the sustainable development criteria, without compromising the competitiveness of nuclear power. The capital cost of the reactors will thus be constrained, as will some options related to the back end of the cycle. Their ability to reach industrial maturity and competitiveness in due course will also depend on the cost of research and development (R&D), on the time taken to complete it and on the international funding of the associated initiative. An option is the creation of a functional "niche" (plutonium burners) to give time for fast breeders to gain operation feedback over several decades before potentially entering the competition as advanced power reactors, combining burning and breeding functions with competitiveness.

The perspective is therefore of gradual transition. The emergence of new issues (and safety is still rising new and challenging questions) results, in the transition from one generation to the next, in a set of choices valid for a limited period of time. Its usual duration in the field of nuclear power is between a quarter of and half a century. The criteria of the sustainable development apply in this framework: the main issues will be weighted as a function of the context and of possible progress over a given period of time with a given (and limited) allocation of resources.

In the past, the efforts made to increase competitiveness and nuclear safety have already contributed to a process of evolution and to the selection of a single reactor type. Light water reactors have succeeded in adapting and have become superior to the other reactor types. This is a result of their intrinsic properties (including the two-phase water-steam physics used from the core to the turbine and beyond in conjunction with an unlimited water heat sink when available such as a river or the sea which can be directly tapped in case of emergency; see the feed and bleed operation with sea water during the Fukushima accident).

The question is how much longer will the LWR type be able to continue to adapt to criteria of increasing stringency, primarily associated with the fuel cycle issues (resources and waste) and hence, to the reactor physics, and how much pressure the LWR will exert on the adaptation of their competitors in the framework of intra-nuclear and inter-energy sources competition, particularly as concerns "moderately clean" coal technology for baseload power generation.

13.3.2. The basic specifications: formulation and discussion

The basic specifications for sustainable nuclear power, as they are currently perceived, are the following.

Competitiveness in base-load nuclear power generation

The same safety objectives as for Gen-III LWR.

Particularly for the reactors:

- A goal of "zero impact" offsite (excluding the residual risk);

- A simplified safety demonstration. The way this simplification is achieved can differ and lead to different economic impact when comparing two types of advanced designs: BWR and HTR, both with passive safety features.

- Defining the best mix of active and passive safety systems, as well as improving and hardening each of them, is on the agenda of advanced BWR as well as PWR. These improvements have to be integrated into the next move, particularly in the aftermath of Fukushima. This affects the safety "software" (including leadership and coordination, procedures) for severe accidents and for their management, with the help of internal and external dedicated resources (pooled skills and tools). This review could also lead to a new ranking table of the coolants, even if the safety issue can be managed (more or less easily) for different types of reactors (using primarily different coolants).

Greater efficiency in the management of natural resources (uranium and thorium).

At the present time, with an open cycle (once-through) and moderate levels of burnup and enrichment, around **0.5% of the mass of extracted uranium leads to a fission** (partly through plutonium).

Hereafter, this ratio (ultimately fissionned mass over natural heavy nuclei – equivalently heavy metal – mass) is referred to as the nuclear efficiency of a system (reactor and fuel cycle). With breeder reactors, almost the entire mass of heavy nuclei could give rise to fissions (with the use of "catalysts" constituted by plutonium for ^{238}U and by ^{233}U for thorium). With an efficiency higher than 50% obtained, for instance, with a fast neutron spectrum and a liquid metal or a gas as coolant, the resources available for feeding a worldwide nuclear fleet of several TWel would be practically inexhaustible, at a cost remaining a low if not insignificant part of the total cost of the kWh.

The resources available at costs in the range of 100 to 500 $/kg Unat could be as high as 20 Mt of natural uranium (NatUr) (see below §12.10.2). This assures the competitiveness of LWR for a transition period of several decades from Gen III to Gen IV systems. This stepwise transition gives time for the development and industrialisation of breeders as "power reactors", building up operation feedback and achieving competitiveness.

At the same time, multi-recycling in high conversion ratio LWR (HCR-LWR) would help build up (from the LWR operation) and save at preserved isotopic quality the first part of the huge plutonium inventory (well beyond 10 000 tons) necessary for the start-up of the "ultimate" world fleet of fast reactors.

This would avoid launching a fast breeder fleet with a massive make up of enriched uranium at a high cost. Building up the second part of the plutonium inventory would be achieved by efficient breeding with a short but reasonable doubling time (hence a high breeding ratio). A gradual transition would thus reduce the global risk of failure of a bullish development of nuclear power, worldwide, after 2025.

A second advantage of a gradual transition would be that, due to increasing nuclear efficiency, the quantity of uranium annually removed would not grow proportionally to the installed nuclear power. This could avoid uranium price spikes, as happened during the 1970's and more recently. Large fluctuations in price could occur periodically well before the natural resources become "economically exhausted" (at a competitive level of cost, for it is primarily a matter of price, as the future of seawater uranium 4 Gt will illustrate).

On the other hand, with breeders, the potential gains in nuclear efficiency (by a factor of 100), are comparable with the potential gains in thermodynamic efficiency since the introduction of thermal machines. History shows that it has been possible to progress from a few percent to more than 50%. However, improvement in technology and system design has extended over a number of generations of machines. In the nuclear domain, generations can last for half a century. It is thus worth considering the next step corresponding to a gradual increase in nuclear efficiency from 0.5% to 2 or 3% at the Gen-III stage. This first step makes it possible to multiply by four to six the nuclear power level without increasing the natural uranium consumption and this is highly motivating. The further gains are important when considering ratios but less in terms of absolute savings.

Can Light Water Reactors meet such specifications without compromising overall competitiveness, particularly without greatly affecting capital costs? Consideration needs to be given to the threshold effect which occurs when the "natural" limit of a technology is approached, and the additional cost associated with a slight gain in performance rapidly increases. The dual role of water as coolant and moderator could result in such a limit with regard to the spectrum in the LWR. On the other hand, the expertise of the designers, helped by the pressure of competition and the emergence of new solutions (particularly in the field of materials), is a powerful resource for a stepwise evolution.

Keeping the potential radiotoxicity and the decay heat (hence the size and cost of the disposal) of the ultimate waste as low as reasonably achievable

It is difficult to determine the best response to this problem without polarising public and expert opinion. However, it seems that the following principles can be laid down.

- A small quantity of long-lived, non-reusable final waste will remain and will need to be isolated from the biosphere for periods (millennia) exceeding historical control timescales.

- Disposal of the waste in a geological formation after effective packaging represents a suitable solution for which there is a broad consensus among the international community of experts.

- As plutonium constitutes the catalyst for virtually complete fission of natural uranium, the goal is optimised multi-recycling of plutonium. This goal is in total harmony with the reduction of the radiotoxicity of the spent fuel which mainly comes from plutonium.

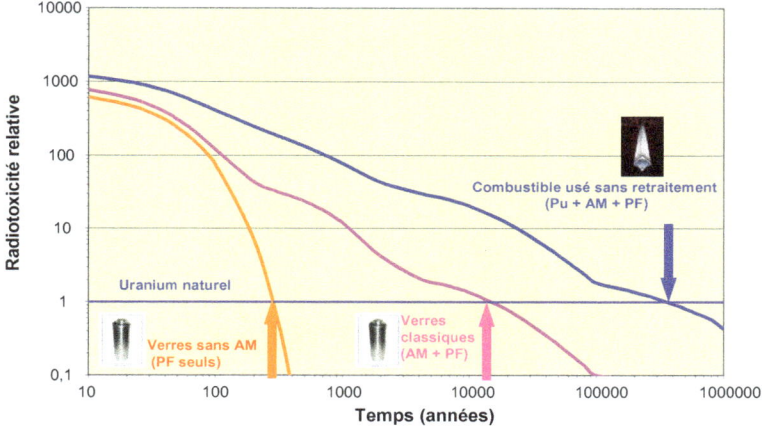

Figure 13.1. Relative radiotoxicity of waste as a function of time after spent fuel removal, depending on the fuel cycle back end option.

- Further improvement in reduction of the potential radiotoxicity can be achieved by the separation and partial transmutation of some minor actinides, or by global fission/transmutation (without separation) of the whole set of actinides (Figure 13.1). In view of the isotopic composition and radioactive half-lives, ^{241}Am seems to be the top priority for transmutation, after plutonium. However, if waste packages are able to withstand degradation to the expected degree during the first few millennia, such transmutation may prove to be pointless, given the long time-to-biosphere from an adequate geological disposal when compared with the half-life of ^{241}Am of around 430 years.

- Optimisation of the disposal – interim storage – recycling/transmutation process may help to reduce still further the scale of disposal, which is of strategic and economic importance. After one century, the ^{241}Am gradually becomes preponderant in terms of residual heat (Figure 13.2). It is therefore tempting:

    o To process and recycle the spent fuel as early as possible (high frequency);
    o Possibly to transmute the ^{241}Am. However, **in-core** recycling of americium incurs a penalty both in terms of fuel fabrication and of neutronics, not only in the thermal but also in the fast spectrum. Moreover, Am transmutation generates as a by-product "fresh" Cm which has to be transmuted in turn, generating still higher elements.

In short, "industrially" optimised recycling of plutonium is a top priority for sustainable nuclear energy

The specifications relating to the thriftiness (saving natural resources and avoiding large fluctuations in the price of natural uranium leading to potential instabilities in the development of nuclear energy), as well as relating to the reduction of potential radiotoxicity

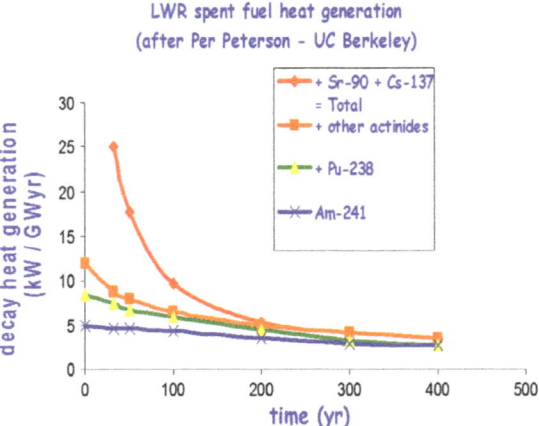

Figure 13.2. LWR spent fuel heat generation and the importance of the ^{241}Am decay heat in the global decay heat released by spent fuel in the time range of 100 to 400 years after removal (after Per Peterson – Berkeley).

and of the volume and cost required for disposal, must be taken into consideration in the framework of a process of overall optimisation. The extent to which the corresponding requirements are integrated in the forthcoming generations of nuclear systems (Gen-III and Gen-IV) remains to be precisely determined, in order to avoid compromising the overall competitiveness of nuclear energy for reasons not clearly substantiated.

The same holds true for economically "lethal" design/safety options taken without considering the whole set of criteria. For instance, the choice of totally passive safety systems for HTRs led to a very small unit power which is not compensated for by modular series effect and this (combined with a low power density) appears to make HTR too costly for power production. The ancient HTRs were not equipped with totally passive safety systems. The solution could well be a mixed passive/active (at least temporarily in case of a severe accident) safety hardware demonstrating the same level of safety; the necessary pumping power for decay heat removal is limited and can even be tapped from the decay heat itself. A – more difficult – transposition to gas cooled fast breeder is under study, helped by innovative device design.

Increasing the resistance to proliferation

Here again, it is convenient to focus on a "zero motivation" goal in using civil nuclear energy as a stepping stone to nuclear weapons. This issue is not returned to below, but is currently the subject of careful thought in terms of Gen-III and Gen-IV systems design.

The same applies to allowance for "new threats" in the design of new nuclear systems and providing them with adequate security, physical protection.

Several other issues influence the choice of future technology. For instance:

- In addition to electricity **Co-production**, of industrial heat or of storable products: petrochemicals, coal to liquid (CTL), synthetic fuels and hydrogen, as well as fresh water.

- The improved **thermo-dynamic efficiency** associated with high temperatures, provided that this is not achieved at the cost of competitiveness.

- The manner in which key innovations are shared with competing technologies and other sectors of industry (e.g. gas turbines, supercritical water, high temperature materials, etc.). The name of the game is **"duality"**.

As for the fuel cycle, an alternative vision has been developed) by MIT, for instance, and could be roughly summarised as follows: the value of destroying toxic radionuclides in SNF is uncertain:

- Destruction of long term radiotoxicity is a very attractive social concept;

- Reduction in radiotoxicity does not necessarily translate into decreased repository risks and could increase total risks from a more complete fuel cycle.

In summary, what are the main objectives and dilemmas related to the back end options?

- Processing is needed as well for waste management as for recycling;

- Recycling is needed for sustainability while taking into account the competitiveness issue;

- Plutonium is the only catalyst for "complete" ^{238}U fissionning

    ○ Thus in the long run, value is on the plutonium side (excepting at the moment the thorium/^{233}U issue);

    ○ "Industrially optimised" plutonium recycling is thus the mean term priority.

Before delving further into back end options, a dilemma has to be dealt with: "Light" waste vs. optimised (back end and front end) fuel cycle facilities and immediate operation related doses, as well as optimised reactor physics and neutron balance management.

Further study of the parameters driving decay heat release, radiotoxicity and neutron balance (and thus breeding capability and performance) is needed in order to "prune" the problem solving "research tree" of this complex issue. Figure 13.2 shows the major influence of ^{241}Am decay heat on the disposal design (and cost) and gives some clues on how to set up a fair and better informed balance between advantages and drawbacks of the main options.

13.4. The physical basis of sustainable nuclear power: high nuclear efficiency and the conditions required to achieve it

Figure 13.3 illustrates the issue of high nuclear efficiency in terms of variation as a function of incident neutron energy (E) of the key parameter characterising fissile isotopes (i): the number of neutrons emitted per neutron absorbed: $\eta(E, i) = \nu \sigma_{fi}/\sigma_{ai}$.

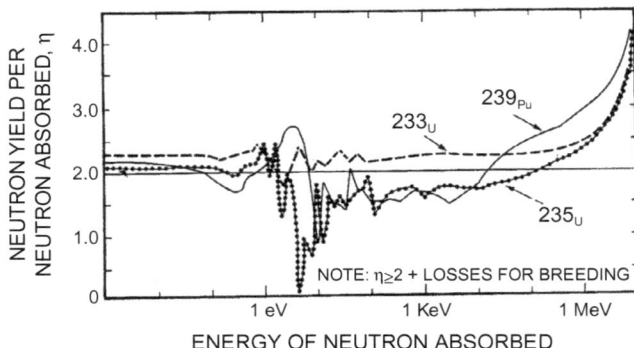

Figure 13.3. Number of neutrons produced per absorption in some heavy nuclei (fissile or fertile) vs. energy: $\eta\,(E, i) = \nu\,\sigma_{fi}/\sigma_{ai}$ (from [3]).

Once $\eta\,(E, i)$ is known, the potential for breeding and for actinide transmutation by fission can be deduced from the spectrum prevailing in the core and from the isotope composition of the fuel.

Globally, the product to be taken into account is close to $\eta\varepsilon$, ε being the fast fission factor related to the "fertile" support nuclide (^{238}U or ^{232}Th). ε depends on the spectrum (fast vs. thermal).

Below 100 keV, with the important exception of ^{233}U, η dips below 2, precluding any possibility of iso-generation of fissile materials in a critical fission system.

Iso-generation means that the conversion ratio c = 1, with:

c = (production of fissile atoms by fertile capture)/(destruction of fissile atoms by absorption),
and with: absorption = fission + capture.

Iso-generation corresponds to the equilibrium; breeding corresponds to a conversion ratio c > 1 and to a net breeding gain: BG = c − 1, enabling an autonomous growth of the breeder reactor fleet thanks to the build-up of an available inventory of fissile material. A breeding doubling time is associated with this autonomous growth rate capability from fertile conversion, requiring a tight neutron economy and a generous production/absorption ratio, thus a high η value.

The ^{239}Pu is the most effective nucleus in the very fast spectrum. It becomes almost the worst between a few tens of keV and a few eV (and the spectrum cannot be customised at will).

In physical terms, nothing is lost: all that is required is to recycle some of the fission energy to produce the missing neutrons by a nuclear reaction caused by impact of accelerated charged particles with diverse nuclei (some heavy for spallation, some light for other types of reactions), or by a fusion reaction, which brings us back to the issue of accelerator-driven systems (ADS) and of fusion-fission hybrids. With 14 MeV neutrons, even the ^{238}U and ^{232}Th η are about 4.5.

If the actual fission spectrum is plotted on the same graph (see Figure 13.4), it is found that the margin is narrow. To guarantee iso-generation, not to mention breeding with a doubling time reduced to a few decades, it is necessary to remain as close as possible to

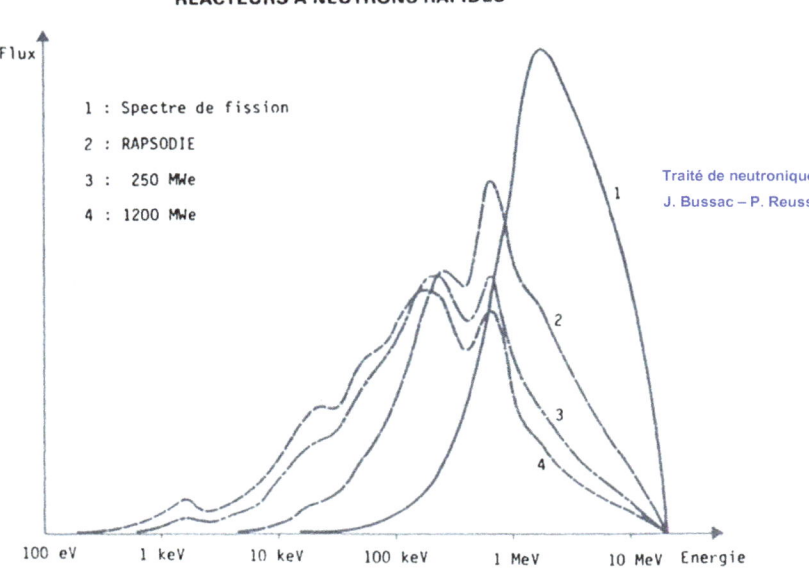

Figure 13.4.

the pure fission spectrum, minimising (elastic and inelastic) slowing down, as well as sterile capture and **leakage**. This latter point means that if outer blankets are prohibited (so as to help proliferation resistance), large cores are advantageous, as are heterogeneous cores with "internal" blankets.

- A high **specific** power P_{sp} (power by unit mass of heavy nuclei) is imperative for the fissile material management in the fast reactor cycle, because fissile material is a rare and costly resource, at the time where one has to launch a large fleet of breeders.

- The fuel concentration N_{fuel} must be high because the dilution would increase elastic and inelastic slowing down by a higher fraction of coolant and of structural materials (excepting gas and lead or lead-bismuth coolants).

- The power density $P_{vol} \propto P_{sp} N_{fuel}$ is thus high.

Figure 13.4 shows that, in reality, it is very difficult to avoid a significant flux level below 100 keV. For instance, Curve 4 shows the spectrum of SuperPhénix (1200 MWe) with a homogeneous core: the conversion ratio is close to 0.85, to be compared with about 0.6 in a conventional LWR fuel lattice. In SPX, breeding is obtained by adding outer blankets (axial and radial).

The slowing down power of hydrogen on neutrons is $\xi = 1$. It is impossible (with the exception of vapour cooling as in the vapour cooled fast reactor concepts of the past) to get a genuine fast spectrum with light water without compromising core cooling effectiveness in normal and accident situations (with a competitive design; otherwise almost anything is possible). Unfortunately, with plutonium, it is necessary to escape from the spectral region where its neutron production performance is mediocre to achieve conversion ratios close to 1.

The only way with LWR is using 3D heterogeneous cores combining "source" zones where the overall neutron production is increased with fertile zones, resembling internal blankets, embedded in the core. A number of advantages (conversion ratio, void coefficient and resistance to proliferation) are offered by this design.

13.5. Fast spectrum: the main constraints and specific issues

13.5.1. The design constraints related to the fast neutron spectrum

To examine to what extent LWR (and BWR in particular) offer potential to increase their thriftiness in the uranium – plutonium cycle without reducing their competitiveness, it is important to **review the main constraints imposed by effective fast neutron spectrum design.**

A high spectral index is required. The indicator is $r = \nu \, \Sigma_f / \xi \, \Sigma_s$, characterising how fast the neutron spectrum is.

This index compares the fast neutron generation rate potential $\nu \, \Sigma_f = \sum N_i \, \nu_i \sigma_{fi}$ with the 'core slowing down power' $\xi \, \Sigma_s$. It is thus necessary to concentrate the fissile material or to only dilute it in a matrix (structures and coolant) that is "transparent" to neutrons (which is more the case in the gas cooled or lead cooled systems).

High specific power. This is related to the management of available plutonium and not to the reduction of the capital costs by increased core compactness. Optimal use needs to be made of the limited inventory of plutonium available at fast reactor commissioning: the first core load (typically around ten tons of plutonium per GWe as for a modern design with reduced positive void coefficient) and the partial reloads necessary to wait for recycled plutonium (typically a "second core" or even more, depending on the out of pile cycle duration for cooling, processing and fabrication).

This is challenging, particularly for countries and utilities which do not perform processing and which send the plutonium to the waste instead, or which consider starting "directly" with fast breeders.

The fissile material inventory may consist:

- either of the plutonium capitalised (and recycled without excessive degradation) for more than fifty years in LWR. The efficiency of this way of generating fissile plutonium is typically of 1:1000 from natural uranium extracted. As a matter of fact, there is about 1.3% of total plutonium left in spent fuel with about two thirds of fissile plutonium still inside, reduced to about 60% after partial decay of ^{241}Pu during the out of pile cycle, before reloading in a core, leading to $\sim 0.8\%$. This has to be combined with a factor of ten typical of the ratio: of natural uranium mass/fuel mass (heavy nuclei), due to enrichment from 0.71% ^{235}U in natural uranium to about 5% in fuel, with a tail enrichment of $\sim 0.2\%$ as for depleted uranium. So the indirect way is long and of limited efficiency, but the fissile plutonium has been paid for (as a material, not as a fuel: the fabrication has still to be paid for) by the UOX burning and power production.

- or of ^{235}U in LEU (low enrichment uranium with e < 20%), with a typical efficiency of about 5:1000 of the natural uranium extracted (once more: from 0.7% in natural uranium to 0.2% in tails). Actually, ^{235}U is not equivalent to **fissile** plutonium in a fast spectrum: an equivalence coefficient has to be used.

Furthermore, in order to achieve short doubling times for the fleet of fast reactors, it is necessary to simultaneously obtain a high breeding ratio, a high specific power as well as short cooling, processing and fabrication times after the SNF (spent fuel) has been unloaded from the core.

A relatively high power density. The power density is the product of two terms: the plutonium concentration (g (Pu)/cm^3) and the specific power expressed for instance in kW-th per gram of plutonium. As previously mentioned, the two terms must be high (except for the concentration in the case of large "diluted and porous" cores without powerful slowing down or capturing coolants and structure materials).

This constraint results in the need for effective and reliable engineered safety features. It requires the best available mix of active and passive systems with a thorough implementation of defense in depth including redundancy, diversification, spatial separation of independent critical safety systems. This has to be integrated in a strong crisis management organisation, including:

- a "hardware" section: pooled, dedicated and rapidly available skills and devices under a strong and unified command at the highest technical and political level;

- a "safety software" crisis management section.

Every type of reactor has its own specific strengths and weaknesses, mainly dependent on the coolant and on the power density, and they must and potentially can be managed by harnessing relevant and consistent "software" (flexible and robust plans for action) and "hardware" solutions. Nevertheless, fast spectrum constraints on reactor design challenge this capability a little more than it is the case with thermal converters.

For instance, in LWR, and specifically in advanced passive or hybrid passive/active BWR designs, two phase flow phenomena make it possible to passively remove a significant decay heat power density (so long as the system is suitably designed to encourage natural convection, powerful condensation in pools, heat pipe operation, etc., with huge amounts of water topping, surrounding the Nuclear Steam Supply System (NSSS), inside the containment). In a gas cooled fast reactor with significant core power density, solutions of the passive HTR type, based on thermal inertia, then radiation, are not adaptable. Natural circulation of a mix of – light and heavy – gases, under an intermediate back-pressure, complemented with innovative decay heat removal systems, is requested (see § 16.6).

A high burnup. This reflects a cost constraint. The proportion of the cost of the kWh due to recycling varies with the sum of the costs of spent fuel processing and of the fabrication of reload fuel, divided by the burnup. Neither of the two numerator terms is directly proportional to the burnup. It is therefore advantageous, at least in the range of 100 to 150 GWdays per ton of heavy metal, to increase the burnup. This results in constraints on fuel behaviour and, as a consequence of the high fast fluence level, in constraints on cladding. In high conversion ratio LWR, the trade-off between the reactivity swing, the reactivity control and the conversion ratio, among other parameters, seems to provide a strong incentive to keep burnups around 50 GWdays/tHM, the consequences of which need to be further analysed.

A high fast flux integral (> 1 MeV). In a thermal spectrum core, most of the fission occurs in the thermal region of Maxwellian form. The corresponding effective cross sections are around one hundred times greater there than those in the fast region. In a fast spectrum, fission reaction rates are the product of low fast effective cross sections (typically 2 barn) and of the fast flux. The integrated fast flux level, for a given burnup, is therefore considerably higher than in a thermal or epithermal spectrum. The result is a strong constraint on the materials, because they must retain their mechanical properties at very high fluence levels.

In short, the specific difficulties with a fast spectrum are:

- High power density, requiring effective and reliable safety features.

- Neutron-related constraints on the fuel, on the stability of its potential micro-structure properties, on the "first barrier", as well as on the internals subjected to flux: low absorption, low slowing down power, low activation, i.e. high neutron "transparency".

- Ability to withstand high levels of fast flux.

- Possibly (given the ambition to combine high thermodynamic efficiency and high nuclear efficiency): excellent thermal conductivity and satisfactory resistance to high temperatures.

- Last but not least, the safety design of the core presents a challenge due to the potential for large reactivity increases from the normal to incidental situations, leading to a partial voiding or collapse (through mechanical compaction, partial melting, boiling …). In this case, depending on the coolant and on the architecture of the core, prompt jump reactivity insertion (several βeff) can occur, (neutron lifetime $\sim \mu$s) providing a large, instantaneous mechanical energy release which must be "practically" avoided by design (frequency reduced below $10^{-7}-10^{-8}$/reactor.year). Currently under study are coolants which are highly transparent to neutrons (for capture as well as for slowing down power), short cores, heterogeneous cores, "diluted" (but still fast) cores, refractory and mechanically robust cores (features which are as important as a high fission gases containment capability – inside the fuel, at high temperature) and dedicated devices for molten core catching and for neutronic decoupling of potentially critical lumps in case of melting.

13.5.2. From the past to the future

This summary leads to three comments:

It illustrates the difficulties encountered from the outset. When the nuclear energy was first introduced, it was anticipated that fissile and fossil resources would be rapidly depleted and plans were put in hand early for developement and commissioning of competitive fast reactors.

Quoting Alvin M. Weinberg in ([4]) – "The collected works of Eugene Paul Wigner" about the spirit of the "forties": "Wigner was much intrigued by the breeder. At the time the total known uranium amounted to thousands rather than millions of tons; and Wigner recognized that with so little uranium, nuclear energy could not be very important unless the breeder were developed. When the values of η (neutrons emitted per neutron absorbed) for ^{233}U and ^{239}Pu as functions of energy were measured it became clear that in principle a thermal breeder based on ^{233}U and a fast breeder based on ^{239}Pu or ^{233}U were possible".

"Wigner was not attracted to the fast breeder; he regarded its engineering problems as formidable. Nevertheless, he and Harry Soodak sketched out the design of a fast breeder. ...Wigner's real love was the thermal breeder, which he regarded as a more practical engineering device than was the fast breeder, even though the breeding ratio was lower than in a fast breeder. In 1944 he sketched out the design of a light-water-moderated converter with plate-type fuel and a thorium blanket for converting ^{239}Pu to ^{233}U...". "Wigner, the chemist, was aware that the breeder would require rapid chemical reprocessing". From E.P. Wigner himself: "Reactors were thought of in terms of structures "that a plumber could put together". In spite of these primitive beginnings, the basic technical pattern of power reactor development was understood at an early date."

Specifically-nuclear technology needs to be developed on its own. Much innovation is called for, which requires a long and expensive validation process. As long as the cost of the fissile resources has not substantially and **definitely** increased, the competitiveness of "newcomers" is difficult to establish. In a world of high discount rates or return on equity (REO) levels, systems with low capital costs are preferred at the expense of systems with an advantage due to their thriftiness later in their life-cycle.

It shows what remains to be done before the transition to Gen-IV advanced fast reactors.

From the point of view of the existing technology and design, sodium-cooled fast reactors, or even GFRs (accepting in both cases limited thermodynamic efficiency as well as nuclear performances, and relying on a robust mix of engineered and intrinsically passive safety features), could lead to early prototyping and "first of a kind" construction and commissioning, but not currently with any guarantee of competitiveness.

It would be nice for a design which is easily convertible into a fast reactor, to first establish a position in the thermal spectrum. An image could be supplied by the (relative) "continuity" of gas-cooled technology, provided the design basis of the HTR is reviewed to make it competitive for power generation (higher unit power and power density, mix of passive and engineered safety systems, temperature derating etc.).

Coming back to Weinberg & Wigner ([4]): "In these early papers Wigner conceded that breeders might develop simply through improvements in non-breeders (e.g. the Shippingport light-water breeder); or they might spring up as a wholly new technology, e.g. the Liquid Metal fast Breeder (LMFBR). He was unwilling to choose between these two possible paths; and although the main line of breeder development has certainly favored the latter path, the very high capital cost of the LMFBR suggests that incrementally improved converters may yet have their day."

This raises an essential question for the future: is there the necessary motivation and any real chance of achieving such objectives in a 3D heterogeneous core, in LWR?

Multi-recycling of plutonium in a hardened spectrum ("smart recycling") would offer to the transition to Gen-IV via the tight management of the fissile resources (largely before any ultimate depletion), two main advantages:

- Preserving the quantity and quality of plutonium inventory considered to constitute a "hard core" for introducing fast reactors;

- Notably reducing the natural uranium make-up for production of a given quantity of energy (typically by a factor of four at equilibrium, compared to that of the current world fleet).

13.6. "Smart" plutonium multi-recycling in LWR: The natural uranium saving context issue

The advanced recycling would represent, remaining within the bounds of well known and tested technology with few modifications (fuel and core structures), the symbiosis of two processes: the conversion from uranium to plutonium and the multi-recycling resulting in the creation of an inventory of plutonium with a stabilised isotopic composition. More than half the energy produced would originate from the recycling. The switchover from "lazy recycling" to "smart recycling" has consequences for techno-economic optimisation (optimal burnup, for instance) which may induce changes in the orientations of the search for technological improvement. Such "smart recycling" could constitute a worthwhile transition pattern "seen from both sides of the river": Gen-III and its fleet of LWR that are here to stay during the century, and hence depend on the management of the natural uranium resources, as well as Gen-IV with the issue of launching the fast reactor fleet with a large (and costly) amount of required fissile nuclei and with very limited (in 2050) worldwide capitalisation of plutonium.

Indeed, the medium term changes in the uranium market, driven by usual processes of cost fluctuations, ramps and plateaux, or even peaks (as in the 1970's), could impose such moves in the LWR fuel cycle by themselves (such as improved conversion ratio and isotopic composition quality).

Yet, provided competitiveness (specifically concerning capital cost) is maintained, this is the first step towards thriftiness that brings the greatest benefit (moreover, it is reached with a limited effort). For instance, reducing the consumption of natural uranium from 220 tons per GWyear to around 50 would avoid a sudden rise in the consumption of natural uranium during a phase (around 2050, probably), when the **installed effective power** (in GWyr/year or GWeff, currently around 300 GWeff) is multiplied by a factor of three to four. It would also assist the transition to fast reactors, paving the way to a **sustainable and robust** development of nuclear energy.

Looking at the natural resources issue, an indicator could be the following ratio:

$$\frac{\text{Affordable uranium resources}}{[(\text{annual worldwide nuclear power production} \times \text{a given period of time}) \times \text{specific uranium consumption}]}$$

i.e.: $M_{NatUr}/(P_{eff\text{-}nuc} \, \Delta t \, Q_{spec\text{-}NatUr})$

The typical time range is related to the XXIth century.

The specific uranium consumption is defined by: the amount of natural uranium extracted to produce a given quantity of energy (electricity), taking into account, depending on the fuel cycle option, the potential recycling of the fissile nuclei of the spent fuel considered as a by-product.

A consistent set of units can be:

- affordable uranium resources in Mt (NatUr);

- annual worldwide nuclear power production in GWyr-el/year;

- specific consumption in t (NatUr)/GWyr-el.

Current figures are:

- affordable NatUr resources: 10 to 30 Mt (depending on the criteria for the affordability assessment, this in turn depends directly on the specific consumtion indicator);

- **annual** worldwide nuclear power production:

    o at the present time: 300 GWyr-el ($\sim$ 2630 TWh-el), with a capacity factor k_p # 80%; higher than 90% in the USA);

    o about 1000 GWyr-el around 2050 and up to 3000 GWyr-el around 2100. In other words, with the competition being with "King Coal" for base-load power production, the annual consumption of coal removed by nuclear power production is about 0.9 Gt presently, and would be about 3 Gt around 2050 and about 9 Gt at the turn of the century, which would be a fantastic achievement, given the total primary energy consumption being presently around 13 to 14 Gtep/year. Note that competition is not with oil for power production nor directly with gas for **base-load** power production – gas being stronger for peak production – although it could happen in the US shale gas case.

- specific consumption which presently stands at 0.22 t (NatUr)/GWyr-el: worldwide: 66 kt/year for 300 GWyr-el/year. This figure can be decreased to 0.05 in "High Conversion Ratio-LWR" with efficient multi-recycling (assuming a correlative sharp decrease in tail enrichment related to an increase in natural uranium cost, which is quite consistent with the context considered in the future).

Comment:

- The installed power is in GW-el unit: it is a power, not an energy;

- The produced electricity is in GWyr-el: it is an energy: power x time.

- The relationship is energy = installed power × period of time (year) × k_p (capacity/loading factor: power mean value/rated power, or: power production/(peak power × elapsed time)).

Let us start with the first term of the denominator, followed by the numerator.

13.7. Energy scenarios and nuclear power worldwide: a prospective framework for the century

Several scenarios have been proposed for the energy growth rate and specifically for the nuclear power growth rate in the current century. These scenarios are a tool for assessing the pressure that would be exerted on the natural uranium resources as well as the requirements that will result from it. They will be used in the short and in the medium term (fifty to seventy years) for the fleet of Gen-III systems and to shape the transition to Gen-IV. In view of the lifespan of the reactors (at least half a century), it is necessary to cover two generations, i.e. approximately a century, which constitutes a major challenge.

A few references are considered, including IIASA-WEC scenarios and primarily an adapted (updated, as of 2005) hybrid "B/C" scenario: see Figure 13.5, with additional comments about the "IEAs Energy Technology Perspectives BLUE Map scenario" (IEA, 2010).

In the wake of combined financial/economic crisis and of the Fukushima events, these data still look robust, taking into account the large proportion of the growth supposed to take place in Asia.

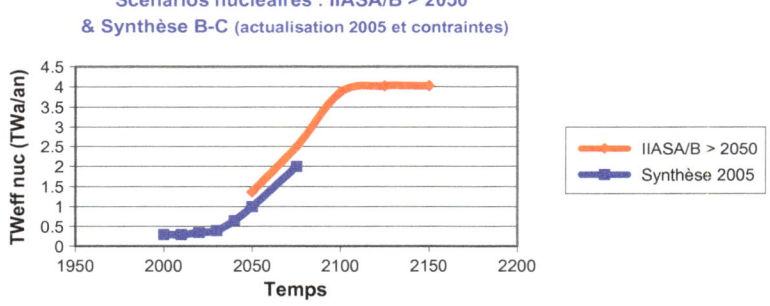

Figure 13.5. Nuclear power growth scenarios: from an IIASA/B-like scenario beyond 2050 to a tentative B/C synthesis, after an update in 2005 taking into account several constraints to nuclear growth.

In the hybrid B/C scenario, nuclear energy, as the primary source of renewable energy without significant greenhouse gas production, gradually supplants oil, if not coal (but not all fossil fuels put together), as the main power (then energy) source worldwide, around the turn of the century.

"The IEA's Energy Technology perspectives 2010 BLUE Map scenario (IEA, 2010) projects an installed capacity of almost 1200 GW-el in 2050, compared to 370 GW-el at the end of 2009. This nuclear capacity would provide 9600 TWh of electricity annually, or around 24% of the worldwide production. In BLUE Map, by 2050, nuclear power becomes the single largest source of electricity, surpassing coal, natural gas, hydro, wind and solar". (From the publication "Technology Roadmap Nuclear Energy" – OECD/NEA & IEA (2010).

9600 TWh annual electricity production in 2050 means 1095 GWyr-el/year, with an improved k_p around 0.9, which is fairly close to the 2005 projection of 1000 GWyr-el/year by the same date (with a k_p around 0.85, leading to about 1200 GW-el of installed nuclear power, the same value as predicted by BLUE Map).

As for IEA's 2010 scenarios, the Baseline scenario 2050 projection is only around 5000 TWh (half the BLUE Map scenario) and the BLUE High Nuclear scenario reaches about 1600 TWh-el. There is still a large scattering, even for a mean term forecast.

In the BLUE Map type scenario, nuclear technologies would be used where possible, safe and profitable, and where it would prepare the solutions for the future from the experience feedback gained in the advanced fuel cycle option field (closed cycle, plutonium management, materials for fast neutrons, etc.).

This could therefore correspond to a **sustainable growth rate** situation.

From a resources-technology pair viewpoint, it is interesting to investigate how to harmoniously achieve the development of nuclear power along a S-curve extending beyond 2100 with the time constants characteristic of the rather slow substitution process of the successive dominant energy sources worldwide. In particular, rational use needs to be made of the resource domain located between $200 and $500 per kg of natural uranium.

At the same time, the vendors will have to answer the question of the utilities on the management of the fuel availability and cost issue after the end of the century for LWR to be commissioned around 2040, which is to come rapidly.

In an uncertain world, the assets of "King Coal" (very abundant in the USA (50% of the power production), in China (83%), in India, in Europe (Germany, Poland, etc.) and in Russia) and of gas, including shale gas (at least for the next 50 years) should not be underestimated.

Despite the problems associated with transport, pollution and carbon dioxide, coal will probably retain its place as a major tool of development in the next few decades, and could even play an increased role in the field of fuel for transports (synthetic fuels, "CTL": Coal to Liquid).

Figure 13.6 shows how the nuclear power generation could be shared between LWR and fast breeders during the century. It will be discussed below in § 13.11.

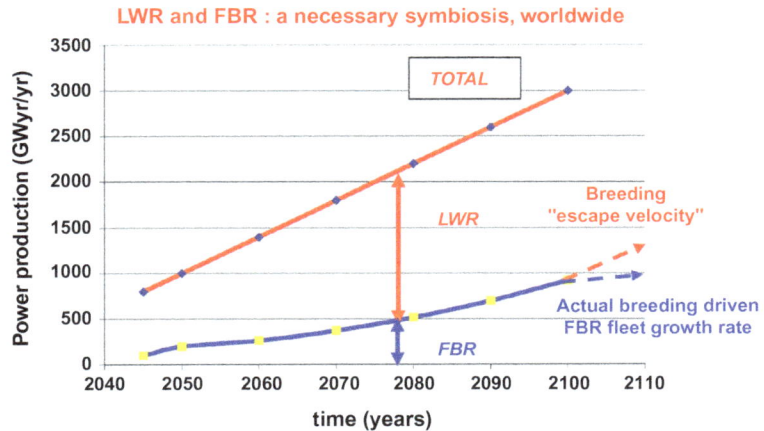

Figure 13.6. LWR and FBR: a necessary symbiosis, worldwide.

13.8. Affordable natural uranium resources

The objective is to distinguish between occasional surges (meaning that the market is waking up and then speculating – sometimes), meaningful mean term ramps and plateaux, and ultimate exhaustion of the resource (at affordable cost for "lazy" fuel cycles; for a breeder fuel cycle, any cost is affordable, well beyond 1000, even 2000 $/kg Unat).

13.8.1. Rising natural uranium prices as ore of decreasing uranium concentration has to be used

The first step of the analysis is of a geochemical nature, as described in the article by Deffeyes and Mac Gregor ([5]) (Figure 13.7). This shows that, in view of the geological history and chemical properties of uranium, it is distributed continuously throughout the earth's crust (mass presence in Mt versus concentration in ppm), in a wide range of concentrations, unlike certain elements which exhibit one or more discrete narrow "peaks". The first three "steps" on the left hand side of the graph show the types of deposits that are currently being investigated and worked: at first vein deposits, then pegmatites, unconformity deposits and fossil placers, sandstones. The product of this first stage is the idea of the elasticity of the resources as a function of the concentration. In the zone they consider to be of practical interest, in the vicinity of the "steps" used, the authors propose a law of the

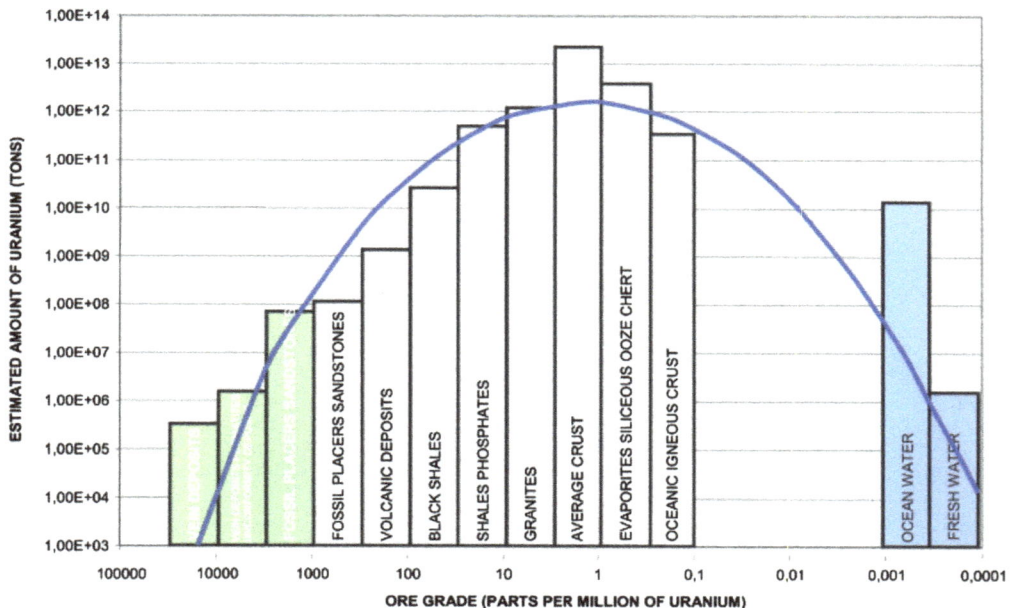

Figure 13.7. Resources in natural uranium: a geochemical approach ([5]).

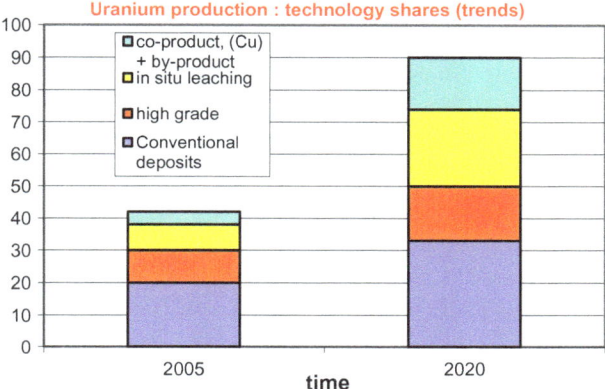

Figure 13.8. A prospective assessment of the share of diverse technologies in the uranium production, in the near future [from Capus *et al.* – SFEN et AREVA].

$M = M_0 (c_0/c)^\varepsilon$ type (where ε characterises the elasticity), with $\varepsilon = 2.48$, which is a rather high value.

Figure 13.8 shows the trends concerning the share of diverse technologies in the near future.

By comparing the data from diverse approaches, the following observations can be made:

- The elasticity exponent related to the concentration can take typical values ranging from 2.48 [5] to 2.35 [Gen-IV cross-cut Working Group], depending on the approach (geochemical or technological), as well as on the considered data set and how they are interpreted. These are actually conceptual values that look very optimistic, even if they are consistent with a general pattern for metals (for example, after the Phillips and Edwards model).

- Using the concentration related elasticity exponent in the following formula:

$$M = M_0 (c_0/c)^\varepsilon$$

One has to find a path to the cost assessment, as a function of the ore concentration (knowing that it strongly depends on the type of deposit). Being interested in decreasing concentration ores, one can take as a rough approximation for the cost p an asymptotic pattern of the type:

$$p \to b/c \; (c \to 0).$$

Combining both approximations leads to:

$$M = M_0 (p/p_0)^\varepsilon$$

Finally, "interpreting" a few points from the Red Book helps set up a simplified formula to be used as a "surrogate model" for further reflection:

$$M \text{ (MtUn)} = M \text{ (p = 130)} (p/130)^{1.6}$$

The second step consists of estimating the cost as a function of the concentration. Basic physical issues lead to an energy related paradigm of the cost, which finally results in making a judgement based on the proportion of energy that has to be recycled to obtain the uranium.

Combining the geochemical analysis with that of the cost as a function of concentration results in a cost-elasticity law: $M = M_0 (p/p_0)^\eta$. The Crosscut Generation IV working group found the value $\eta = 2.35$, with a technological approach weighing the characteristics of various processes, including in situ leaching. This approach is different from that of Deffeyes but the results are consistent. Doubling the cost results in the quantity accessible being multiplied by five. This value appears to be optimistic, specifically when taking into account the increasing role of in Situ Leaching (ISL) with a lower exponent. **In general, the law is used to deduce the mass of natural uranium available at a given cost**, by extrapolation from *the third foundation of the analysis in OECD-IAEA Red Book,* which constitutes the source of reference for the evaluation of resources up to a cost of $130 per kg of natural uranium (and, since 2009, up to 260 $/kg Unat). Such an extrapolation requires utilisation of resources with a concentration below 1000 ppm.

The case of resources below 100 ppm will not be considered (apart in exceptional co-production cases), for energetic-economical reasons (effect on cost of production) as well as for environmental reasons.

To establish orders of magnitude, use can be made in this region (extrapolating beyond the Red Book) of a surrogate model:

$M = M_{130} (p/130)^{1.6}$, with $\varepsilon = \eta = 1.6$ as a result of the $p \propto 1/c$ hypothesis. M, the accessible mass, is given in Mt-NatUr for p, the cost, in $/Kg-NatUr.

The curve passes through the point: M_{130} (Mt-NatUr) for p = 130 $/kg.

On the other hand, 260 $/kg NatUr (equivalently 100 $/lb U_3O_8, the new category introduced in the 2009 Red Book issue) still means affordable uranium with a slightly advanced once-through cycle.

For instance, 20 g NatUr/MWh-el (175 t NatUr/GWyr-el) is a modest performance (with tail enrichment of depleted uranium consistent with a "high" uranium cost). The cost of production of 1 MWh-el is close to 60 $ ($\sim$ 45 €). A fraction of 10% of the power cost dedicated to natural uranium leads to 6 $ for 20g thus 300 $ kg (NatUr). The present situation parameters are close to: 25 g (NatUr)/MWh-el (worldwide), 130 $/kg (NatUr) (long term contracts), thus 3.3 $/MWh-el or about 5% of the power cost. Dividing the consumption by a factor of two with efficient multi-recycling would be a big advantage. The 500 $/kg (NatUr) class of resources would then be considered part of the affordable set of resources.

It seems possible to extract around 20 to 30 Mt of natural uranium at profitable/affordable costs/prices, keeping nuclear competitive in LWR beyond the end of the century, and supporting the future deployment of the breeders.

The role and the limits of uranium co-production merit further study, including the issue of the adjustment of the consumption of the co-products. Olympic Dam (copper, gold and uranium) produces more than 10 Mt ore every year, providing 220 kt of copper, 4.5 kt of uranium, 90 000 ounces of gold and 900 000 ounces of silver. Uranium is a by-product of copper, which generates more than 70% of the income. The concentrations are not very high: copper: 1.1%, uranium: 600 ppm, gold: 0.5 ppb. The potential of Olympic Dam is

4.4 Gt ore, including 600 kt low cost uranium (< 40 $/kg). The updated potential as for uranium resources (2007) is up to 1.9 MtU.

The option of 4 Gt of uranium from sea water (concentration 3 ppb) of which the cost seems to be situated well above $1000 per kg (estimates around 2000 $/kg at the present time) and which, for low levels of annual consumption (with breeders), is the ultimate and practically inexhaustible resource. In 2009, Japanese experts issued a paper regarding an innovation which would potentially lead to a cost of around 250 $/kg NatUr. But a real world operation could well lead to over 1000 $/kg, which is still not discouraging (at least as far as quasi-breeders are concerned).

Finally, the prediction, for years of virtually stable resources at a cost level expressed in **current $** per kg, probably reflects, as in the case of many other metals, technological progress compensating for the increase in cost due to the gradual drop in the concentrations, as well as the result of exploration being resumed. See the interview given by Colin Macdonald, vice president of exploration at Cameco, to Nuclear Engineering International in January 2004 and the document issued in the framework of the Blue Ribbon Commission: "Advanced Fuel Cycle Cost Base" [6].

To summarise, the increasing cost of uranium due to the reducing concentrations of the workings is inevitable (even if allowance is made for technical progress). A smooth dynamic excepting large fluctuations due to a lack of anticipation (which would be damaging for every player, even the promoters of specific advanced systems (**because the prices could fall down again**)), is necessary to avoid the threat of getting stuck, while reserves of increased thriftiness exist for LWR.

13.8.2. The strategic risk of preclusion of access to natural uranium is latent and may take form for a number of reasons

Historically, the American leaders had this risk in mind when they launched the Atoms for Peace initiative in 1953. Currently, one can observe that:

- Some countries with substantial resources take fluctuating positions regarding exploitation of their deposits and selling uranium;

- Concerns associated with proliferation, combined with the risk of progressive dissemination of the enrichment technology could be the driver to try to control and limit the supply of natural uranium.

13.8.3. Shortages and price fluctuations in the short and long term uranium market

If the breeder technology is an insurance against long term increase of uranium cost due to decreasing grade in ore deposits, stepwise "convertible" core LWR of increased thriftiness can be seen as an insurance against uranium price fluctuations which would be due to lack of dynamic adaptation of exploration/operation investment to the variation in rate of natural uranium consumption (Q) and of its derivative (dQ/dt). Furthermore, only a few countries have enough plutonium available for the rapid commissioning of a fleet of fast power reactors (France could launch a dozen commercial reactors starting from 2040).

After a rapid review of the main strengths and limitations of LWR, we will consider their potential for adaptation so as to optimise the transition period towards a U/Pu fuel cycle with breeding capability (while omitting the Th/U-3 option).

13.9. Light Water Reactors, the current situation: Strengths, Weaknesses, Opportunities, Threats

13.9.1. Current situation

At the present time, nearly 90% of nuclear power worldwide is generated in LWR.

Over half a century, a large amount of experience has been accumulated. Much progress has been made under the pressure of increasingly stringent requirements. New criteria and constant improvements in design and operation have however resulted in somewhat disorderly adaptation, by successive additions (for example in terms of nuclear safety) with no fundamental design changes expected before the next generation of plants, because of the advantage generally given to "evolutionary" design. Yet there have been few new generations so far. This is the reason why the new specifications emerging for Gen-III and the changes that will be implemented through "Darwinian pressure" before the actual transition to Gen-IV, are extremely important. Care must be taken to avoid design dead ends that might compromise the chances of ultimate adaptation of Gen-III LWR to the requirements that will inevitably apply (and this would probably threaten the Gen-IV systems for the same reasons). Let us therefore consider a few criteria in order to assess the current situation and to consider the margin for useful and credible progress.

Econonomics and competitiveness (disregarding at this stage the resource issue)

In order to be able to compare economic performance levels, it is necessary to establish the context, particularly with regards to investment. Furthermore, even with common rules, it would appear that costs are difficult to interrelate between different regions of the world.

Using a levelised electricity cost analysis based on a specific "discount rate" option as carried out in France by the Department of the French Ministry for Industry and by the OECD and using typical rates of 8%, the MWhel (for base-load operation) can be estimated to be around 45 €, with the capital cost representing around 60% (see an interpretation of the 2003 and 2008 studies in Figure 13.9). The "taux d'actualisation" concept, primarily used in the framework of public investment, is different for instance from the ROE (Return On Equity) calculation used in the framework of a conventional business plan.

The proportion of the capital cost (capex) in the total levelised kWh cost is so high that it is essential to maximise the capacity factor (k_p) value (the ratio of the amount of energy produced annually over the product of the rated power by 8766 h; this indicator combines availability with plant grid call). Increasing the k_p from 75% to 85% has a major effect on competitiveness. In comparison, increasing the lifespan, although extremely profitable for a utility operating an "inherited" plant, appears as being of secondary importance in

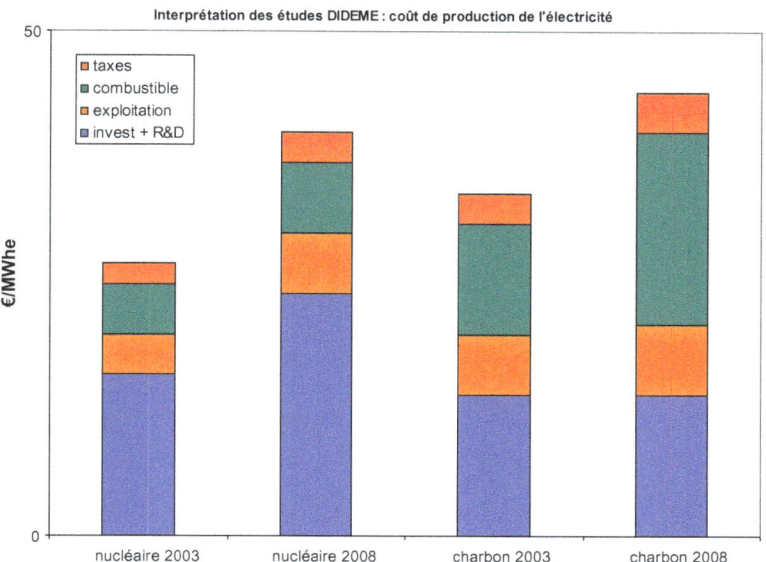

Figure 13.9. An *interpretation* of the 2003 and 2008 studies from French DGEC.

such analysis, as results of the weight of the discount rate. The same applies to the gradual growth in the cost of the enriched uranium. The determination of the provisional k_p for a "newcomer" (for instance the first series of a new type of reactors) would thus be thoroughly scrutinised in the process of assessing the economic balance of the investment. The same applies to the cost of the first core and next partial reloads. This is detrimental to newcomers of an innovative reactor system family. This is favourable to the SMR concept.

In the USA, k_p usually exceeds 90%, a high value currently threatened by the growing fraction of power produced by solar PV and wind intermittent renewables, without adequate storage (which, in some cases, would be out of reach technically and economically), forcing a significant back-up power to be available and reducing its k_p. Keeping refuelling and maintenance outages as short and as infrequent as possible (every two years for instance) is a constant concern. It tends to be somewhat detrimental to optimisation of the use of uranium obtained by increasing the subdivision on reloading the core. This option reduces the cycle duration at fixed burnup (see below), leading to lower k_p. Reconciling both constraints leads to increased enrichment and burnup, up to a given limit (due to fuel cycle facility design, and to fuel technology qualification). With a 5% ^{235}U enrichment and a good neutron economy (few captures, low leakage and well subdivided reloading), a burnup of 65 if not 70 GWdays per ton of heavy metal can be achieved. This example shows the interplay between diverse parameters.

Reduction of capital cost is a major priority. Currently, with moderate discount rates common to diverse financing schemes, but not universal, and for "typical" Gen-III reactors, the capital cost seems to be in the region of 5000 $ (2014) per kWe (at least for the FOAK – First Of A Kind – and maybe a few NOAK – Nth Of A Kind). This is dependent on local boundary conditions (in some cases a factor of two on capital cost – capex as capital expenditure – between some Western countries and ASIA). It is fair to note that construction costs generally soared over the 2000–2010 timeframe. Estimates for gas and coal stations increased by 60–70% before entering a period of fluctuations, due to a combination of rising raw material costs, more stringent specifications and regulations, along with a global boom in orders, during a period of time.

Reducing the time spent in construction and commissioning is also essential. The goal is a duration of less than four years. Here again, from the investor point of view, a reduction of one year may be preferable to a substantial increase in lifespan, depending on the discount rate.

The financing scheme issue and its effects on capital cost created a gap between studies that were issued on the one hand in France or by the OECD and, on the other hand, in the USA (MIT, Harvard or Chicago).

In the French Ministry study, the discount rates considered are around 8%.

In the US studies, investment is assumed to be financed, in the context of a conventional business plan, from equity seeking a high rate of return (typically 15% after tax) and from debt, loans, with a reduced rate of return, but specifically increased to cover the "nuclear risk" (nuclear premium). Another way of covering the nuclear risk: the proportion of equity capital is raised to 50%, instead of a quarter as is typically the case in other projects.

When drawing equivalences with a mode of calculation based on a single discount rate, a rate of 13.5% is obtained. At this level, the investment ends up representing as much as 80% of the cost of the kWh, and imposes options that serve to reduce, in any possible way, the cost arising before commissioning. The first conclusion to be drawn from these studies is that nuclear power in a market economy has no chance of attracting investment without seeking government assistance, at least for the first series of plants.

In the USA, in a second stage, this assistance has been obtained in the 2005 Energy Bill, in the form of government loans, but this is still limited to a handful of lucky newcomers.

The techno-economic vision underlying these choices can be summarised as follows, following recent contributions from MIT which roughly conclude that:

- reactor capital costs drive busbar costs ($\sim$ 80%):
    - uranium cost is not an economic constraint for most of the century;
    - once-through fuel cycles are the most economic;
    - fuel cycle choices are constraint by reactor capital costs;
- economic conclusions:
    - LWRs with a once-through fuel cycle are likely to be the primary nuclear system for a long time;
    - etc.

At the same time, state withdrawal seems to occur, at least partially, in Europe. Nuclear power is considered to be a commodity and not as an enabler to build a dynamic economy on a dependable low-cost basis. Trapped between fossil fuels for which the investments are lower (with the notable and expensive exception of "clean coal" which is the only competitor on a long term worldwide basis for massive "low carbon" baseload production) and assisted renewable energy, nuclear energy appears to suffer a "sustainable" disadvantage.

However, this disregards the increase or at least the large fluctuations (see the oil case) of fossil fuel costs which justifies regained interest in nuclear power. Currently, and for several decades to come, these increasing fossil fuel costs do not include the US shale gas, at least in the USA where the priority has been put on a domestic competitive use, pushing more US coal to the export trade.

At the same time, in the post-Fukushima context, in China the CPR1000 starting from the 1970-vintage French PWR design has been developed into a "generation 2.5" design (CPR1000+).

An optimized "CPR1000+" under construction at Yangjiang costs around $1800 per kilowatt installed. A replicate CPR1000 unit costs less than $1500/kW, after CGNPC, quoted by Nucleonics Week.

On the other hand, for any type of reactor, "The cost of a major nuclear accident can dwarf the extra cost of third-generation nuclear plant safety features."

Typical Gen-III PWR would include a core catcher and double-reinforced containment capable of withstanding a large aircraft crash, among other improvements.

The harmonisation of safety principles and requirements will surely be a powerful driver. This pressure could lead to increased costs and it will be mitigated by international cooperation between vendors and the main utility companies. This cooperation will lead to an optimisation of the overall cost of new reactors proposed worldwide.

Nuclear safety: the trend towards zero impact on health and environment beyond the site boundary (excepting ultra-low frequency "residual risk")

The review of modern designs (EPR, KERENA, ABWR and ESBWR, AP1000) in terms of probabilistic studies shows that the design goals have been reached, these being more stringent that the explicit safety goals (with core meltdown frequency lower than 10^{-5} per reactor per year and early massive releases being "practically eliminated").

On the other hand, more defense in depth must be respected in the design and in the crisis management and resource allocation (skills, tools (software and dedicated hardware), leadership and coordination). It is particularly important to re-assess the contribution of extreme and combined external aggressions, of crisis duration and of the number of plants hit during potential events.

But nothing is specifically threatening the LWR concept and implementation, neither PWR nor BWR.

Safety case simplicity is another important issue associated with the deterministic-probabilistic evaluation. Progress in combined numerical and experimental simulation has made it possible to guarantee safe and effective design bases. However, limits are determined by the extreme non-linearity, the difficulties of multi-physics and multi-scale simulation, and the virtual impossibility of conducting predictive deterministic calculations in certain cases. The control of the rapidly growing uncertainties leads to the provision of substantial explicit or implicit margins, and the safety case can also be handicapped by procedural complexity. Simplicity can be achieved by "decoupling" the physics leading to failures (thermal-hydraulics, thermo-mechanics and chemistry, for instance), by reducing the transients to a succession of "simple" and fast transitions and of slow, virtually stationary kinetics, dominated by competition between antagonist driving forces (decay heat vs. heat removal to an external heat sink). This can be easily calculated in the form of a balance sheet and is related to "analyse par état" ("state oriented approach"). For instance: feed and bleed (LWR); adiabatic heating (decay heat) versus thermal inertia, then radiation (HTR). This could greatly assist a well-founded assessment/agreement.

The context favours the integration of an increasing number of "passive" devices.

Classification of coolants and of their performances is also influenced by these trends. The merits of a two-phase water-steam system in arrangements **originally** designed to exploit them (for instance designs of the KERENA and of the ESBWR) become apparent in the effectiveness, in terms of removal and transfer of energy, offered by latent heat (vaporisation, condensation), by the difference of density transformed into motive power in normal and accident operation (natural convection, "heat pipe" (condensing and reflux) operation), at all pressures. This is clearly in contrast to gases, the rival industrial coolant, for instance.

Once more, with their advanced designs, the Gen-III LWR appear to be well placed.

Fukushima: a few additional comments

- The primary goal for Gen-III LWRs is to avoid health and environment damage and detriment beyond the site boundary. This goal has not been reached in Fukushima, partly due to the reactor design, despite some very interesting features of this 1965-old design, to the lack of upgrade enforced by a safety authority, to organisation issues.

- There will certainly be an increased contribution by passive safety systems for decay heat removal, with a very long term capability and autonomy (weeks vs. days). Some of them will possibly be chosen because they run during normal operation in order to reduce the risk of failure in fulfilling the "pre-conditions" at the launching time. Such powerful assets can be implemented on any type of reactor but the quality of implementation, the resilience and the "hardening" must be improved. For instance, even for old BWR, the isolation condenser basic principle was adequate, but did not have the full set of required features, and unfortunately, there is no such thing as a sufficient condition but only a set of necessary conditions to fulfil. For instance, active and passive systems must withstand the hard environment that results from an accident, without incurring or provoking breaks, bypass, etc.

- In order to get a cooling-down flow for a 1000 MW-el (thus 3000 MW-th power plant), at a decay power rapidly reduced to 2% of nominal power (i.e. 60 MW-th), in monophasic liquid flow, $Q \times cp \times \Delta T$ around ambient conditions = 6×10^7 W – thus: Q min # $6 \times 10^7 / 2.4 \times 10^5$ = 250 kg/s.

- The powerful heat removal capacity in feed and bleed mode resulting from the high latent heat value around 2.25 MJ/kg, leads to a reduced flow of: $6 \times 10^7 / 2.5 \times 10^6$ = 24 kg/s; an order of magnitude lower than if one-phase cooling down mode was used.

- Batteries are a powerful tool for long term DC and AC power delivery. Too many batteries have a short capacity of a few hours. This must – and can – be significantly improved. At 100 Wh/kg specific capacity, which is a quite typical performance coupled with a satisfying specific power capacity (below 1 c), 1000 kWh results in a 10 t battery for 10 hour full power operation at 100 kW power. Batteries can be put in floating mode on the grid. These mere "order of magnitude" figures give an insight into the simple – but resilient – means to be combined. Among the necessary upgrade decisions are: Hydrogen venting and recombiners and filters, even if scrubbing in the torus-shaped wett-well has led to a great reduction in the release of elements like

Iodine and Cesium, etc. apart from the noble gases Xe and Kr, when compared to the molten core source term.

- Innovation: Core catcher, enhanced containment withstanding external aggressions including plane crash.

13.9.2. LWR strengths: robust options, wealth of experience

The initial advantages of LWR were the following:

- use of an industrial coolant;

- compactness;

- applicability of the experience feedback gathered from the military naval propulsion for the PWR (Westinghouse) plus the intrinsic advantages expected from the transition to civil steam supply systems with the BWR design (General Electric);

- availability of financially-competitive industrial-scale enrichment for civil power generating applications, making it possible to overcome the obstacle of capture by the light water moderator and to achieve burnups substantially higher than with natural uranium. This trend in increasing burnup has continued. In 1985, the following prediction was made illustrating general opinion at the time, already based on techno-economic analyses: "it is probable that there is an optimum corresponding to an average burnup of 30,000 MWdays per ton (40,000 MWdays per ton at a maximum)".

In some twenty years, the nature of the problems has evolved. There is a gradual trend towards burnups of 70 GWdays per ton of heavy metal and, for exploratory reasons and as a function of considerations associated with the back end of the cycle, strategies to achieve 100 GWdays per ton are being studied. Techno-economic analysis alone is not discouraging such strategies, even though technological progress and "robust" qualification of such advances may result in deferring or even abandoning them.

The capacity to adapt

Their abundant operation feedback places LWR in a situation comparable to that of the piston engine in the automotive concept.
Improvement in operation, nuclear safety and the fuel cycle (cost, innovations (particularly relating to BWR) and experience with recycling) has been continuous.
As concerns the safety issue, the adaptation to justifiably more stringent requirements for a nuclear industry with the ambition to be both robust and sustainable, has been completed without changing the basic architecture. This adaptation has consumed economic margins. Gen-III and Gen-III+ designs need to exploit some simplifications of principle and a few shortcuts, while making reasonable allowance for the "new threats". As such, the ESBWR and the KERENA concepts are appealing.

13.9.3. Weaknesses

Can the LWR be said to have reached a point of maturity and stability? This conclusion would fail to take into account the new challenges appearing in the context of the cycle (and related to reactor core physics), as a function of the criterion of thriftiness and of the less focussed criterion of "cleanness", without missing the long term common trend towards high thermodynamic efficiency, provided that it does not compromise competitiveness (something which is shared with fossil fuel plants). On the other hand, before the end of the century, the quest for high nuclear efficiency would probably overtake the quest for high temperatures, if a technological contradiction between the two should appear. The same also applies to gas-cooled and sodium-cooled technologies.

The limits of the LWR relate to the spectrum and to the temperature. They are linked to the **coupling** of water (an excellent coolant, even in the provision of passive cooling by two phase means) with neutron slowing-down and with chemistry. On the other hand, (pure) helium is also "virtually transparent" to neutrons and it is chemically inert.

13.9.4. Opportunities

The LWR has not yet realised its full potential. Any progress with LWR would enable a more gradual transition to a worldwide fleet of fast reactors (in uranium – plutonium cycle, before considering hybrid cycles associating in addition, in a broader range of spectra, thorium and ^{233}U), and consolidate its preparation (fuel cycle options, materials, design, experience feedback of prototypes and FOAK).

Building up and keeping a strategic inventory of plutonium in the useful and profitable process of multi-recycling in LWR without degrading its quality, could be combined with substantial savings in natural uranium. This could even free up a "reserve of natural uranium", which may help in constructing fast reactors in countries where recycling was not previously practised, or countries that would directly start nuclear power using fast reactors without being able to "purchase" the necessary amount of plutonium (there is no such thing as a plutonium "market"...). This help from a plutonium generating LWR fleet will be necessary until the full development and full usefulness of breeding with "reasonable" doubling times can break up the dependence of the FBR growth on feeding from an external source of fissile nuclei, i.e. **in the second part of the fast reactor population S-shaped growth curve**, namely the XXIInd century.

How would a LWR project be oriented towards increased performance and a harder spectrum for optimised recycling?

To avoid breaking the chain of competitiveness, in a Gen-III+ perspective, an "evolutionary" approach would be adopted for the nuclear steam supply system (NSSS). The potential major changes, for instance use of supercritical water, will be examined in the Gen-IV context. Consideration is therefore not given here to the issue of high temperatures which changes chemistry-related aspects (corrosion). It is the core and the fuel that are focussed on (including the topic of structure and cladding materials), with minimal changes in operation, only the ones due to optimisation under safety-related constraints.

The steam-water coolant (and certain gases in the combustion turbines and elsewhere) corresponds, to standard practice in terms of thermal machines (provided steam of adequate quality is delivered). Furthermore, in nuclear operation, two specific qualities become apparent:

- optical transparency, favourable to ISIR and during refuelling operations;
- relying on water, during the same operations and more generally during outages, to fulfill the functions of natural convection cooling, biological shielding (making it possible to fully profit from its optical transparency), avoiding the need for a complex bulky and heavy handling machine (defense in depth against melting of a fuel assembly during transfer), and allowing direct access to all parts of the core.

The particularities of water as a moderator play a role in the design of advanced cores. One of them is the migration area. In liquid water, with a standard lattice pitch and a core of conventional composition (uranium enriched by less than 5%, single-pass recycling of plutonium), the migration area is a few tens of centimetres. This corresponds to the capacity of hydrogen of thermalising neutrons in one – single – shot ($\xi = 1$). The spectral heterogeneity of a core can thus be structured over rather short distances, of the order of magnitude of the mean chord of an assembly, (but with the related power peak problems).

Finally, the race to Gen-IV gives to LWR a double opportunity:

- they become the necessary "missing link" to produce more plutonium for launching the fleet of fast reactors; their capacity to improve as converters (up to multi-recycling and to accepting in addition thorium and ^{233}U?) is thus challenged;
- by the way, they need to stay in operation well beyond the end of the century; their ability to enter the world of "sustainable" reactor type is also challenged, including a better efficiency in using natural uranium and thorium by smart recycling (in the "fuel states", the "reactor states" being supplied by the fuel states with fresh UOX for once-through cycle, but "obviously" for a limited installed nuclear power).

These challenges are new opportunities for improvement. They could be of the, "market pull" type if the natural uranium cost increases by steps, leading to a continuous optimisation process setting the fuel cycle parameters (tail enrichment, conversion ratio) in accordance with "incremental" shift of the power cost decomposition working as cost driver analysis.

13.9.5. Threats

- "Clean coal" and gas, including emerging shale gas. In the longer term, as soon as their part in the power production on the grid increases substantially, the intermittent technologies (wind turbines, solar power plants, PV or technologies based on concentration and on thermodynamic conversion) can be viewed as developing a necessary synergy with flexible base-load power sources, such as nuclear, coal fired plants and gas fired power plants.
- An accident, anywhere in the world, on any reactor type. The severe accident that occurred in Japan does not represent a threat that is specific to LWRs, given that they proved to be globally compatible with their natural environment where it became necessary to use ultimate, universal tools to cool down the cores and the spent fuel

pools placed "upstairs". These tools included temporarily sea water in "feed and bleed" operation and (universal and ultimate remedy) fire-engines and concrete-lift trucks. No criticality problems were confirmed during the periods of injection of water devoid of boron poisoning.

Finally, it is the case that both the above threats are occurring gradually or through disruptive events, yet LWRs are still showing potential for improvement in terms of resources, competitiveness, survivability and adaptability.

13.10. LWR: further improvements in fuel cycle efficiency by spectral hardening

13.10.1. LWR: an overview of the present fuel cycle performances, of the trends and of some possible improvements

The overall "nuclear efficiency"

Let us call nuclear efficiency the mass fraction of natural uranium giving rise to fission before the end of the fuel cycle, the meaning of this "end" being: waste generation, i.e. materials without further use for energy production. The fraction giving rise to fission depends on the fuel cycle strategic options: once-through cycle, mono-recycling (plutonium or plutonium plus the amount of ^{235}U left in the spent fuel) or even multi-recycling, should this be possible, either in innovative LWR cores or in other reactor types (FBR, PHWR). What is the current level of "nuclear efficiency" in the dominant LWR fuel cycle?

Basic data

- Fissioning 1 g of heavy nuclei delivers 1 MWd-th.

  The accurate result depends on the element and isotope as well as on the spectrum, through the variable values of κ (around 200 MeV per fission) and of the atomic number of the isotope (A), from ^{233}U ^{235}U and ^{238}U to the fissile plutonium isotopes. The elementary calculation steps are the following:

- Basic data lead to the first result: the fission of 1 g of heavy nuclei delivers about 1 MWd-th

  200 MeV = 3.2×10^{-11} J and 1 MWd-th = 8.64×10^4 J, so 1 MWd-th means 2.7×10^{21} fissions.

  The related mass is M = N_fA/"Avogadro" # $2.7 \times 10^{21} \times 235/6 \times 10^{23}$ = 1.05 g/fission

  # 1g/MWd-th from fission.

- 1 t of fission → 1 GWyr-el in modern LWRs

  With a net efficiency close to 0.36, 1 GWyr-el = 1 MWd-th $\times 10^3 \times$ 365 days/efficiency (0.36)

  # 10^6 MWd-th ← → 1 t of heavy nuclei (fissile and fertile) fissioned.

- 1.25 t of "fissile nuclei" absorbed/consumed in a thermal spectrum ← → 1 GWyr-el.

  1 t of fission of any type of heavy nuclei (fissile or "fertile") means 1 t/ε of fission of "fissile nuclei" (namely ^{235}U and ^{239}Pu – ^{241}Pu in a U-Pu fuel cyle), ε being the fast fission factor of the four factor formula for k∞.

  This means (1 t/ε)(1 + α) t of fissile nuclei absorption, with absorption = fission + capture. The global result is close to 1.25 t/GWyr-el (for instance, one can start from (1.05 g fission/GWyr-el/1.06) × 1.28, for a mix of fissile nuclei in a thermal neutron spectrum.

- Equivalently, 1 GWd-th of burn-up means a decrease of around 0.125% in the concentration of fissile nuclei (^{235}U + ^{239}Pu + ^{241}Pu), by absorption (fission + capture), without taking into account the 0.125 × conversion ratio (c) produced by fertile capture and partially compensating for this destruction (see below).

What are the (worldwide) typical current performances of a LWR fuel cycle?

Presently, the LWRs deliver about 90% of the worldwide annual nuclear power production which is close to 300 GWyr-el (or equivalently 300 t of heavy metal fissionned, see Basic data). The dominant option as for fuel cycle is the once-through option.

The mass of natural uranium consumed each year is about 66 kt NatUr, and most of this amount will not be recycled in the short term, so the assumption will be to calculate the "nuclear efficiency" without any further power production. This leads to a "specific consumption rate" of: 66000/300 = 220 t NatUr/GWyr-el. and to a "nuclear efficiency" defined as the ratio of the number of atoms fissioned over the number of atoms extracted from the deposit, in this case: η_v = 300/66000 = 0.5% of fission from natural uranium. This is a very low value when compared to the potential of breeders (requiring multi-recycling) which can be a hundred times higher or even more. This value of 0.5% in once-through can be compared to the difference between the concentration of ^{235}U in natural uranium (0.711%) and the enrichment left in the tails after the isotopic separation process, with an enrichment close to 0.2% (typically) in so called depleted uranium. But this difference which leads to the correct order of magnitude must be interpreted in a more subtle way.

- At the EOC (end of cycle), the core reactivity is close to 1.0/1.01, requiring a significant concentration of residual fissile nuclei in the core and even in the spent fuel assemblies which will be unloaded. The ^{235}U loaded in the fresh fuel has not been totally burned.

- On the other hand, a massive production of plutonium has occurred and a large part of this plutonium has been burnt. This plutonium contributes to the two terms completing the global balance: residual fissile nuclei, and fission up to the final burn-up of the spent fuel.

- All these values can be easily calculated from a small amount of data (main figures and orders of magnitude) given by an assembly calculation (with APOLLO, from instance ([6])), from the fresh fuel to the final burn-up, under a critical buckling calculation assumption. The main data can also be derived from Table 13.1 by simple interpolations. These calculations are the subject of a tutorial, the results of which are presented in the Annex 1 at the end of the chapter.

Table 13.1. Spent nuclear fuel composition.

System (loading)	PWR (UOX)	PWR (UOX)	PWR (UOX)	PWR (MOX)	PWR (UOX)
Discharge burnup (MWd/t HM)	33000	55000	60000	36500	40000
Initial ^{235}U concentration %	3.25	4.5	4.95	0.25	3.55
Initial Pu concentration %	–	–	–	5.3	–
Final Pu concentration %	0.95	1.28	1.29	3.97	0.9
Fissile Pu (%)	70	64	64	59	60
% ^{238}Pu	2	4	4	3	2
% ^{239}Pu	56	49	49	41	50
% ^{240}Pu	23	23	24	28	29
% ^{241}Pu	14	15	15	18	11
% ^{242}Pu	5	8	9	11	8

Let us now consider how, as soon as the actual cost of uranium would definitely increase to levels justifying a related increase in thriftiness, one can decrease the specific consumption by a factor of two (when compared with the **present worldwide** mean value), thereby multiplying the "nuclear efficiency" by two. This can even be achieved keeping the traditional 17 × 17 lattice in PWR, with an optimised in core fuel cycle management, reduced depleted uranium enrichment and mono-recycling,

Improving (stepwise) the use of the natural uranium resources and the nuclear efficiency in LWRs without core modification (lattice pitch and moderation ratio, clads, structures), with the exception of a more robust fuel and a safe reactivity control at an increased burn-up.

- Reducing the tail enrichment in the front end of the fuel cycle will be a direct consequence of the natural uranium cost increase. The way it works is shown in the Figure 13.10: the optimal tail enrichment is calculated from the ratio of Natural Uranium cost in $/kg (UF6) over Separation Working Unit cost in $/SWU.

During a long period of time, the ratio was about 0.5 (40$/ kg Unat after conversion into UF6, for about 80 to 100 $/SWU), and the related optimal tail enrichment was about 0.3%.
At the present time, and for known long term contract cost, the ratio is closer to 1.5 (around 130 $/kg U (as UF6) and about 100 $/SWU), so the optimum enrichment has moved below 0.20%. In the long run, the ratio could go as high as 2 or a little higher due to

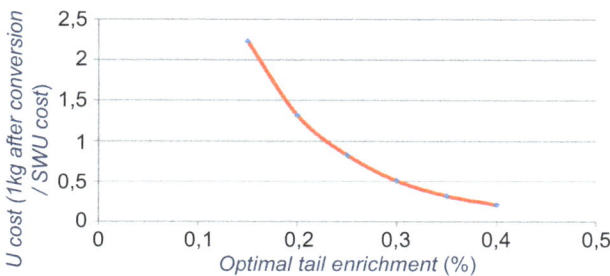

Figure 13.10. Optimal tail enrichment calculation.

the increase in cost of uranium, without jeopardizing the global nuclear competitiveness, and a low 0.15% value could be reached.

Then, between 0.3% and 0.15%, starting from 0.71%, the decrease in natural uranium consumption for the same power production could be 25%, for typical enrichment fuel.

- With a slightly reduced specific power (the EPR trend for instance, as for PWR), and with a typical enrichment value of around 4.5% to 5%, a reasonable cycle length can be combined with a fractioning by 1/4 at the expense of a higher mass of fuel.

- A further reduction of leakage and of sterile capture can be achieved in a very large core with a thick reflector, and with low capture clad and structural materials, but retaining plenty of sterile capture in diverse reactivity control poisons. In BWR with pumps, a spectral shift can substitute fertile capture to sterile capture. There is no equivalent solution in PWR at the present time. It would require to move heavy and bulky structures through the core and this raises a big challenge with limited incentive.

The global result of these first steps, in one through cycle, would be close to 135 t NatUr/GWyr-el, with: $e = 5\%$, $\tau_s(4) = 64$ GWd-th/t, plant efficiency: 36%, e_r(tails) = 0.15%.

- The last improvement – the (mono)-recycling of the plutonium – is currently used in France and Japan. It can be completed with the recycling of the residual ^{235}U left in the spent fuel.

For instance, in the projected 5%/64GWd-th/t fuel cycle, the final plutonium content is about 1.29% total Pu with 64% fissile content, including 15% of ^{241}Pu. During out of pile cycle (let us say about seven years, for a 14 year half-life, leading to a 30% decay and consequently 10.5% ^{241}Pu left and globally 59.5% of 1.29% = 0.77% fissile plutonium. By the same final burn-up, there is still 0.75% ^{235}U in the spent fuel. This value must be downgraded, because the ^{235}U is poisoned by other isotopes. So, globally, when using more accurate equivalence between fissile nuclei in a thermal neutron spectrum, the fissile nuclei to be recycled (when compared to the make-up of pure ^{235}U at 5% enrichment) mean natural uranium savings of about 20%. This leads to a net consumption of 135 × 0.8 = 108 t NatUr/GWyr-el; and a "nuclear efficiency" close to 1%.

All the previous improvements can be implemented around 2030 in a renewed fleet of modern LWRs and with conventional fuel cycle processes and facilities (with the important exception of new processing and MOX fabrication facilities for plutonium recycling).

13.10.2. The last decades: fluctuations in the objectives, shooting on a mobile target

Spectral shift in conventional lattice BWR

In a BWR, the transformation of sterile captures needed for the cycle reactivity swing control into fertile captures leading to additional plutonium production usable at the end of cycle, has been routinely achieved by varying the pumping power, hence the mean void fraction, instead of moving the solid absorbers through the core. This "spectral shift" makes elegant use of the two phase flow regime in the core, saving natural uranium and SWU (see the chapter dedicated to the BWR specific features).

If the search of a harder spectrum is easy to understand today, it was more the AMR **(Augmented Moderation Ratio)** that was a trendy thing during the late 90's.

APA advanced fuel for reduction or even minimisation of the amount of plutonium in PWRs.

At the end of the 1990's, in a relatively uncertain environment concerning nuclear power and its sustainability, particularly in the European context, a persistent question remained relating to plutonium: was there, with standard PWR, a way to avoid the build-up of unused plutonium, or even to reduce the global Pu inventory (Figure 13.11)? Several solutions were compared, from multi-recycling on slightly enriched uranium (the most industrially mature way), to more sophisticated CORAIL type assemblies using conventional fuel rods.

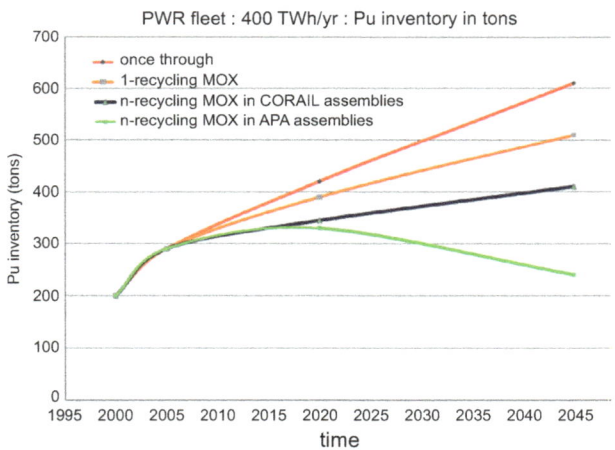

Figure 13.11. Some Plutonium management strategies considered around 2000.

The most innovative concept, called APA, reflected the desire to concentrate the plutonium in a minimum number of objects to be fabricated, as well as in a minimum number of reactors. The principle is illustrated in Figure 13.12. The annular rods containing the

Figure 13.12. APA fuel assembly.

plutonium occupy a "supercell" equivalent to 2 × 2 standard cells, in an *Augmented Moderation Ratio* zone.

Recent trends in BWR fuel. In the single phase region of the core, the BWR moderation ratio is greater than in a PWR. In the upper regions, as a result of the high void ratio (typically 65% at the outlet), the situation inside the casing is radically different. Reference was earlier made to the use of the spectral shift by variation of the void ratio, in order to compensate for the extra cost of BWR fuel through improved utilisation of the natural uranium.

Currently, the mean burn-up and enrichment are increasing (average enrichment 4.6% corresponding to a maximum enrichment of 5% and to necessary zoning to limit power peaks near water gaps and to control the shape of the power distribution in 3D during the cycle). Correlatively, diverse designs (hollow cross or square atrium at the centre of the casing) provide a single phase tube lane from the bottom to the top of the core. This increases the moderation ratio in the upper part of the core and distributes a thermal neutron source at the centre of the fuel assembly. It thus avoids supply of thermal neutrons **only** from periphery (intercasing space for control cross circulation) which causes sharp power peaks which are overcome by enrichment zoning. Water rods had already been used previously in a similar role. This design provides greater benefit, in an open cycle or in "lazy" recycling, where the make up of fissile material eventually produces most of the energy, enabling the offset of the over-costs of the BWR fuel fabrication at the front end of the cycle.

Flexibility. It can thus be seen that optimisation of the moderation ratio depends on the context. To date, thermalhydraulic-neutronic equilibrium in a LWR is established on the basis of moderation ratios in the region of 2 in single phase or equivalent states. Opportunistic fluctuations have been observed, ranging from an AMR tendency to spectral shift. A more radical trend towards reduced moderation ratios is conceivable in the context of multi-recycling with due regard to preserving the quality of the plutonium and to greater thriftiness. The consequences for capital and operational costs need to be minimised.

13.10.3. The state of the art regarding the limits and the trends for the burn-up and for the recycling of plutonium

Increasing the burn-up beyond 70 GWd/t: a challenge without decisive fuel cycle incentive?

At the present time, increased burn-up of uranium oxide fuel still appears to be a long term trend, but substantial technological problems remain to be overcome, particularly as concerns the "robustness" of the qualification of fuel assemblies at high burn-up. As the PWR fuel assemblies are open (no casing) and subjected to lateral flows, severe thermal-hydraulic and thermomechanical compatibility conditions are imposed and new failure modes may emerge with time. Qualification through gradually stretched operation feedback and semi-empirical improvements have made it possible to rapidly and economically achieve 30 then 40 Gwdays per ton of heavy metal. The successful semi-empirical method becomes a longer, more unpredictable and costly process beyond 50/60 GWd/t. Numerical and experimental simulation of the fuel **and** the fuel **assembly** will need to be improved to meet this challenge (the "numerical core" target). In BWRs, the presence of casings that divide up the core and isolate the fuel assemblies avoids the need to meet the whole set of compatibility conditions. This is one of the reasons for the faster and daring development of BWR fuel, as is necessary for competitiveness in view of the high fabrication cost. Neighbouring fuel assemblies may exhibit very different axial profiles, with fuel rods of partial length ("Manhattan skyline" profile), water rods or internal hollow crosses or atriums, and even different numbers of rods and rods of different sizes, all within a common boundary condition: the core pressure drop Δp. Furthermore, the small size of fuel assemblies (which is an advantage for fine tuning of fuel shuffling, but contributes to complexity and cost) makes it possible to perform, particularly with two phase thermalhydraulic conditions (the essential issue in view of the strong neutronic coupling), parametric experiments on "full scale" representative thermal-hydraulic loops with a fuel assembly of a given design (unlike in PWRs). However, the long-term behaviour of the casing, the guaranteed gaps for moving the control crosses, constitute delicate issues.

To summarise, increasing mean burn-up around 70 Gwdays per ton of heavy metal ($\geq 5\%$ enrichment, then, above 5%, efficiency of further increasing the enrichment would max out) remains a medium-term perspective. Such a level of spent fuel burn-up would not discourage competitive single-pass recycling in thermal spectrum of the fissile material contained in spent fuel.

Plutonium (and residual ^{235}U) mono-recycling

Beyond the concerns expressed relating to proliferation, the recycling of plutonium has been the subject of criticism in several studies (e.g. Harvard), unless costs of natural uranium above $ 200 per kg are reached. The French analyses, and that of the OECD, agree on lower natural uranium "cost indifference" level.

At around 5% enriched UOX, the savings in natural uranium represent around 20%.

13.10.4. What could be the next step?

As for plutonium isotopic composition and for the saving of natural uranium, **multi-recycling in the thermal spectrum is not helpful. The plutonium is degraded and ultimate energy production rapidly peaks.**

It is necessary to change over to another mode of multi-recycling where most of the energy is produced by the recycled materials, where the quality as well as quantity of the plutonium is preserved, once the strategic inventory necessary for efficient starting of a fleet of fast reactors has been built up, with the plutonium remaining in the useful loop and continuing to produce revenue and energy. This advanced or "smart recycling" assumes availability, at least in some places in the core, of a harder spectrum.

A short analysis of an advanced multi-recycling mode in LWRs leads to the following results:

- *Strengths, advantages*:
    - Actually closing the fuel cycle with multiple plutonium recycling: no more "available on the shelf" plutonium and a flexible management of the plutonium inventory and of its isotopic quality; correlative reduction of potential radiotoxicity of the spent fuel as an ultimate waste (plutonium being the main source of it, in the U-Pu cycle).
    - Preserving the quality and quantity of plutonium through seamless use and valorisation (preparing the transition to Gen-IV);
    - Reducing the cycle reactivity swing with less sterile capture control: safety and operation advantages (reducing the role and the related risk of boron as well as of mobile absorbers); plus gains in neutron balance and conversion ratio;
    - Increasing the conversion ratio, reducing the natural uranium consumption.

- *Drawbacks, weaknesses*: they are mainly related to the following issues:
    - Reactivity coefficients (void coefficient); solutions are at hand but the issue must be kept in mind;
    - Large amount of initial fissile mass required: cost and out of pile fuel cycle issues, as well as reactivity and criticality management issues, similar to those for FBR (solutions start from the same principles, too).
    - Acceptation of active industrial and commercial plutonium management, including transport, spent fuel processing and fabrication, fissile material swap. Same as for FBR.

- *Opportunities*
    - In an uncertain world, and with a growing pressure of some newcomers on uranium market and resources, nuclear plant vendors will have to explain to the utilities how they will operate competitively well beyond the turn of the century their future LWRs built and commissioned around 2050.
    A reduced uranium consumption will be an asset, the best solution being the convertible reactor, able to start operating with a "light" (in terms of fissile material amount) first core and produce plutonium, then to shift to a higher conversion

ratio, thriftier operation around mid-life, only modifying internal core structures, and with a complementary make-up of fissile material (namely of ^{235}U).

- A question will arise, worldwide, about what would be the "missing link" between Gen-III and Gen-IV, in order to strengthen the credibility and the sustainability of the overall process. The thrifty LWR would take a medium term insurance or option value.
- The convergence with operation and safety driven innovation (fuel assembly innovation, reducing the risk associated with the control of the reactivity swing, advanced clad material) could help.

- **Threats**

  - There is still no "market pull" effect, just the interest of a few utilities and vendors.
  - A closed fuel cycle is not favoured by many powerful players. But it could be a roadblock for any type of breeder and for Gen-IV strategy as a whole, being a common mode failure.
  - The cost and the industrial management of the first core could quench the potential interest, apart from utilities supported by "having" countries (countries that have access to spent fuel processing and plutonium management).

Useful conversion ratio performance targets

The first priority is avoiding a substantial degradation of the plutonium isotopic quality. This requires a reduction in the moderation ratio from about 2 to about 1

The assessment needs a relevant criterion. It can be taken from the comparison of multi-recycling with the decay of spent fuel plutonium waiting for **further** use over 30 more years. At the starting point (t = 0), used fuel plutonium is unloaded from an equilibrium cycle in a tight lattice LWR, and the comparison begins after about seven years of out of pile cycle. Figure 13.13 shows that rapid processing and reuse are better than pure decay, because the isotopic quality is "rectified" during the in-core burning, while it declines monotonically during the out of pile phase, due to the ^{241}Pu decay. Recurrent recycling therefore has the beneficial effect of periodically "rectifying" the plutonium composition.

As for assessing the fuel cycle gain from efficient multi-recycling, one can start from the point previously defined in the primary requirement (limiting strongly the plutonium degradation) and check the added value from the natural uranium saving viewpoint.

To do this, one needs a curve relating the natural uranium consumption at the equilibrium cycle, to the conversion ratio typical of a given moderation ratio. One must take into account the fissile material management from the beginning, including the first core and reloads, in the same way as in fast breeders. The quantity of fissile atoms necessary to start a high conversion ratio LWR is comparable to the quantity required for a FBR, which means about four times more than for a conventional core LWR, assuming simplified equivalences between different fissile isotopes in diverse spectra.

For example, the mass of heavy nuclei in a typical, loosely packed standard lattice, 1300 MWe (installed) PWR is 104 t, giving an order of magnitude of 100 tons of heavy

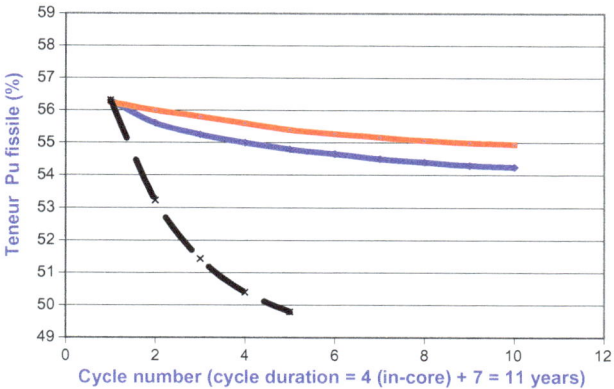

Figure 13.13. Fissile plutonium concentration at "BOC" and "EOC" as a function of the cycle number during multi-recycling in reduced moderation ratio LWR, compared with on the shelf degradation due to ^{241}Pu decay. The "cycle" includes the in-core "rectifying" period under burn-up, as well as the out-of-pile cooling down, processing and fabrication (plus transportation) pure decay period.

nuclei for 1 GWeff (1 GWyr/yr) capacity. In EPR, the linear power is slightly reduced, so the mass of heavy nuclei is slightly increased when compared to the 1300 MW-el situation. In a BWR, one finds about 150 t/GWeff. The relevant data are the mass of fissile nuclei invested and, finally, the equivalent natural uranium mass required. For instance, for the 1300 MW-el PWR (1 GWeff), a first core and the first reloads would mean around 1000 t NatUr.

As for the equilibrium cycle natural uranium consumption rate, a simple approximation relates the average conversion ratio during the cycle (c) to the make-up of fissile nuclei necessary to produce a given quantity of energy (Q), according to: $Q \propto (1 - c + \varepsilon)$, where ε represents loss and degradation (including decay of ^{241}Pu into ^{241}Am) and is arbitrarily estimated as being in the region of a few % for typical out of pile cycle.

The formula assumes that a handful of cycles are completed, so that the fissile material introduced in the first cycle can be recycled and finally absorbed, which does not necessarily occur during the first cycle (there is a significant quantity of initial loading left in spent fuel) and so on for the next quantities produced by conversion, with a reasonable cut-off.

Under this assumption, loading one atom of fissile material, it will finally be absorbed and thus will produce $1 \times \varepsilon/(1 + \alpha)$ fission as well as, by definition of the conversion ratio, c atoms with similar $(1 + \alpha)$ ratio, entering the same process of absorption, fission, production, along the many cycles. There will be an amplification effect: $1 + c + c^2 + \ldots + c^n + \ldots$ leading to $1/(1 - c)$. Actually, taking into account the losses, the out of pile fuel cycle returns only $c - \varepsilon$, to be reloaded in the core, and finally the effective value of c is reduced to $c_{eff} = c - \varepsilon$.

The corresponding chart is displayed in Figure 13.14.

The shape of the curve having been established, the next question is to determine where the simple approximations make it possible to fix an "entry" point on the curve.

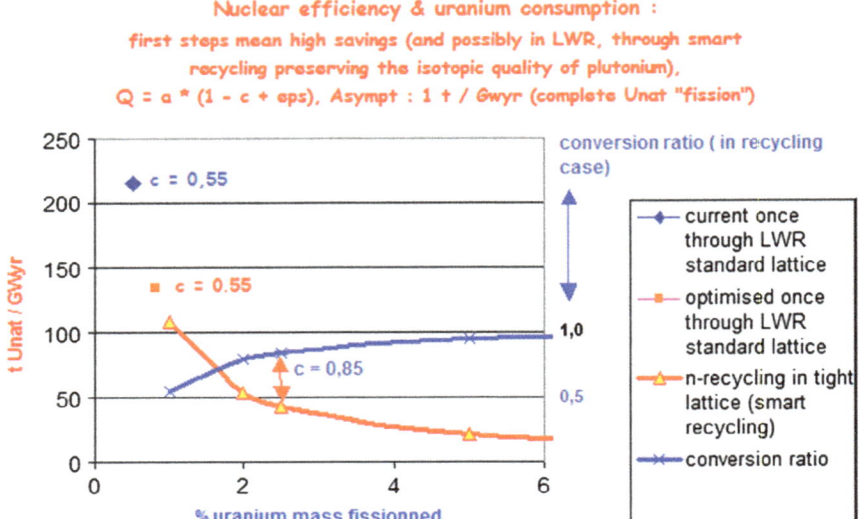

Figure 13.14. Nuclear efficiency and uranium consumption as a function of the conversion ratio.

Obviously, the well-known reference points in once-through are not located on the curve. They just set up a loose framework.

In a LWR with a standard lattice, c/c_{eff} is generally estimated to be around 0.6/0.55. Applying this formula, taking 0.5% of ^{235}U from the extracted mass of natural uranium by the enrichment process leads to $0.5/(1 - 0.55 + 0.05) = 1\%$, which corresponds to the rough assessment of the limit of multi-recycling with a standard lattice (approximately 100 tons of natural uranium per GWyear) – see above.

When c = 0.80 (typical PWR target?), 2% is obtained, i.e. 50 kt/GWyear).
When c = 0.85 (typical BWR target?), 2.5% or 40 kt/GWyear.

The last part of the analysis is to set up a scenario under the (debatable) following assumptions: mixing the continuing construction of LWR and the As Quick As Reasonably Achievable development of a Fast Breeder Reactor fleet, worldwide.

In any case, the limits in applying the formula are related to the fulfillment of the necessary conditions:

- the mix of fissile nuclei must be known; generally, it associates a steady recycling of plutonium with a constant make-up of fresh enriched uranium, and with used uranium recycling;

- there must be no quick degradation of plutonium with successive recycling. This is a necessary condition for the development of the series of conversion productions: c, c^2, etc. This is why the entry point on the curve can't be calculated from the UOX cycle in conventional pitch, loosely packed lattices. Limiting the degradation of plutonium quality with recurrent recycling is thus the very criterion for efficient multi-recycling. It requires a moderation ratio close to 1 (vs. 2 for conventional LWR lattices) and guarantees a conversion ratio of at least 0.75 to 0.8;

- the α value of the fissile nuclei must be similar, which is not too far from the real situation.

These figures can be compared with those from other reactor types, typically:

- natural uranium gas cooled graphite moderated reactor: c # 0.8;
- standard lattice PWR: 0.6;
- tight lattice PWR (RSM) (45 GWj/t, 10% Pu): 0.8;
- fast breeder without blanket (RNR-Na, 100 GWj/t, 20% Pu): 0.85;
- fast breeder with external radial and axial blankets: 1.2.

Updated FBR design leads to higher – close to 1 – conversion ratio in heterogeneous core and no blankets (at least, no radial blankets, but the heterogeneous concept means "inner blankets").

The next issues to be addressed by designers concern safety, performances and competitiveness.

13.11. A stepwise transition, a synergistic cohabitation: defining a flexible scheme for a sustainable nuclear fleet growth rate, worldwide, and transferring fissile material to the future through continuous valorisation

13.11.1. Introduction

As for the transition from Gen III to Gen IV when considered as a whole, several questions arise, depending on the players and on their specific viewpoints and interests: utilities, fuel cycle companies, vendors, investors, fuel states and reactor states. These players/"actors" have to take decisions on a day to day basis (keeping a plant operating, securing short term uranium and fuel supply), but also on a longer term perspective (nuclear plant investment, fuel cycle option choice) and finally from a global vision. For most countries and companies, entering the nuclear energy game around 2020 will depend on their confidence in "sustainable" acceptability of nuclear power, along with access to natural resources and to the availability of "enablers" like breeders as well as fuel cycle processing and fabrication plants, or alternatively to an indefinite supply of cheap nuclear fuel and of adequate back-end fuel cycle services, while not alienating their political independence.

Aside from the safety considerations, three conditions have to be satisfied for nuclear energy sustainability:

- authorised, regulated and controlled worldwide fissile material management and trade (including tens of kt of plutonium), in the framework of a closed fuel cycle;

- adapted (thrifty fuel cycle options), acceptable and competitive reactor types;
- fuel cycle facilities for a closed cycle with efficient and "clean" processes.

Competitiveness has to be continuously ensured, whatever the time scale.

In an oligopolistic market, a strong interplay between major players integrating several links of the value chain can lead to powerful alliances and strategies. These connections may well prove to be mandatory on the long term, isolating and discouraging many potential players and so reducing the pressure on natural resources by reducing the pace of the (desired and useful) nuclear power growth.

The last part of the chapter will address a few typical issues related to diverse scenarios and strategies.

13.11.2. How to manage, from the uranium extraction rate viewpoint and from the nuclear plant type viewpoint, a strong nuclear energy growth after 2025/2030?

Growth scenario

Assuming a linear shape (preferred to an exponential one), a typical "slope" would correspond to commissioning more than 1000 GWeff (GWyr/yr) – or equivalently 1200 GWel – newpower capacity every quarter of a century, in order to get an **additional** capacity of 40 GWeff (50 GWel) every year (because the old reactors have to be replaced).

Natural uranium cost: the estimates still strongly vary with time; a few examples

In order to define the nuclear plant type strategy, a pattern of natural uranium cost depending on the total uranium mass extracted is required. Figure 13.15 shows two models.

The first is the simplified law previously defined (in 2004) from general considerations about geoscience and extraction technologies, consistent with the Harvard and Gen IV Crosscut Group reflections: $M = 10 \times (c/130)^{1.6}$ with M (natural uranium mass extracted at the time t, in Mtons) and c (cost, in $/Unatkg). The curve is extrapolated up to 40 Mtons, with its smooth pattern.

An updated (in 2010), simpler – linear – law could be: cost ($/kg Unat) = 50 + 8 M, with M in Mt Unat extracted, leading to 210 $/kg Unat for M = 20 Mt.

The second law has been used in techno-economic studies led by the ITESE (Institute for Techno-Economic studies of Energy Systems) in CEA. This law shows a yield, around 35/40 Mtons, with a plateau at 1000/1100 $/kg, tentatively representing the extraction from sea water. So, over a range of a few Mtons, the cost is supposed to jump from 300 to more than 1000 $/kg.

What could be the consequences of such a jump on the decisions about building a new type of reactor (namely a fast breeder), assuming that such a choice would be authorised all around the world, or at least to the players able to contribute to the largest part of the nuclear power growth?

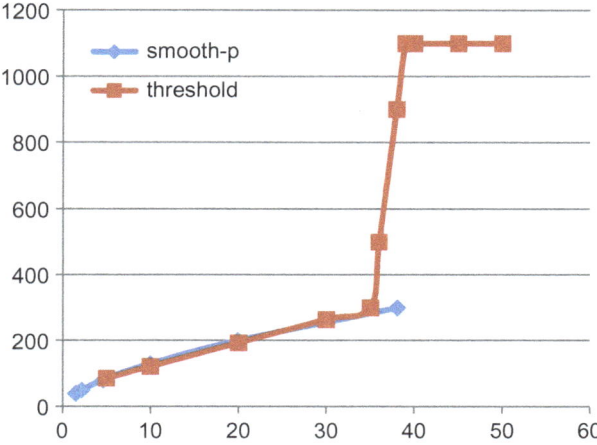

Figure 13.15. Natural uranium cost pattern: a smooth "toy" model *vs.* a "threshold" model: cost of uranium in $/kg U as a function of the total uranium mass already extracted, starting around 2010.

As for the near future, J. Hinze, from Ux Consulting, quoted by Platts, *"said there is still a lot of uranium in the ground around the world, but asked, "will it flow fast enough ?"*. The cost ceiling will be determined by the balance between speculation and the threat of a loss of competitiveness of nuclear power versus "king coal" for baseload electricity generation.

Some remedies can slow down the process. In the long run, more powerful drugs are required: **the breeder belongs to this class of antidotes and it has a tremendous value as an option, which justifies a large R&D effort, including an "ASAP" build up of operation feedback.** Which leads back to the question: what will be the right time to decide to only commission breeders as power reactors? Or will the "ultimate" fleet be symbiotic, with an optimized support ratio between fast breeders and advanced converters (mainly LWR, maybe gas cooled reactors)?

One must keep in mind that a general trend – as for raw materials – is to shift to the advanced technological alternative only after the primary resources have been exhausted and the related profit has been taken. The same holds true for the competition between two materials when the depletion of the first one leads to a substitution by the other.

Recent studies carried out by ITESE give an updated, parametric assessment of the evolution of the natural uranium cost as a function of the total mass already extracted (Figure 13.16). The main improvements are:

- the parametric assessment of the thresholds and of the slopes, related to several assumptions covering the most probable range of variation of the driving parameters;
- an updated assessment of the cost of uranium extraction from the sea.

The main consequences of the update, partly due to the time elapsed (about a decade) between both assessments, in terms of current $ value as well as of conventional starting year, are:

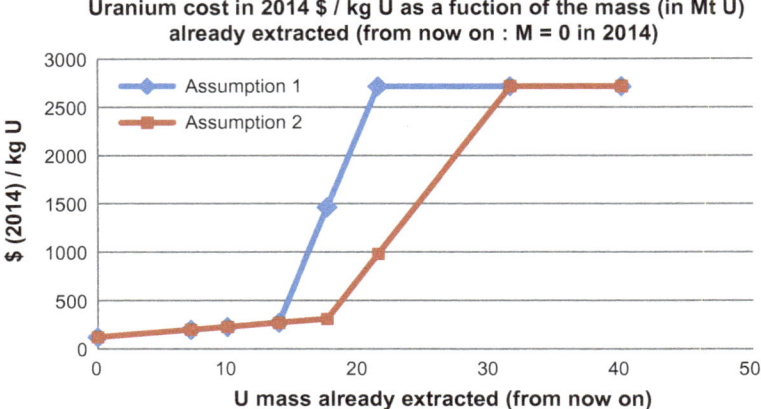

Figure 13.16. Uranium cost (in 2014 $) as a function of the mass already extracted (from now on): from a parametric analysis by CEA/ITESE.

A faster increase in cost (now in 2014 $) and a "cap" put in place by the extraction from the sea of around 2700 $/kg rather than 1100.

In any case, at the present time, two figures seem to play a key role, at least for the purpose of brainstorming:

- the extracted mass related to a uranium cost of 500 $/kg, which means entering into an uncomfortable zone at the strategy planning level; this threshold would be crossed at around 15 Mt under assumption 1, and at around 20 under assumption 2. Incidentally, it gives an incentive to prepare a reduction in the consumption of the future Gen-3$^+$ reactors (belonging by definition to the Gen-4 reactor fleet, after 2040, as competitive "power reactors"), by a factor of two to three, first reducing the tail enrichment and preparing an efficient multi-recyling capability (see "smart recycling");

- crossing the 1000 $/kg line: beyond this line, even for fully amortized plants, after 20 to 30 years of baseload (or close to baseload) operation, the competitiveness of the once-through cycle is questionable. At that time, the busbar cost would no more be driven by the reactor capital cost alone.

On the other hand, the geochemistry and the steady progress of the extraction industry strongly support the assumption 2 pattern which illustrates, roughly speaking, a higher mean value of the elasticity exponent in the considered extracted mass range, at least up to 30 Mt U, shifting away from the conventional Red Book accounting standards interpretation.

Last but not least, the cost of the uranium extracted from the sea soared from 1100 to 2720 (2014 $). This figure has to be balanced by the fact that only favourable locations would be exploited (with a strong current and in "warm" marine waters). These necessary conditions limit at the same time the cost of uranium extraction and the annual flow of worldwide production from this type of process. A detailed assessment has still to be carried out taking into account the related trade-off.

In any case, engineered systems working as plutonium or ^{233}U factories by implementing the nuclear energy synergetics in hybrids (ADS or fusion – fission hybrids, see below) would put a cap at a lower cost level. Neither are they likely to be needed in the current century.

A reduced uranium consumption vs. an increased first core inventory: a difficult trade-off

A conversion ratio of around 0.8 could reduce the (equilibrium) specific consumption to about 50 t NatUr/GWyr-el.

On the other hand, a significant drawback is the high fissile mass for the first core and then the first reloads, waiting for the recycled plutonium and uranium to be available. This mass is comparable with the mass for a breeder of similar installed power. It could be a blend of fresh enriched uranium (LEU) and of available plutonium.

If the plutonium is available "for free" (as a fissile material, not as fuel ready to use), it can be a good opportunity. In other cases, this is a large over-cost, and closely related to capital cost before the revenues from the first kWh of power production.

13.11.3. Competing options around 2040–2050 for the utilities and for the countries launching a large fleet of nuclear reactors

A first constraint is related to the fuel cycle front end, namely to the maximum natural uranium annual extraction rate, Q, and to its derivative, representing fluctuations in production and demand.

Around 2030, several scenarios (see WNA) are considered, including worldwide productions close to 100 kt/year. But what about doubling this rate? The data released concerning some Namibian projects show what could mean a rush to launching plenty of low concentration, low unit capacity uranium mines intended for a short operation time, due to high uranium prices making them competitive. A kind of industrial and financing "overheating", related to an increased rate of opening and closing deposits and probably also related to cost fluctuations could be generated by a sort of "Hubbert" peak uranium similar to the peak oil. It could be wise to put in place a cap in order to prevent that risk and to define arbitrarily Qmax = 200 kt/yr: at any time t, $Q \leq Qmax$.

The strategic option around 2050 is largely open to recent (but powerful) newcomers such as China, which has not yet have time to accumulate plutonium (more than 50 years of LWR operation and plutonium storage (or single recycling, like in France)). These newcomers could start new reactors with LEU plus the amount of available Pu. The breeding performances will be altered but the fleet would be started.

It is thus necessary to compare the strategies based on early introduction of fast breeders and on a "conventional buffer" of a first generation of converters (likely to be LWR), taking into account the uncertainties concerning the potential instability of the uranium cost, and specifically:

- when the annual rate of extraction increases over a "uranium Hubbert-like peak" for affordable costs (up to 500 $/kg Unat without **major** adaptation of the fuel cycle);

- when the cost of available uranium **"forever"** climbs the cliff of Figure 13.16 (at least over 1000 $/kg) and reaches the cost corresponding to seawater uranium (or from engineered system processes, for instance fusion – fission hybrids – see below – if the seawater uranium has definitely been taken out of the play, apart for feeding a pure breeder fleet).

On the other hand, if the valorisation of the plutonium is not immediate, its value suffers two drawbacks: physical degradation by ^{241}Pu decay ($T_{1/2}$ = 14 years) and discount rate effects, as a dormant asset.

The larger amount of fissile material necessary to start the fast reactor and the tight lattice LWR are detrimental to their competitiveness, unless these materials can be considered as having been given "for free" (which is usually the case for fast reactors in "having" countries). If this advantage is not ensured, the benefit of reduced consumption comes too late to compensate for:

- the higher capital cost incurred from the "heavier" core initial loading in the tight lattice LWR;

- the same cause plus the higher capital cost of the plant for the same electricity production capacity in the fast reactor case, at least with the current design, as well as with its extrapolation.

Benefits coming at a time when the plant investment is amortised, a long time from the financial and risk taking viewpoint, makes the early "power reactor" option risky for the breeder, as long as the uranium cost is not **definitely established** well above 500 $/kg.

For instance, what happens to the owner of a LWR 30 years after plant commissioning, if the uranium cost jumps to 1000 $/kg for the second half of the plant lifetime?

The competition is with baseload coal plant, which is assumed to be also amortised, as for its (significantly high) investment cost. In this case, the comparison is limited to the sum of O&M and of fuel cycle cost. The O&M cost of the coal plant is lower than that of the nuclear plant. It is therefore necessary for the latter to have the lower fuel cycle cost. It will be assumed that the coal cost is 70 $/ton, with a low 30 $/t (C) carbon tax (less than 10 $/t CO_2), or equivalently 100 $/ton without any carbon tax (nobody wants to hurt; neither coal vendors nor core users, see the European case). This leads globally to 100 $/tce (tonne of coal equivalent) which is an affordable cost.

1 GWyr ~ 3 Mtce (300 M$), and 130/150 t NatUr – thus 130/150 M$ – in an optimised once-through cycle with a low tail enrichment for depleted uranium.

Even taking into account the other fuel cycle cost components (conversion, **enrichment**, fabrication and back end), the nuclear once-through option is competitive, after the investment amortisation has been achieved on both sides: coal and nuclear.

A convertible reactor running at an almost preserved power rating during the second half of the plant lifetime would be a very attractive concept in this perspective. The adaptation would involve the fuel elements, the core internal structures and the reactivity control means, with a minimum impact on the power rating and the operation envelope. Due to the presence of a casing separating the fuel elements (which only share the core pressure drop), the BWR allow different types of fuel elements to be loaded in the same core, which would likely enable a smoother transition.

The adaptable LWR would combine the following advantages:

- a low initial fissile inventory,

- an increased thriftiness during the second half of its lifetime,

- the (self)-build up of additional amount of fissile materials required to switch from the first to the second mode of operation (from standard lattice once-through to smart recycling in tight lattice) during the first half of its lifetime. Nevertheless, a complementary make up of fissile nuclei (^{235}U from natural uranium) would be necessary at the "conversion" time. It could be anticipated and fairly provisioned through at least two decades of operation.

The adaptable LWR would be the best compromise between a low initial fissile inventory and an improved utilisation of uranium resources, encouraging some plant owners who do not have an initial stockpile of plutonium or uranium to enter into the nuclear power business with a first generation of LWR, and initially with a standard lattice and a low amount of fissile material.

For "having" countries and utilities, the tight lattice LWR can help to wait more securely and flexibly for operation feedback and improvements in the competitiveness of fast reactors, while preserving the quality and the quantity of already produced plutonium. Moreover, following this path, there is no plutonium left available outside the active fuel cycle loop.

A possible transition scenario from advanced LWR to FBR is shown in Figure 13.6, respecting a sustainable growth rate and simultaneously a maximum uranium extraction rate.

On the other hand, and from a uranium resource viewpoint, it would be possible to generate and maintain a growth rate of 40 GWeff/yr for pure breeder construction, as soon as they are mature, competitive, and backed by a significant operation feedback (even in a "burner mode" instead of a "power reactor mode"). This growth rate would be achieved in a "uranium driven" mode. The slow "breeding driven" component of growth has to be added. This option leaves a few dozens kt/year for existing LWRs and it is still possible to launch a minority of conventional LWR in parallel.

A first conclusion from this simplified analysis is that both alternative "pure" options are viable ways for a significant nuclear power growth rate worldwide. Additionally, any combination of these options would hold. It makes every individual player free to take an independent option without jeopardising the global perspective. The predominant issue will be to comply with the international nuclear order, still in slow progress amid sharp conflicts of doctrine and interests.

Another issue concerns the value of the fissile materials owned by the utilities at the end of the "transition plant" lifetime. A "dormant asset" strategic stockpile waiting over many decades for the emergence of the breeder wave is probably not the best choice. Furthermore multi-recycling in **standard lattice** LWR is rather inefficient as for thriftiness (requiring an increasing ^{235}U make-up) and it rapidly degrades the Pu quality.

Figure 13.17 shows a sketch of a long term growth strategy based on three means (initial Pu inventory, conversion in LWR and final breeding driven phase) and on a smooth transition. The direct use of ^{235}U in FBR has an efficiency of about 0.5% (0.5% ^{235}U from 0.711% in natural uranium), which is higher than the efficiency of the conversion strategy (about 0.1% Pu from the same amount of natural uranium), but in the latter case, the resulting plutonium is considered as being available for free, apart from the processing and fuel fabrication cost.

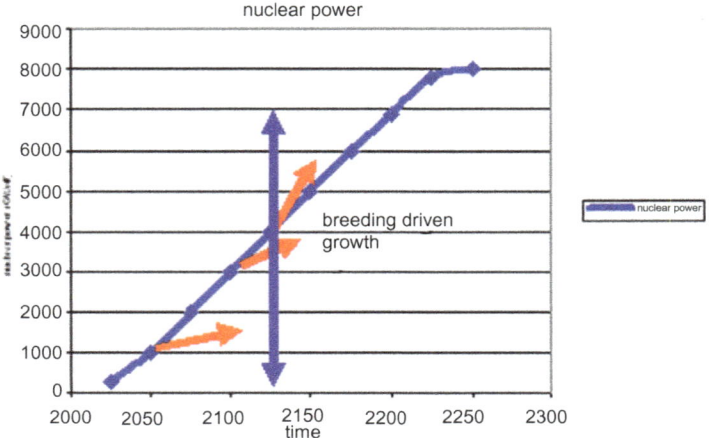

Figure 13.17. The boundary between the ^{235}U (from natural uranium) driven growth and the purely breeding driven growth periods, where the "escape velocity" of the installed fleet for a given nuclear power growth slope is obtained by breeding alone.

13.11.4. Best available technologies for "thrifty" Gen-3⁺ NSSS

There is a structural difference between PWR and BWR potential for improvement, the two phase flow bringing new opportunities:

- the two-phase flow with high void fraction in a large part of the BWR core creates the opportunity to more easily design heterogeneous 3D cores.

- the two-phase flow requires compartmentalisation (achieved by the casing surrounding every fuel element bundle), which makes it possible to load into the core very different types of assemblies (sharing only the core pressure drop, with a further tuning by individual orifice plates), opening the way to the "adaptable" concept.

Nevertheless, both PWR and BWR are candidates to "smart – multi – recycling", entailing a significant increase in conversion ratio.

The goal of smart recycling is to fulfil the following specifications:

- **significantly reducing the degradation of the isotopic composition of plutonium and enabling multi-recycling** (during the lifetime of a plant, about half a century i.e. a handful of cycles).

- **reducing the consumption of natural uranium for a given energy production**, or increasing the plutonium production in converter mode, or keeping a constant inventory of plutonium (quantity and isotopic quality) and minimising the uranium make up. The key word is **flexibility**, depending on the constraints and opportunities that can hardly be predicted at present.

A basic requirement is to avoid any modification of the plant, except for the fuel, the internal structures and the reactivity control means, as well as to avoid plant power de-rating. Taking into account the structural differences between PWR and BWR, the conversion ratio objectives are slightly different, leading to a difference in performances, which can be balanced by other advantages of the PWR.

The PWR objective seems to be within reach, as well in square as in hexagonal (densely packed triangular lattice) assemblies, with moderation ratios close to unity.

Figure 13.18 shows that hexagonal assemblies currently exist in the VVER (with a conventional moderation ratio).

Figure 13.18. VVER fuel hexagonal assemblies. ©TVEL.

As for BWR, Figure 13.19 shows some examples from Japanese studies. It seems that, for a fissile Pu stabilisation objective (which is one option among several), a realistic objective in terms of safety, plant rating and competitiveness would be a global conversion ratio close to 0.85. A 3D heterogeneous core design is a key to success. More ambitious concepts aim at a conversion ratio close to one. The usefulness of such performances seems questionable for a long time to come.

A high level of precision is required when calculating c (uranium make-up varies with 1–c), checking the power capacity, the ability to control the power distribution, the acceptability of the reactivity coefficients, the criticality analysis and finally the confirmation of the overall realism of industrial scale operation, from the point of view of a techno-economic balance integrating reactor and cycle.

Benchmarks are in progress concerning these different issues in order to initiate more detailed design studies, related in particular to materials, operation and optimisation (burn-up, reactivity control).

Finally, one has to keep in mind that a large negative void coefficient is currently a key feature for current BWR operation flexibility.

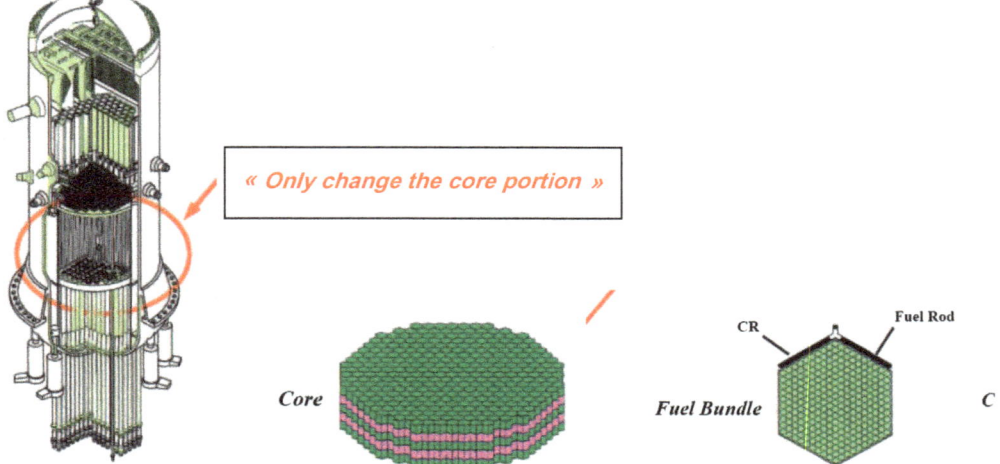

Figure 13.19. A possible mean term evolution path for BWR (after GE-HITACHI).

13.11.5. *Thorium and related strategies (basically, it is a ^{233}U issue)*

The ^{233}U η being rather insensitive to the epithermal spectrum (at least less sensitive than the ^{239}Pu or ^{235}U η) is an advantage that makes it possible to design symbiotic optimised "cross cycles" using uranium, thorium and plutonium in several types of reactors and lattices. But the optimisation is somewhat complicated and testing many inefficient options can be frustrating. Up to now, history has drawn a logical path to nuclear energy:

- **Fissile nuclei, thus natural uranium** giving rise to the build up of a huge amount of depleted uranium stored (waiting for future use; or as waste);

- Energy production (in thermal spectrum), thus **plutonium build up** and the powerful U8 – Pu synergy in fast spectrum.

If in the future it looks easier to keep going with thermal or intermediate spectrum (intermediate could mean a softened fast spectrum, hardly compatible with full Pu efficiency (η)), then U3-Th would be a powerful tool in order to:

- get an overall Breeding Ratio BR (or equivalently conversion ratio) ~1.0;

- help burning existing plutonium and some of the depleted uranium.

In this case:

- LWR would be strong and proven competitors as reactor types (and as contenders gas cooled reactors) for implementing this strategy: "incrementally improved converters may yet have their day" (E. Wigner);

- gas cooled reactors with a harder spectrum (but not a genuine fast spectrum) could do the job, if only they had operation feedback with a thermal spectrum application, for instance;

- more natural uranium would be needed in order to support the transition.

13.11.6. An "exotic" enabler from "Nuclear Energy Synergetics": fusion-fission hybrid as fissile plutonium (and ^{233}U) factories

These concepts were assessed (and compared to ADS alternatives) in the late seventies, in the wake of the INFCEP initiative launched by the US president Jimmy Carter. The objective was to offer a (conceptual) alternative to the breeders, and also to enable an early introduction of fusion technologies into the field of energy dedicated systems, being aware of the specific drawbacks delaying their maturity and competitiveness (as "power reactors") in what would be a far-distant and uncertain future. A systematisation was undertaken in "Nuclear Energy Synergetics" from Harms and Heindler, issued in 1982.

The improvements in fusion technology achieved in the last decades make it possible to update the image and the assessment of fusion-fission hybrid Pu (or ^{233}U) factories, clearly reserved for "fuel states".

One has first to determine the order of magnitude of achievable performance targets, starting from the nuclear data for 14 MeV neutrons, from the Weale experiment (Table 13.2), and from a simple image of a two zone blanket surrounding the plasma and producing energy, tritium and fissile nuclei.

Table 13.2. Main data from the Weale experiment with 14 MeV neutrons.

Nuclear data for 14 MeV neutrons			Weale experiment: **one** fusion neutron on a finite volume **natural uranium** block (reaction rate)	
	^{238}U	^{232}Th		
σ (n, f)	1.15	0.35	^{235}U (n, f)	0.28 ± 0.02
ν	4.45	3.9	^{238}U (n, f)	1.18 ± 0.06
$\nu\sigma$	5.12	1.37	^{238}U (n, γ)	4.08 ± 2.4
σ (n, 2n)	0.8	1.2	^{238}U (n, 2n)	0.28 ± 0.01
σ (n, 3n)	0.5	0.8	^{238}U (n, 2n)	0.33 ± 0.05
			Leakage	0.42 ± 0.02

A simplified model of the hybrid blanket has to take into account a limited "cover ratio" around the plasma, defined as: "useful" available blanket surface facing the toroidal plasma, behind the first wall, over the total surface. With modern, less bulky heating devices (when compared to massive heating by injection of "neutrals" considered more than three decades ago), one is led to the following performance **target**, for an improved "ITER-class" fusion device (500 MWth fusion, $Q \geq 5$), associated to fast fission blankets, with $k_p \sim 0.6$ and a thermodynamic efficiency of the energy conversion system close to 0.4 : a net annual power production around 500 MWyr-el (thanks to the generous, yet not so welcome fission reactions in the blanket) and around 1 ton of Pu per year.

The fission blankets would operate in complicated geometries and a hard environment, intertwined with the tritium process systems. The tritium issue is critical.

At the turn of the century, the Pu factory is a conceptual alternative (at a cost of between 1000 and 2000 $/kg of equivalent natural uranium for use in fast spectrum) to the extraction of natural uranium from the ground or from the seawater. There is a structural difference between the two ways of fissile material extraction/production. Hybrid Pu factories (including ADS based ones) are engineered systems, with a limited environmental footprint. They benefit from the formidable energy concentration of nuclear physics and lead to "cyclopean" but conventional industrial scale facilities, limiting the amount of re-circulated energy required by the process in a deterministic way. At the opposite end of the spectrum, with decreasing ore concentrations, or through tapping the seawater resources, one faces an increasing ratio of recirculated energy to feed the process to energy from the extracted materials, with a high sensitivity to local, natural parameters. Moreover, in the case of seawater, a large annual production rate would lead to numerous areas of the oceans being dedicated to this activity, occupying the best suited locations and stimulating environmentally based opposition. Seawater facilities, as well as hybrid factories, have stable locations. This is not the case with mines and rapid turnover will be aggravated by a Hubbert-like syndrome, leading to volatility and potential overheating in the mining industry.

Finally, both ways (low concentration ore and seawater extraction on the one hand and hybrid Pu factories on the other hand) could deliver a few more dozens of kt Pu or a few more Mt equivalent Unat.

This is an unlikely scenario but it gives an idea of the flexibility from basic physics and from technology improvements, taking a reference at 30 years ago and making a rough update.

13.11.7. FBR fleet breeding doubling time: estimates and sensitivity analysis

The strategy of growth of the worldwide FBR fleet is generally based, after the bootstrap phase where this option is too slow (whatever the doubling time), on a fleet doubling time which is a parameter of utmost importance and very sensitive to many primary parameters: physical, technological, industrial and political (proliferation resistance and transports, etc.). *A short doubling time is related to high specific power, high global breeding gain (GBG), and a short out-of-pile cycle.*

A short presentation of the main concepts and values is given in the Annex 2 at the end of the chapter. It follows the conventions and definitions presented in [3].

Quoting George Vendryes reflections in 1990 after calculating a RLDT (necessary time to the net production by breeding of the plutonium inventory of the same FR core, see Annex 2 below) of 40 years as an example ([1]): "Such figures show that the introduction and deployment of FNR is a very progressive and rather slow process".

The fleet doubling time is highly sensitive to several parameters. Safety, design optimisation and proliferation resistance modify Psp, GBG, T_{out}. The values taken in the current example are typical ones, but it is very difficult to determine what would be a worldwide FBR fleet effective doubling time around 2050–2100 and to give an estimate of the related uncertainties. For instance, if fertile blankets are prohibited, even with a high IBG (internal breeding gain involving fertile core portions) close to zero, one cannot bet on any doubling time. This is why it is important to be equipped with alternatives and possible additions to the pure FBR breeding doubling time strategy, at least for the first part of the S-shaped growth curve.

13.11.8. Conclusion

Around 2040/2050, when FBR will be ready to use, several choices will be available for utilities and for countries entering nuclear power and aiming at a large growth rate of installed capacity.

A few players (including France), having accumulated by fertile conversion in LWR a large inventory of plutonium kept in a useful and profitable closed cycle, will be able to start a fleet of FBR, directly in breeding mode or perhaps, in a dedicated transmutation mode ("burner reactor", starting with Pu: the ancient CAPRA concept) for a transition generation kept free from competition with LWR considered as "power reactors".

Among the newcomers, and specifically the "having not" (countries without fuel cycle industry or significant plutonium inventory), many will probably make the choice (if this choice is internationally accepted, being dependent on the recycling perspective) of a LWR transition generation, in order to build up a plutonium inventory by conversion, following in the tracks of their forerunners. The few remaining countries, should the initial investment (probable plant overcost plus larger amount of natural uranium for the first core and the first reloads) be affordable, could launch their FBR fleet with enriched uranium (LEU) and with a limited amount of Pu and, if they are authorised to do so, at that time could take the closed cycle path..

Considering that by 2040, a potential jump in the price of uranium to an ultimate "plateau" associated with seawater uranium or symbiotic hybrids will not be realistic (except for the hypothetical case of a strategic and forced blockage of the market), the LWR transition generation could be a profitable option for the following reasons:

- a low initial fissile inventory;

- a continuous competitiveness vs. coal, even for the end of life with a potential large jump in uranium cost;

- a continuous valorisation of the fissile material inventory running into closed cycle, combining energy production and plutonium build-up by conversion (with a fine tuning capability).

A mix of the aforementioned options, as well as each in its own right, lead to a sustainable and high rate of nuclear power growth. This is a very robust perspective in a game where powerful players will make unpredictable inputs, forcing their partners to adapt. Every individual player can follow its way without jeopardising the whole nuclear progress.

The open question is how to join without collective failure the last phase of the growth. This phase will be breeding driven (well beyond 1 TWeff), and this requires a FBR fleet of more than 1 TWeff and more than 20 kt Pu of final inventory. The extraction of 10 Mt additional natural uranium may be acceptable if the overheating of the global market and the large related fluctuations can be dampened, on the way to the "sustainable" low cost resource amplifier (breeding). The main issue is to find a path to the future for the diverse players, with different optimisations in terms of what could be their best "plutonium mine", starting (rather blindly) in the first decades of the millennium.

In any case, the enabler of nuclear power survival is **quick** fissile nuclei **recycling**, if (and only if) it is internationally authorised, possibly in regional centres.

Improvement in thriftiness of Gen-III LWR in parallel to FBR development is therefore a mandatory task for the maintenance of an improved competitiveness to any LWR launched between now and 2050–2075 and up until the end of their lives, well beyond 2100.

Last but not least, for several countries, the thorium issue will be a strategic one. This fact, connected to the lower sensitivity of the nuclear performance of ^{233}U to the neutron spectrum, could lead to symbiotic cycles based on U, Pu and Th, in diverse reactor types. This flexibility would be an advantage for naturally adaptable (up to a significant but limited spectrum index r) reactor types like LWR and gas cooled reactors. One comes back to the early reflections of nuclear energy pioneers, suggesting that "incrementally improved converters may yet have their day". The ^{233}U η curve could play a role in this competition in relation to safety design and technology challenges associated with very fast breeders and with the fissile mass inventory required in a fast breeder core.

Further studies with a broader scope will be necessary to explore these paths to a future which might be open to a large spectrum of nuclear energy users worldwide.

The rationale of the related approach (schematically illustrated in Figure 13.20) is as follows, in two steps:

- respecting two global constraints, one dynamic and one integral:

    o limiting Q (Unat), for instance Q < 200 kt/year,

    o keeping a 30 Mt extracted Unat limit at a distance, for as long as the knowledge regarding uranium resources and about technology improvement and acceptability is not strongly established and as long as the breeding solutions have not been further demonstrated (safety, operation, performances and cost);

- allowing "individual" players to make local choices (an option still open thanks to the respect of the aforementioned limits), under "individual" constraints (techno-economy, strategy, etc., as well as under the constraints of the international nuclear order still in slow progress).

This leads (**hopefully**) to an ASARA (*As Soon As Reasonably Achievable*), low risk transition to a worldwide sustainable fleet composed of breeders (fast or thermal) or of symbiotic systems (breeders + efficient converters, both with multi-recycling).

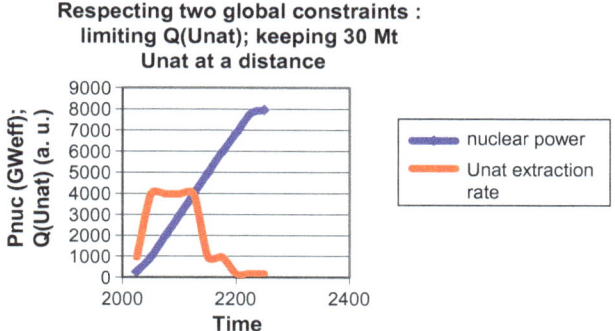

Figure 13.20. A path towards the future respecting inventory as well as flow constraints.

References

[1] G. Vendryes, *The science of fast reactors and why it has been studied*. Philos. Trans. Roy. Soc. Lond. A **331**, 293–300, 1990.
[2] J.B. Thomas, *Nuclear power plant types and the management of plutonium and minor actinides in search of fuel cycle flexibility*. C.R. Acad. Sci. Appl. Phys. **3**, 783–796, 2002.
[3] A.E. Waltar, A.B. Reynolds, *Fast Breeder Reactors*. Pergamon Press, 1981.
[4] *The Collected Works of Eugene Paul Wigner*, edited and annotated by A.M. Weinberg with the Assistance of Alfred M. Perry. Springer Verlag, Part A: "The scientific Papers" – Volume V: Nuclear Energy.
[5] K. Deffeyes, I. MacGregor, *World Uranium resources*. Scientific American **242** (1), 66–76, 1980.
[6] D.E. Shropshire et al., *Advanced Fuel Cycle Cost Basis*. INL/EXT-07-12107 Rev. 2, December 2009.
[7] R. Sanchez, I. Zmijarevic, M. Coste-Delclaux, E. Maziell, S. Santandrea, E. Martinolli, L. Villatte, N. Schwartz, N. Guler, *APOLLO2 Year 2010*, Nucl. Eng. Technol. **42** (5), 474–499, 2010.

Annex 1

A typical once-through cycle: simple calculations using basic physical data and a conversion ratio approximation

From Table 13.1 (see above, §13.10.1), taking the second column: e = 4.5%, τ_R = 55 GWd/t, τ_R being the spent fuel burn-up.

55 GWd/t mean about 7% destruction/consumption of fissile heavy nuclei (55 × 0.125) by absorption. In-between, 4.2% fresh fissile nuclei (^{239}Pu + ^{241}Pu) have been generated

by fertile capture on ^{238}U and ^{240}Pu, with a conversion ratio equal to 0.6 as a mean value from fresh fuel to spent fuel ultimately unloaded from the core (including the slow (when compared with the in-core residence time under flux) decay of ^{241}Pu into ^{241}Am (half-life around 14 years)). So the final balance should be approximately 4.5 – 7 + 4.2 = 1.7% fissile nuclei concentration in the spent fuel at 55 GWd/t.

This is consistent with the balance of residual fissile nuclei: around 0.9% of ^{235}U concentration in the spent fuel; 1.28% plutonium with 64% fissile fraction giving 0.82% fissile plutonium. 0.9 + 0.82 = 1.72%. This gives a flavour of the driving phenomena in the fuel cycle and of their order of magnitude.

Figure 13.21 shows the variation of the concentration of ^{235}U and plutonium isotopes (fissile and fertile) with burn-up, in a 4% enriched UOX.

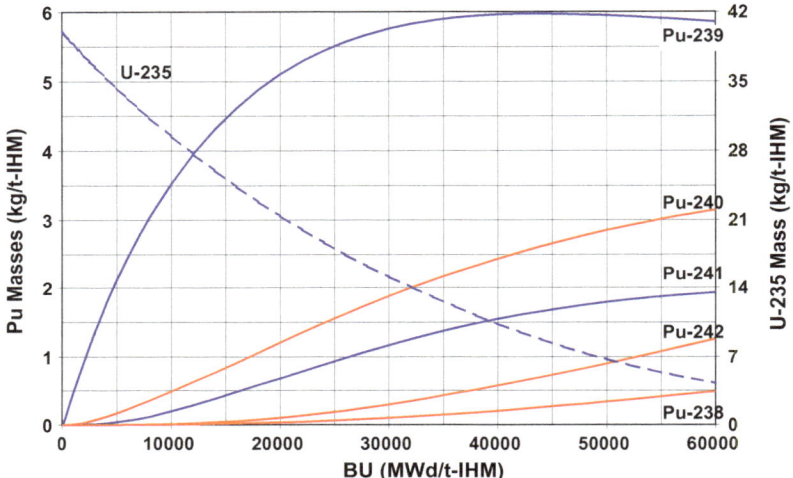

Figure 13.21. Variation of the ^{235}U concentration in kg/t(HM), on the right side scale, as well as of the plutonium isotopes (same unit, on the left side scale) with burn-up.

Simple modelling/interpretation

The following calculations are related to Figure 13.21, and therefore to a 4% enrichment in the fresh UOX fuel. The main idea is that the evolution of the fissile nuclei concentration can be represented by the following for formula (the s index being for "spent", for instance: τ_s being the discharge burnup) :

$N_{fiss}(s) = e - A (1 - C_{average(0 > \tau_s)} \to \tau_s)$, with a linear variation of c (the conversion factor) with τ (the fuel burn-up).

With A = 0.128%/GWd-th/tHN and with a linear variation of $c(\tau)$, corresponding to a mean value of 0.6 through the fuel cycle, the comparison with the $N_{fiss}(\tau)$ from an APOLLO-2 calculation ([7]) is quite good (see Figure 13.22).

The same holds true for a simplified calculation of the concentration of fissile Pu isotopes (see Figure 13.23). Then, the U-5 variation can be calculated by difference (see Figure 13.24). This type of approximation can be extended to the calculation of $k = f(\tau)$.

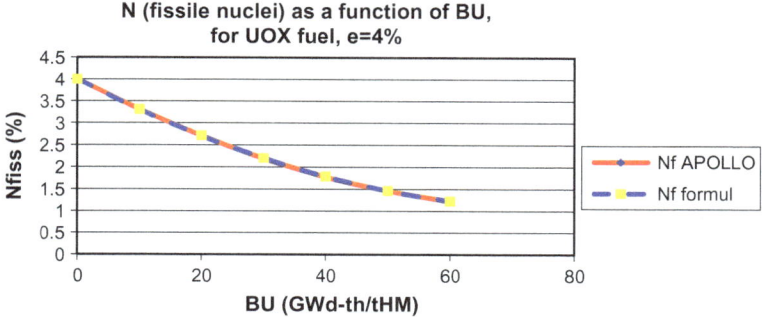

Figure 13.22. Fissile nuclei concentration in % (HM) as a function of the burn-up, in a 4% enriched UOX: a simple approximation.

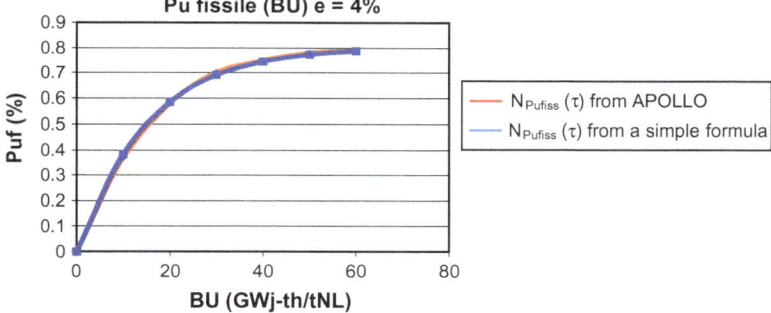

Figure 13.23. Variation of fissile Pu concentration with burn-up: a simple approximation.

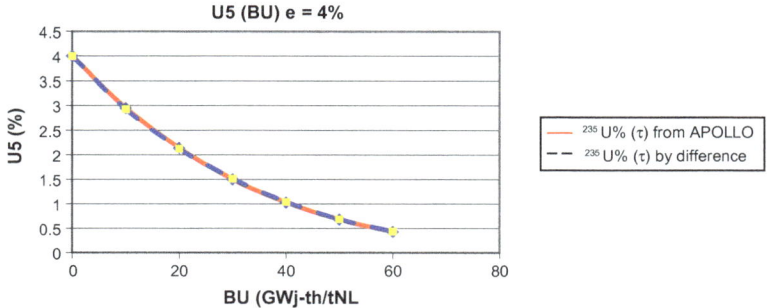

Figure 13.24. Variation of ^{235}U enrichment with burn-up: a simple approximation.

Another useful approximation is given by the linear core neutronics approach of the fuel burn-up and of the cycle duration dependence on enrichment as well as on reloading fractioning, of the order n, meaning that the core is loaded by 1/n (rather than at once, an option called batch loading).

The end of life of a core in batch cycle determines at the same time the ultimate burn-up of the spent fuel, called $\tau_s(1)$ (s for spent or used fuel) and the fuel cycle length $\tau_C(1)$. In this case, $\tau_s(1) = \tau_C(1)$. An approximation is given by 10 (e − 1), with τ in GWd-th/t-HN and e in % (^{235}U). This very crude approximation holds between 3 and 5% with a reasonable accuracy, in a regular lattice of traditional moderation ratio (17 x 17 in PWR, for instance), with a large core ($\geq$ 1300 GW-el), a good reflector, low capture clads and structural materials and, finally, with a classical loading pattern.

Now, if one adopts the assumptions of linear core neutronics, the core reactivity is roughly driven by the mean burn-up of the core represented by $\tau_{eff}(n)$, for a given core management strategy. Thus, for a core loaded 1/n by 1/n, the situation is as follows, when equilibrium has been reached:

BOC: $\tau_{eff}(n)@BOC = 1/n\ [0 + \tau_C(n) + 2\tau_C(n) + ... + (n − 1)\tau_C(n)]$
EOC: $\tau_{eff}(n)@EOC = 1/n\ [\tau_C(n) + 2\tau_C(n) + ... + n\tau_C(n)] = [n(n + 1)/2]\tau_C(n)/n$
$= (n + 1)\ \tau_C(n)/2$
With $\tau_C(n) = (1/n)\ \tau_s(n) = \tau_{eff}(n)@EOC − \tau_{eff}(n)@BOC$
For the same criticality condition at EOC: $\tau_{eff}(n)@EOC = (n + 1)\ \tau_s(n)/2n = \tau_s(1)$
$\rightarrow \tau_s(n) = [2n/(n + 1)]\ \tau_s(1)$
And $\tau_C(n) = (2/(n + 1))\ \tau_C(1)$

The consequences of those formulae are:

at constant enrichment and thus at constant $\tau_s(1) = \tau_C(1)$, $\tau_s(n)$ increases with increasing fractioning. This increase has an asymptotic behaviour towards the limit: x2 (continuous reloading during operation).

Conversely, $\tau_C(n)$ decreases with increasing n. This is a significant drawback from the viewpoint of the best possible use of the high capital cost, due to the associated fall in the − necessarily high − value of k_d, the availability factor.

The solution of this dilemma is of a dialectic nature: increasing simultaneously the enrichment and the loading order of core reload fractioning keeps the "competitive" level of fuel cycle length (around one and a half year up to two years) and increases the fuel burnup at discharge according to: 2n/(n + 1).

The gain is still significant, from 1/3 to 1/4 for instance (a factor of 1.5 to 1.6).

The simple rules and calculations described above are translated into a network of curves in Figure 13.25. Nevertheless, the assumptions taken for the figure are slightly different from those taken in the simple calculations, so the results can differ substantially in some places. Moreover, the simplified rules hold only between e = 3% and e = 5%.

Annex 2

Definition and simplified calculation of a few concepts leading stepwise to the evaluation of the breeding doubling time of a FBR fleet ([1.3])

In a **FBR** and therefore in a fast spectrum, a heavy metal (mainly fissile plutonium) fission releases about 213 MeV, leading to: **1 MWd$_{th}$** $\leftarrow \rightarrow$ **1 g of heavy nuclei fission**.

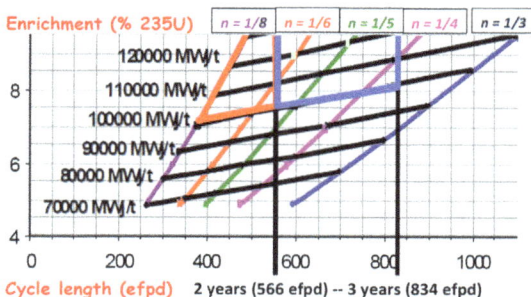

Figure 13.25. Interdependency between spent fuel burn-up (in MWd/t(HM), cycle length (in full power equivalent operation days), initial enrichment (in % of 235U) and reload fractioning.

With a typical thermodynamic efficiency close to 40%, this means fissioning 920 kg of heavy nuclei per GWyr$_{el}$. Taking into account the fast fission contribution of ^{238}U (ε term of the $\varepsilon\eta$pf formula) in a fast spectrum, 800 kg of fissile Pu are fissioned.

The amount of absorbed fissile Pu (fissioned and captured) is 800 (1 + α) with α being the (fast) capture/fission ratio (about 1.25). The approximate result is thus:

$$\text{1 ton of fissile Pu absorbed per GWyr}_{el}$$

The Global Breeding Gain (GBG) = Breeding Ratio (fissile core fast conversion ratio plus "internal" and external fertile blanket contribution) minus 1 = BR − 1.

Taking into account the whole set of approximations concerning space and time mean value calculation, this leads to:

$$\text{Net fissile Pu production} = \text{« GBG »} \times \text{1 ton/GWyr}_{el}$$

For instance, for GBG = 0.2, the net fissile Pu production is $\sim$ 200 kg/GWyrel.

This result can be compared with the fissile Pu production in a standard lattice, once-through fuel cycle PWR # 140 kg/GWyr$_{el}$.

The commissioning Pu(total) inventory for a 1 GWe(installed) FBR – sodium cooled – modern core with lower void coefficient is about 9.6 t, meaning about 7.2 t of fissile Pu. With k_p = 0.75, this leads to 9.6 t of fissile Pu for a 1 GWeff core. This amount can be produced in: 9.6/1 × "GBG" years of operation. With GBG = 0.2, for instance, the "Reactor Linear Doubling Time" or RLDT (time necessary to the net production by breeding of the plutonium inventory of the same FR core) is:

$$\text{RLDT} = 9.6/0.2 = 48 \text{ years.}$$

However in practice it is necessary to provide twice this amount at commissioning, because the first reloads are necessary in order for the plant to run while waiting for the return of the bred Pu from the out of pile cycle in the fuel factories dedicated to spent fuel processing and fabrication. This means about a "second" core available, pushing the doubling time to 48 × 2 = 96 years.

The same result can be obtained from the "System Doubling Time" or SDT concept: plutonium availability requires waiting for: Tcore (in pile cycle) + Tout (out of pile fuel cycle), thus:

SDT = RLDT [(Tcore + Tout)/Tcore](1 + L). L is for losses occurring through decay during the whole fuel cycle, by material losses including irremediable process rejects; L, in a range of a few %, will be neglected below.

Neglecting the losses and taking Tcore = Tout = 6 years leads to SDT = RLDT × 2. The word "system" refers to the set of processes: reactor (core, then cooling down) + transportation + processing + fabrication + return to the reactor and waiting for loading.

Finally, for a large fleet of breeders or converters (EPR-like, for instance) providing a correspondingly large influx of plutonium, the relevant parameter is the "Compound System Doubling Time" or "Fleet Doubling Time" (FDT): plutonium can be loaded almost as soon as it is available, without waiting too long for a build up of the inventory of a new core. The reduction factor related to this process acceleration is Ln (2) = 0.693.

$$CSDT = FDT = SDT \times Ln (2)$$

With the same data, the FDT is reduced to 96 × 0.693 = 66.5 years.

The FDT doubling is rather optimistic, specifically for the "first" doubling time in a rising fleet of newcomers.

Obviously, with GBG = 0.4, which can probably be considered as an upper limit with realistic core and fuel design (while the $\varepsilon\eta$ values from Figure 13.3 provide a GBG potential up to 0.8), the previous figures can be divided by two, leading respectively to:

RLDT = 9.6/0.4 = 24 years; SDT = 24 × 2 = 48 years and finally, for large fleets, CSDT = 33.3 years.

14 Nuclear fusion

J.-B. Thomas

14.1. Introduction

The R&D issue with fusion is complicated by the tight coupling of two challenges: a basic research challenge regarding plasma engineering (magnetic or inertial) on the one hand and a system design challenge on the other. Moreover, the system design issue cannot be dealt with without first solving the basic research problems (at least for the basic choice of the physical/technological means to be used). For instance, the role(s), the nature and the power level of the plasma heating determine not only the plasma confinement capability and mode, but many more design characteristics, including a critical one: the high availability required from a power production system.

That is the point where ITER will play (as for the Tokamak line) an essential and dual role and will represent a major step on the way to a DEMO, except for two major nuclear issues (the materials and the fuel cycle). It will make it possible in two steps to firstly demonstrate and to improve the physics of burning plasma (i.e. mainly heated by the fusion reaction energy) as well as the technologies required to reach this goal, and then to tackle the integration issue of all the components and devices involved in a reliable tritium and power generation perspective.

The current chapter will not enter into the details of the two paths to achieving fusion. It will not describe the ITER project (see "The ITER Design" in [1]) either. Its objective is to give an overview:

- Of the principles and of the order of magnitude of the main physical requirements;

- Of the two classes of issues to be dealt with: the ones focused on plasma confinement and the system design ones. The second class shows, by comparison with the fission system case, the distance to industrial maturity for power generation fusion systems and the main ways that will need to be taken in order to reduce this distance, stepwise, starting from the ITER achievements expected during the next decades.

14.2. Principles and basic data

14.2.1. General

The principal fusion reactions are:

$$D + T \rightarrow {}^4He(3.56\ MeV) + n(14.03\ MeV)[17.59\ MeV\ in\ total]$$

And the deuterium-deuterium reactions:

$$D + D \rightarrow {}^3He(0.82\ MeV) + n(2.45\ MeV)$$
$$D + D \rightarrow T(1.01\ MeV) + p(3.02\ MeV)$$

Obviously, the reaction: $p + {}^{11}B \rightarrow 3\ {}^4He$ (8.7 MeV) without n, would be very attractive… Figure 14.1 shows the effective cross sections in these processes.

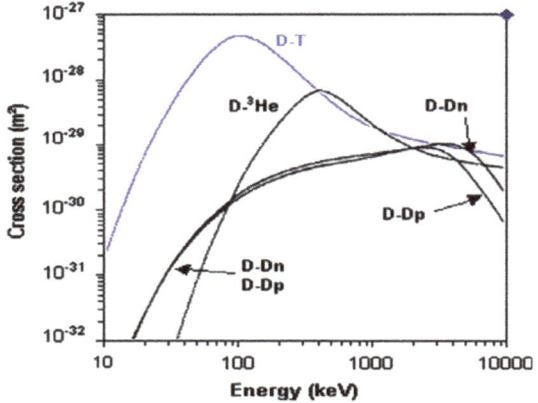

Figure 14.1. Typical fusion reaction cross section as a function of ion energy.

In other words, 17 MeV per reaction for D-T fusion corresponds to the consumption of 1 kg of fuel per day, in a plant generating 1000 MWe with an efficiency of 30%. In practical terms, this means that with fusion, each litre of seawater from which 3.3 mg of deuterium would be taken could produce as much energy as 250 litres of oil, with no negative effects on the environment.

It must be borne in mind that one gram of tritium (the product of a D-D reaction but, above all, fuel for a D-T reaction) represents around 5 kCi ($1.85*10^{14}$ Bq – 185 TBQ), and that the management of the high flux levels and inventories needed by industrial-scale reactors imparts certain constraints.

The difficulties inherent in fusion include the following:

- the low effective cross sections, which are in the order of 10^{-28} m², in competition with other concurrent phenomena such ionisation, Coulomb scattering etc.;

- the necessity to raise the energy level of the particle to at least 100 keV (l eV = 11,600 K, or about 10^4 K) in order to pass the Coulomb barrier and reach the nuclear force level. The D-T reaction is in this regard the most readily accessible. It is this reaction which is considered below.

The need to work at relatively high energy levels complicates the application of a projectile-target model as a result of loss of the energy beam to competing parallel processes. To be efficient in terms of energy, fusion must occur in a medium that is such that collisions are an almost neutral phenomenon. This is the case in a hot gas where the primary role of collision is to preserve Maxwellian distribution of the velocity between the particles. The temperature values that give the best results are somewhat above 10 keV of average energy. In such a gas, there are particles whose energy is well above the average temperature. It is this small proportion of particles that give rise to "thermonuclear" fusion reactions.

This plasma model has two implications:

- the energy that must be spent to raise a quantity of gas to temperature T is:

$$U \text{ (in J/g)} = 1.8 \times 10^8 \times T \text{ (with T in keV)}$$

U can be compared to the energy that can be derived from one gram of fuel, i.e. 3.3×10^{11} J, giving a factor of 300. Care must be taken to ensure that the efficiency is acceptable by avoiding unwanted dissipation;

- plasma is easily cooled by any phenomenon that disturbs the situation. This guarantees that the reaction will rapidly stop in the event of an incident. Although this does not simplify operations, this feature makes the process intrinsically safe, and in addition there is no decay heat to contend with as a result of the disintegration of fission products. It is only activation of the surrounding structure by neutron bombardment that causes any residual nuclear activity, but this is very slight.

14.2.2. More on physical principles and basic data

Jumping over the Coulomb barrier

The height of the Coulomb barrier can be written as: $E_c = Z_1 * Z_2 * e^2 / 4\pi * \varepsilon_0 * r_m$,

ε_0 being the vacuum permittivity ($8.25 * 10^{-12}$ C^2/m.J), and r_m the distance between the nuclei where the nuclear force emerges then takes over, as shown in Figure 14.2. One can take $r_m \sim 10^{-14}$ m for (D, T).

For example, in the D, T case, with $Z_1 = Z_2 = 1$, $E_c \sim 0.14$ MeV = 140 keV, a rather high value.

Taking a slightly different expression by separating the nuclear radii R1 and R2 leads to yet a higher value.

The corresponding temperatures amount to several billion Kelvin, while the ignition temperature in the magnetic confinement fusion will be around 100 million Kelvin (10 keV). The tunnelling effect makes sub-barrier reaction probability significant, even well below the barrier height. Depending on the centre of mass energy it gives for (D,T)

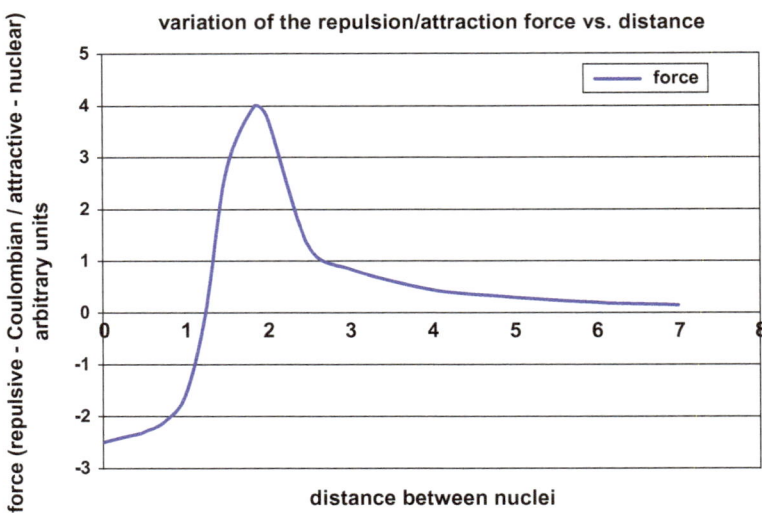

Figure 14.2. Variation of the repulsion/attraction force as a function of the distance between nuclei.

and (D,D), the following barrier transmissions at two different temperatures:

	10 keV (~ 120 million K)	5 keV (~ 60 million K)
(D,T)	2×10^{-5}	2×10^{-7}
(D,D)	8×10^{-7}	2.5×10^{-9}

The Gamow's tunnelling formula gives an approximation to the transmission coefficient, T(E), giving the probability for the particle to surpass the barrier:

$$T(E) = \exp(-2/h \int_{x1(E)}^{x2(E)} \sqrt{2m(U(x) - E)} dx \text{ (at x1 and x2, } U(x) = E)$$

The variation of the cross sections with E_i

The measurement of the cross sections vs. the energy of the ions leads to the curves shown in Figure 14.1. For the (D,T) reaction the peak is obtained around 120 keV which is close to the approximation of the barrier height at 140 keV. The peak value is 5 barn. At Ei = 10 keV, the cross section is dramatically reduced to 10^{-3} barn. The quantum "tunnelling" effect makes it possible for ion energies lower than the "wall height" to generate fusion reactions.

Therefore on has to shift from a ballistic one shot inefficient process to a statistical method in a collisional regime inside a hot ionised gas i.e. a plasma. The reason for this is to reduce the required energy, introducing the confinement as an enabler (first a self-confinement (pinch), then a strong external confinement (toroidal field with huge and powerful coils)).

The temperature distribution of the ions inside the plasma

- The temperature distribution obeys the Maxwell-Boltzmann law, with k (Boltzmann) = 1.38 x 10^{-23} J/K).

- To E = 1 eV is related T = E/k = 11600 K # 10^4 K.

The temperature associated with the Coulomb barrier height or, equivalently, with the (D,T) cross section peak around 100 keV is thus 1 Billion K.

- The fusion dedicated plasma (neutral ionised gas) is composed of ions with n_i ion density and of electrons with $n_e = n_i$ density. It is kept neutral if the MHD forces do not lead to a divergent drift of both populations (see section below describing centrifugal drift exerted on a circular trajectory in a purely toroidal field, without geometrical (Stellarator) or poloidal field compensation).

This neutrality assumption is valid above a given scale. The scale boundary is related to a characteristic minimal length named Debye length: $\lambda_D = \sqrt{\varepsilon 0 * kT / ne * e2}$

In a typical fusion plasma, the Debye length is around a few tens of μm. Actually, plasma particles can collapse in bunches in phase space and lead to density fluctuations. If, locally, but on a significant scale, an increased concentration of electrons (for example) occurs, it generates a fluctuation of electric and magnetic fields, disturbing the confinement. The particle kinetics is a key issue, as well as being a sensible representation by clusters in numerical simulation.

Strong internal turbulence phenomena can feed the development of instabilities up to chaotic bifurcations. The main problem with quasi-steady magnetic confinement is that (beyond fulfilling pure MHD conditions for a perfect "quiescent" plasma) instabilities can develop (if the system is deprived of any active control) over a large range of frequencies. The most spectacular plasma confinement breaking effect is related to a violent expulsion of a part of the plasma towards the first wall. It is called an (external) disruption. Some "internal", "core" disruptions also occur.

The negative (or potentially positive, see "H mode" (H for High, sustained confinement)) influence of the external heating of the plasma, as well as the strongly modified situation generated by the approach of ignition (the regime by which the plasma is self-heated (and then "poisoned") by its He - α fusion-emitted charged particles), will be dealt with below (see "the main challenges and the main R&D trends" in § 14.10.2).

- The Maxwell-Boltzmann energy distribution in the plasma means that if the "local" (at a scale well above the Debye length) mean temperature in a quiet region of the plasma is slightly below 100 MK, a small percentage of the ion population reach a higher temperature and can work as a "flame" for fusion reactions, in a "milder" environment.

Reaction rate and power density calculation

The reaction rate is: $R = n_{i1} \times n_{i2} \times <\sigma \times v>$.

Calculating the $<\sigma \times v>$ variation with the ion energy leads to a peak value around 70 keV for a (D,T) plasma. At 10 keV, $<\sigma \times v> \sim 10^{-16}$ cm^3/s, about one tenth of the peak value.

In the hot, dense and disruption-protected region where fusion is supposed to run in long pulses (1000 s?), the power density can be calculated by multiplying the reaction rate R with the energy released by the fusion reaction (17.6 MeV in the (D,T) case). With ion densities typical of a magnetic confinement plasma around 10 keV, the core plasma power density would be around 1 MWth./m^3. This is a rather low value, when compared with fission core values. Furthermore, in fission, the core is the place where everything is concentrated. In fusion, it is the most internal part of the global heat and tritium releasing device. Confinement tools, heating, heat removal and tritium recovery and transport are located outside the first wall containing the plasma torus. The global dilution and spreading of massive, sophisticated and costly components, starting with the giant superconducting coils, is thus a major issue. This point has far reaching system engineering and design consequences (see below).

14.2.3. Plasma

Very diverse plasma densities and temperatures exist in the universe. For plasmas in fusion, at around 10 keV, relativistic corrections mainly apply to electrons. Magnetic fusion plasmas are of very low density whereas inertial fusion ones are of high density and may require quantum correction.

The following two properties are important in the present context:

- wave-plasma interaction where:
    - electromagnetic radiation is used to heat the plasma,
    - the thermal energy of the plasma feeds some waves, tending to make the plasma unstable,
    - the waves permanently present in the plasma make it turbulent, affecting its characteristics (particle scattering, heat transport etc.);

- radiation from plasma where as a result of the thermal agitation and of the induced impacts, the particles (mainly electrons) emit bremsstrahlung radiation (consisting of x-rays). At a temperature Te, radiated power is:

$$P_B = 4.8 \times 10^{-37} Z^2 n_i n_e \sqrt{T_e}$$

With P_B in W/m^3, Z the average charge number of the ions, n_i & n_e the density of the ions and of the electrons and T_e the temperature in keV.

This formula calls for the following remarks:

- when n_i and n_e are greater than 10^{20}, the radiated power is considerable and strongly cools the plasma

○ term Z^2 gives an incentive to minimise the presence of impurities such as iron or (worse) tungsten and molybdenum, constituting grounds for choosing certain materials over others for construction. Despite this, W (Tungsten/Wolfram) is considered for replacement of carbon in the structures facing the plasma (first wall, divertor) in the current design of TOKAMAKS, including ITER. The issue of dust (plasma poisoning, risk of explosion in the case of a LOVA – Loss of Vacuum Accident) is thus raised with this option.

14.2.4. The ignition criterion

Figure 14.3 shows the key components driving the balance of power of the hot confined plasma, with the following definitions:

$P_{fusion} = P_\alpha + P_n$; only $P\alpha$ contributes to plasma heating, the neutrons escaping outside the plasma and being used for tritium breeding and heat removal inside the external blankets;

in the plasma: $P_{sources} = P\alpha + P_{ext.}$ (external power coupled to the plasma – for heating and current generation, excepting the fraction injected but escaping outside);

P_{losses}: the sum of all losses

$P_{inj.}$: Power injected into the plasma in order to create and heat it, then to help sustaining the temperature, as well as driving the fusion reaction.

One has to be careful when working out the global thermodynamical balance and efficiency performance of the plant and be aware that Pinj. is not the relevant term: the whole set of auxiliary systems (including the cryogenic ones, for instance) have to be taken into account. Moreover, the efficiency of transforming the grid power into injected energy depends on the nature of the physical process considered, an example of which is heating, such as the radiofrequency heating from klystrons, magnetrons, but also injection of neutrals – accelerated/neutralized ions -, as well as ohmic heating by induction. The efficiency cannot be calculated directly from what is used as a physical indicator of the progress towards controlled fusion: the amplification ratio $Q = P_{fusion}/P_{inj.}$.

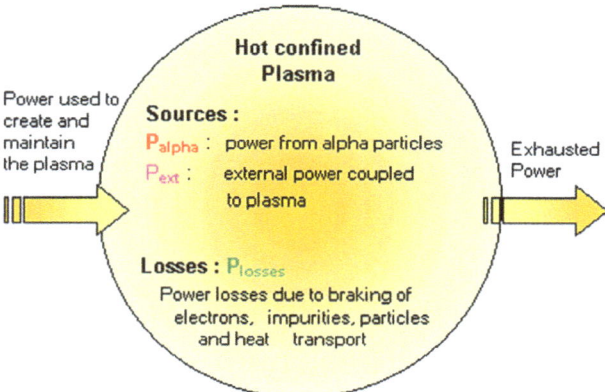

Figure 14.3. Key components of the hot confined plasma balance of power.

By definition:
The scientific break-even is related to Q = 1 and the ignition to Q "infinite" (Pinj. # 0).
The confined plasma power balance can be expressed as follows, with E_{pl} being the plasma energy:

$dE_{pl}/dt = P_{sources} - P_{losses}$

It is difficult to model all forms of losses from plasma (radiation, conduction and convection). A time τ is formally defined to characterise the duration of the spontaneous loss of the plasma energy and is referred to as the energy confinement time. Gas kinetics theory gives the following for the energy of a plasma:

$$E = 3nkT$$

where n is the density of the ions and electrons, k is the Boltzmann constant and T is the temperature. Instantaneous power loss can then be expressed as follows:

$P_l = 3nkT/\tau$ (with P_l representing the global level of the losses)

These losses should at least be compensated for by the fusion power, which can be written:

$$P_f = n_D n_T <\sigma V> E_{fus} \qquad \text{where } n_D = n_T = n/2$$
$$P_f > P_l \rightarrow (n^2/4) <\sigma V> E_{fus} > 3nkT/\tau, \quad \text{thus } n\tau > 12kT/<\sigma V> E_{fus}$$

This condition depends only on T (for a given fuel).
Heating of plasma by the fusion power it releases can take place by:

- recirculation of the energy extracted which is converted into electricity and returned to the plasma by an appropriate heating process (see below), where a series of efficiencies intervenes,
- self-sustaining, with particles produced by the fusion that are hotter than the plasma itself thereby maintaining its temperature. When it is capable of working autonomously, without any further external heating, this process is referred to as ignition of the plasma.

In the D-T case, the neutrons produced, being neutral particles, escape without giving up their energy. They interact further on with the first wall and the blankets. As for the helium ions, they interact with other charged particles, particularly electrons, and release their energy in the plasma.

To assess the conditions of ignition, the following approach can be used. By substituting E_α for value E_{fusion}, the ignition criterion becomes:

$$n\tau = 12kT/<\sigma V> E_\alpha$$

When E_α = 3.56 MeV, $<\sigma V>$ = 10^{-22} m²/S and T ≈ 10 keV, we obtain:

$$n\tau = 10^{20} m^{-3}.s \text{ (Lawson criterion)}$$

As derived, this criterion is local and instantaneous.
It appears to leave great scope for choice. In reality, only two approaches seem to be worthy of interest and are being investigated. They differ in the duration of confinement, the energy loss time, designated τ:

- either it is not sought to retain the plasma, which burns and disperses and only the inertia of the particles avoids $\tau = 0$. τ is nevertheless well below 10^{-10} s. Densities are therefore necessarily high ($>10^{30}/m^3$ if the criterion is applied), in which case the situation is referred to as inertial confinement,

- or a magnetic field is created to retain the plasma and prevent it from dispersing. In this latter case, confinement times can and indeed must be of the order of one second. This is because the densities can only be very low ($10^{20}/m^3$ and less than that of the ambient gas). This is referred to as magnetic confinement.

What the two methods have in common is their fuel (D-T) and the temperature of the medium (10-20 keV). All the other properties are completely different.

Figure 14.4 shows the evolution of the ignition Lawson criterion L as a function of the temperature

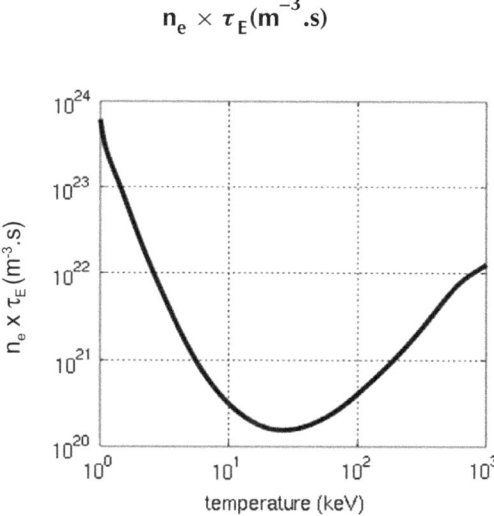

Figure 14.4. Evolution of the ignition Lawson criterion L as a function of the temperature (from Wikipedia).

For (D,T), $n_e \times \tau_E$ minimizes near 25 keV (around 300 million Kelvin).

14.3. Fusion by magnetic confinement

14.3.1. Principles

In a uniform magnetic field B, a particle of charge e, of mass m, with a speed v considered to be in a plane perpendicular to the field is subjected to a force

$$F = e\, v \wedge B$$

that imposes a uniform circular movement of radius R = mv/eB (Larmor radius).

With a field of 5 Tesla and an energy of 10 keV, the radius of gyration of the ions is less than one centimetre. The radius of gyration of electrons is far less, and even helium ions with an initial energy of 3.56 MeV have a radius of gyration that is less than 10 cm. This can be considered to constitute confinement in the plane perpendicular to the field line.

Furthermore, the component of the speed parallel to the field is unchanged. The particle adopts a corkscrew trajectory around the lines of force. As a result of certain mechanisms (collisions and collective effects), transverse scatter persists and limits the imposed transverse confinement.

Taking a fluid mechanics approach, it can be demonstrated that the situation is equivalent to one where the magnetic field is exerting "pressure" amounting to $B^2/2\mu_0$.

Yet experience shows that a magnetic field can resist a gas pressure of around one tenth that of the magnetic pressure. For a field of 5 Tesla $B^2/2\mu_0$ amounts to 100 bar, and therefore the maximum kinetic pressures are 10 bar (1 MPa). The perfect gas formula indicates that:

$$P = nk_B T \text{ where } k_B = 1.4 \times 10^{-23} \text{ and } T = 10^8 \text{ K}.$$

For p < 1 MPa, n must not exceed 5×10^{20} particles per cubic metre. Comparison with the density of the particles of a gas at room temperature (around 10^{25}) shows there is already a reasonable vacuum. With the density ($10^{20}/m^3$) and the ignition criterion ($n\tau > 10^{20}$), minimum confinement time τ can be found (around one second).

From a technical and economic viewpoint, some rough figures can already be determined, for instance the power density, expressed in MW_{th}/m^3. Taking the fusion power formulae and substituting in the values calculated earlier, a value of a few MW/m^3 is found. This is the reactive core power density. The power density of the complete nuclear installation is even lower (the plasma ring, constituting the core of a large and complex electromagnetic device which is described below).

For comparison, orders of magnitude for fission cores are 1 MW/m^3 in Natural Uranium Gas Cooled Reactors, 7 MW/m^3 in HTRs, 50 MW/m^3 in BWRs, 100 MW/m^3 in PWRs, 300 MW/m^3 in Sodium Cooled Fast Breeder Reactors and up to 10 000 MW/m^3 in NERVA type space propulsion engines. Moreover this power density comprises the fuel, the structures and the volume dedicated to the coolant flow in the case of fission, whereas in the case of fusion this is remote, with the neutrons being handled by blankets outside the first wall. In addition, fission does not require any massive devices (like coils) outside the core to foster the chain reaction, making it more compact.

These considerations are not neutral as concerns the overall economics of fusion as a source of power.

Furthermore, in view of the effects of scaling and non-linearity, the minimum unit power for a fusion reactor with magnetic confinement of industrial scale is very large (typically equivalent to that of a modern fission power station). This is not only a handicap for the transition from R&D to commercial use (as compared to the rapid advance from Fermi's CP1 critical experiment (0.5 W) in December 1942 to the first plutonium production reactor commissioning in the USA, in Hanford (250 MWth), started in September 1944), but also for large-scale commercial introduction, as the capital costs (and the risks) cannot be broken down in either time or space.

14.3.2. Confinement and the Tokamak principle

The effect of transverse confinement associated with the presence of a magnetic field is described above. The problem of confinement along the field lines must now be considered.

The basic principle consists in closing the field lines. One of the first ideas was to arrange electromagnets to form a toroid (thus the terms "toroidal coil" and "toroidal magnetic field"). The lines of force are then endless, as they form circles. But this does not suffice to solve the problem. Study of the individual movement of the particles shows drift. This has to be combined with the effects of the electric field associated with charge separation of electrons and of ions.

The toroidal field is curved and decreases in strength moving away from the axis of rotation (with a related pressure gradient profile). The ions and the electrons move parallel to the axis, but in opposite directions. The charge separation leads to an electric field and an additional drift, in this case outward (away from the axis of rotation) for both ions and electrons. To counter the drift, a current J is added to the configuration in the plasma ring. J gives rise to a poloidally orientated field component (see Figure 14.5) and the lines of force become helices that coil round the toroidal surface surrounding the plasma (see Figure 14.5, which shows an exaggeration of the effect). The consequence of creating a twist around the toroidal field lines is an alternate circulation on the outside and then on the inside of the related toroidal flux surface. Some of the outward drift is thus compensated by an inward drift on the same flux surface and there is a macroscopic equilibrium with much improved confinement. Closed configurations always have magnetically nested surfaces, the greatest pressure being at the centre, with both density and temperature decreasing outwards. A fine tuning of the shape of the plasma (stretching it vertically) is obtained by combining several controlled fields in a sophisticated way.

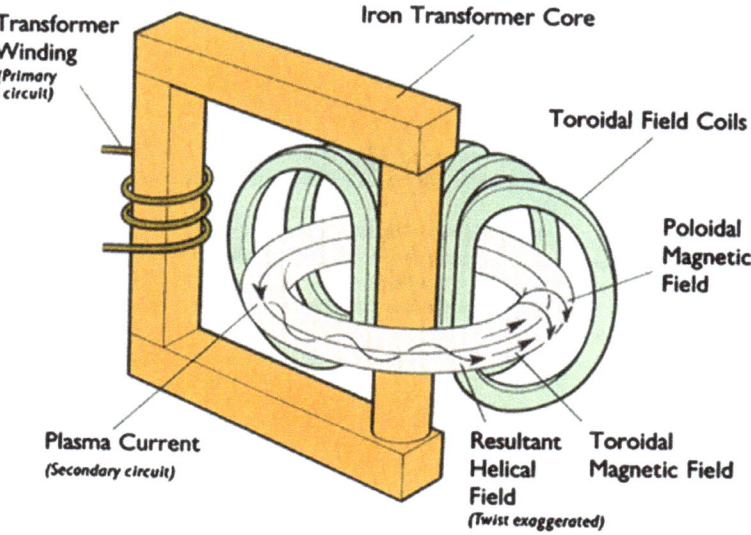

Figure 14.5. Magnetic confinement: toroidal (+ current and poloidal component) configuration.

The above diagram showing a mainly toroidal field and a current circulating in the plasma corresponds to a Tokamak configuration. This is the type that has been the most comprehensively studied and is the most effective at the present time. Other configurations exist, such as the closed configuration Stellarator which avoids the presence of a current in the plasma at the cost of external coils of a more complex design that are delicate to construct (Figures 14.6 and 14.7), as well as a linear configuration with a magnetic "plug" at each end (Figure 14.8), giving rise to the Tandem Mirror Experiment in the Lawrence Livermore Laboratory.

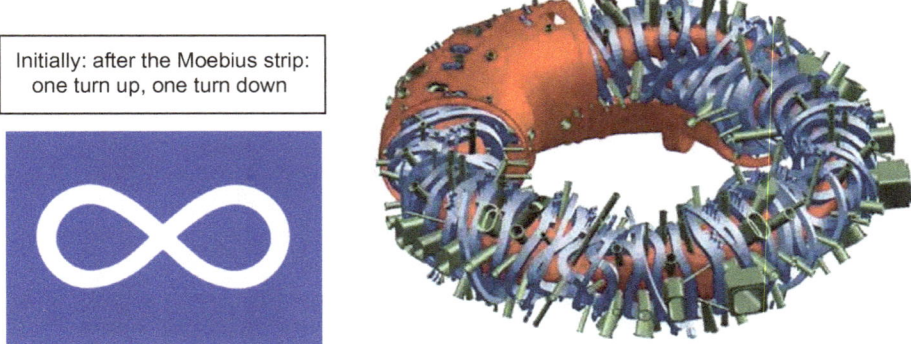

Figure 14.6. The stellarator: from first principles to modern implementation Wendelstein: after Max Planck Institute for Plasma Physics.

Figure 14.7. From Spitzer Stellarator experiment to the coils for the German Wendelstein W7-X.

By conservation:
  of the magnetic moment $\mu = mV_\perp^2/2B$;
  of the the particle energy $E = (mV_{//}^2/2) + (mV_\perp^2/2) = mV_{//}^2/2 + \mu B$;
$V_{//}$ could vanish if: $V_{//0}/V_{\perp 0} < \{(B_{max} - B_{min})/B_{min}\}^{1/2}$.

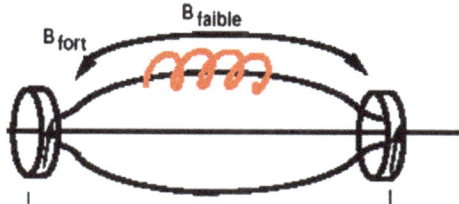

Figure 14.8. Magnetic confinement: magnetic mirror – tandem – effect.

The main difficulties with the Tokamak configuration are the following:

- **Generation of the plasma current**

    The basic approach is to initiate it by induction. A coil consisting of horizontal layers is placed in the hole in the middle of the torus (see Figure 14.9). By varying the current (an upward ramp), the plasma current is created by induction, as in an electrical transformer. The duration of the pulse is necessarily limited as the plasma current is sustained, and the current in the coils must continually increase.

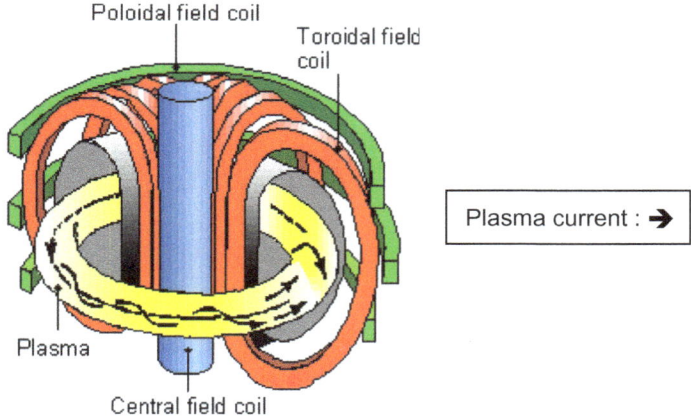

Figure 14.9. Plasma current generation in TOKAMAK: the inductive way and the role of the current in generating the poloidal field.

When the primary circuit is exhausted, then comes the end of the confinement pulse with vanishing poloidal control. It is therefore of utmost importance to develop non inductive ways of current generation. On the other hand, as the resistivity of the plasma decreases with temperature, the heating by the current alone is limited to a very few tens of million K and other means are mandatory (see below).

- **Disruption**

    The magnetic configuration that confines the plasma is established by the current carried by the plasma itself. There are limits on the various physical variables (density, plasma current etc.). If a disturbance causes the limits of stability to be reached, disruption occurs. The plasma immediately disperses towards the walls with its energy content (of GJ) and the process stops. This problem has not yet been completely solved.

Figure 14.9 shows a diagram of a Tokamak:

- the set of coils placed at the meridian planes (toroidal) creates the strongest magnetic field;
- the "horizontal" coils that create the poloidal field are interconnected:
  - ▶ the first stack is the core situated at the centre of the configuration, used for creating current J in the plasma, which in turn creates the poloidal magnetic field,
  - ▶ the horizontal coils create a weak vertical field intended to stabilize and control J.

14.3.3. Heating of magnetized plasma

Three basic principles are used (Figure 14.10).

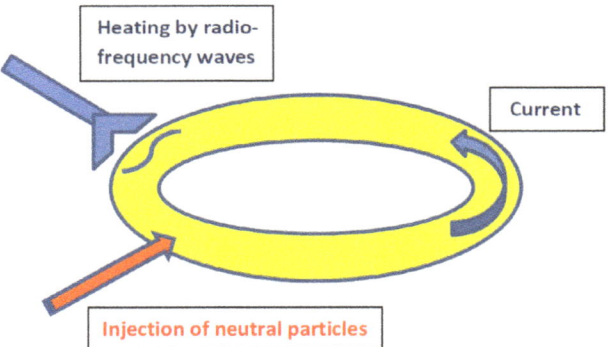

Figure 14.10. Plasma heating.

Resistance heating

Plasma constitutes a good conductor of electricity (its resistivity being well below that of copper at normal temperatures).

On the basis that $P = RI^2$, there is an incentive to keep R and I high to heat the plasma. However, increasing I raises stability problems for the confinement configuration, and R decreases rapidly as the temperature increases, making the heating less effective.

Electron and ion temperatures of several keV are obtained, but this process is not used alone to achieve ignition. Other means of energy thermalisation and transport are used in the plasma, and these are described below.

Heating by neutrals

Energy can be carried by energetic ions, but these have to be "neutralised" to be able to enter the plasma. The process is as follows:

- ions are created with a source then accelerated (100 to 200 keV),

- the high-energy ions are then neutralised,

- on reaching the plasma, these particles become ionised, and are then trapped in the magnetic confinement and gradually give up their energy to the surrounding medium which becomes hotter,

- such systems have been used with H+, D+ and T+ up to 1 to 2 MW at 150 keV per particle. For a reactor, energies above 1 MeV need to be envisaged, resulting in the use of negative ions for a better neutralisation efficiency. There are also difficulties in achieving high currents.

Heating by high frequencies

To ensure efficient energy transfer and coupling, it is important to remain close to the resonance levels. Two resonance levels in plasma give promising results that could be extrapolated to a reactor; they are related to the cyclotronic frequencies of the:

- ions: between 40 and 90 MHz,

- electrons: of which a typical frequency is slightly below 200 GHz.

In current experiments, such systems are delivering more that 10 MW; for a reactor, 100 MW or more needs to be emitted.

Figure 14.11 shows an implementation of antennas in the framework of the WEST program in the TORE SUPRA facility in CEA.

Other applications of heating systems

A smart use of beams of particles and waves can supply vectored energy momentum. Transfer can affect the plasma as a whole (which can be put in rotation), or only one class of particles.

It has been possible to demonstrate (on the Tore-Supra installation in France) generation capacities of the order of 1 MA for several minutes. This paves the way for continuous operation of a Tokamak by providing the necessary current in addition to that which it generates itself (bootstrap current).

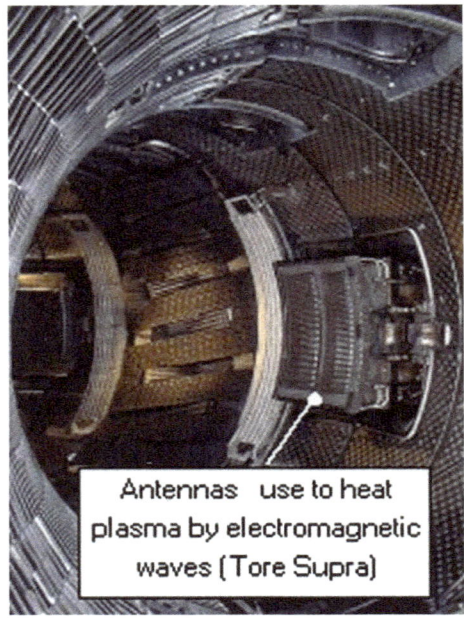

Figure 14.11. RF antennas: TORE SUPRA implementation (CEA).

14.3.4. Findings: principles and noteworthy facts

Scaling and extrapolation of performance

To calibrate the performance levels, to determine the amount of progress made and to extrapolate to power generating, laws have been sought that govern the key parameters. For instance, for a confinement time τ, a law of the following type is needed:

τ = f (R, a, B etc.) as a function of the geometrical and mechanical properties etc. of the machine

No such law currently exists in which theory and experiment both concur. Simulation is a big challenge in this complex field, which is highly non-linear and in which instability is prevalent, whereas for fission reactors the simplicity of the linear Boltzmann equation for their neutronics has greatly contributed to the development of the corresponding technology. A semi-empirical monomial law (involving a number of variables and supported by adimensional theoretical scaling laws) has been devised by combining the findings of experiments. The experimental parameters range over two decades, in various Tokamaks and under different conditions, yet the deviation between law and observation is less than 15%, in the overall field of experiments. This excellent agreement is shown in Figure 14.12.

The tritium experiments: from JET to ITER

If consideration is limited to JET in Europe (Figures 14.13 and 14.14) with reference to work conducted elsewhere (such as in the USA (TFTR)), two key dates show the current progress, which is still continuing:

14 – Nuclear fusion

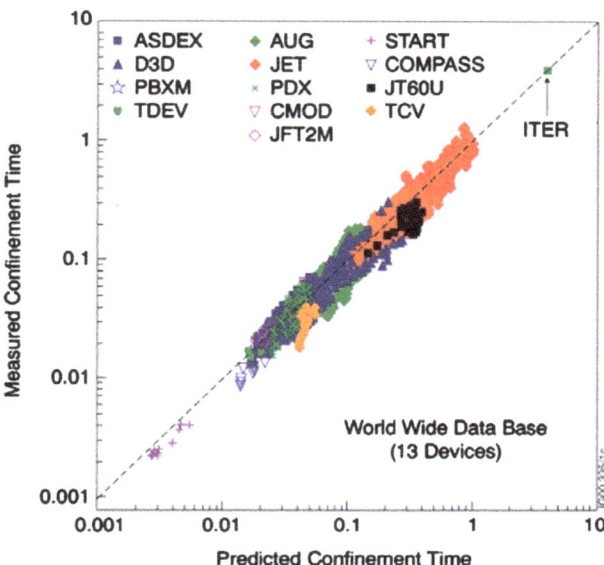

Figure 14.12. Scaling law agreement with diverse experiments; extrapolation to ITER.

Figure 14.13. Joint European Torus (JET) at Culham (UK): 1973: start of the design; 1983: beginning of the operating period.

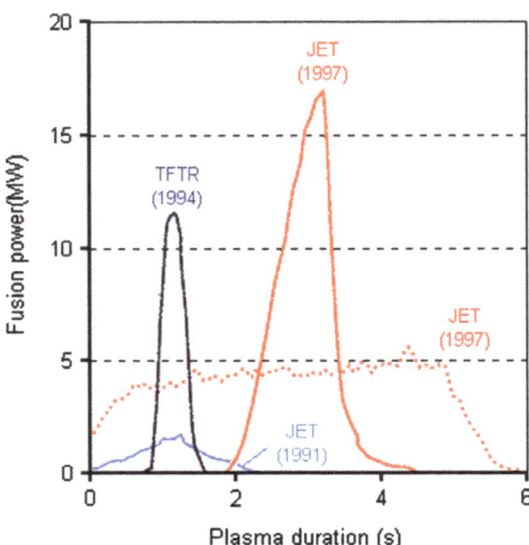

Figure 14.14. JET Joint Undertaking. 1991 - Preliminary Tritium Experiment (PTE): 1.7 MW fusion peak power; 2 MJ fusion energy; 1997 - deuterium-tririum experiments: 16 MW fusion power; 13.8 MJ fusion energy; 4 MW fusion power sustained for 4 s.

- 1991: first introduction of tritium into deuterium plasma, resulting in the production of 2 MJ by fusion, i.e. 10^{18} neutrons,

- 1997: 22 MJ in a single experiment, bringing the Q ratio (between the amount of thermal energy produced and the amount of electrical energy injected) to 0.6. The duration of the experiment was close to one second. The correctness of the models used was confirmed. For the first time, the effect of α heating was observed.

The way to the ITER Project

After many experimental facilities (for instance in France, TFR (Figure 14.15) and TORE SUPRA – now dedicated to the WEST program – (Figure 14.16)), the latest research initiative is the ITER Project (Figures 14.17 and 14.18), with the goal of building a Tokamak capable of producing hundreds of megawatts for many hundreds of seconds, with a Q value close to 10. Should the results be satisfactory, this will constitute a first step towards the subsequent construction of a demonstrator (DEMO). Physics, heavy-component technology and operations are all among the sectors that should be able to benefit the most from this type of overall programme (Figure 14.19).

18 superconductive coils generate the toroidal field of 5.3 Tesla, shaping the global configuration.

The central solenoid generates a variable magnetic flux inducing, inside the plasma, a current of about 15 MA. The magnetic field produced in this way combines with the field from the coils to set up the tokamak dynamic confinement, compensating for the drift of

Figure 14.15. General view of TFR (Tokamak de Fontenay-aux-Roses, CEA/FRANCE); designed as a slight extrapolation of the best performing Russian Tokamak T4, it has been the most powerful experimental facility worldwide from 1973 to 1976. Temperatures up to 2 keV were obtained and important results on confinement and plasma heating were achieved.

Figure 14.16. TORE SUPRA, in Cadarache – CEA/FRANCE.

the charged particle trajectories stemming from a mere toroidal field: the field lines are coiling themselves around toroidal surfaces.

The plasma (about 800 m^3 for about 500 MWth – about 1 MWth/m^3 plasma) is contained inside a vacuum chamber with a first wall. The heating is achieved by RF waves and charged particle injection, for about up to 50 MWth altogether, as well as by the interaction of α from fusion (about 100 MW for 500 MWth fusion energy release). The ^{4}He must be removed as well as the impurities: this is achieved with use of the divertor.

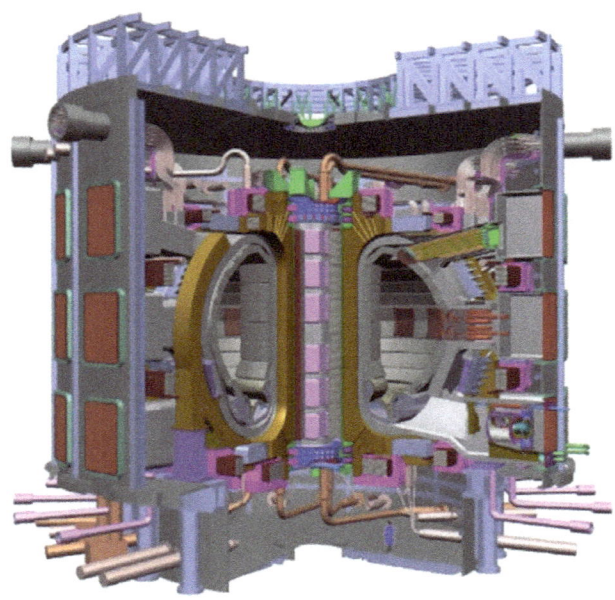

Figure 14.17. International Thermonuclear Experimental Reactor (ITER).

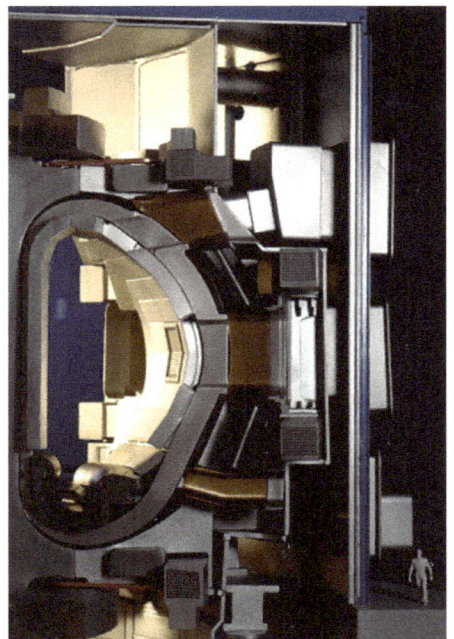

Main parameters

R: 6.2 m (the scale is represented by the man walking on the ground on the right side of the image)

a: 2 m

Toroïdal field B_T: 5.3 T

Current I_p: 15 MA

P_{fusion} : 410 MW

Neutron power flux (wall): 0.5 MW/m2

Q: 10 (ignition – marginally - possible)

Figure 14.18. ITER main design parameters.

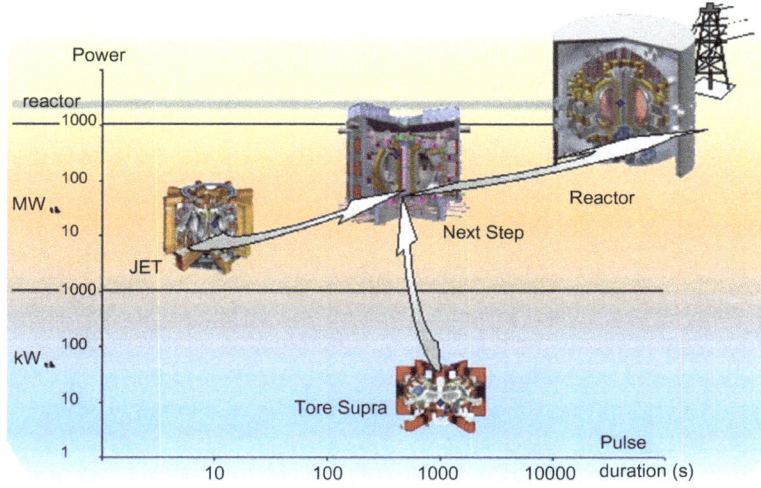

Figure 14.19. Mapping the way towards the fusion reactor.

The rest of the fusion energy (400 MWth) will escape as neutral particles from the plasma and will be stopped and used of in the blanket (thickness: about 50 cm) surrounding the torus, just beyond the first wall. A few TBM (tritium test blanket modules) will make it possible, during the second operation stage, to test tritium production and removal capability. The blanket works as a radiation shield protecting the coils and the vacuum chamber (28 m diameter, 25 m height). The neutron flux through the first wall is only 0.5 MW/m^2 but should increase significantly when moving to the DEMO. The same holds true for the divertor challenge.

14.4. Fusion by inertial confinement

14.4.1. Introduction: orders of magnitude

The fusion by inertial confinement was studied for weapon applications (NIF and MEGA-JOULE facilities belonging to the USA and France respectively) and an important civil effort was achieved with (more or less) powerful dedicated lasers, including for example:

- GEKKO XII (10 kJ) in OSAKA, being currently upgraded with the addition of a second "side-by-side" laser, the **LFEX** (*Laser for Fast Ignition Experiment*), in order to deliver a 10 kJ pulse of energy to a target in 10 picoseconds as a further exploration of the fast ignition regime.

- The OMEGA laser at the LLE (**Laboratory for Laser Energetics:** a scientific research facility of the University of Rochester), capable of delivering 30 kJ at up to 60 TW onto a target less than 1 millimeter in diameter.

- In France and in Europe, the HIPER project aims at preparing the study of energy generation, and The PETAL laser, located in the Aquitaine region of France, would be a fore-runner to the HiPER facility to address physics and technology issues of strategic relevance to HiPER.

- Besides these tools, the LULI laboratory located on the Ecole Polytechnique and Université Pierre et Marie Curie campus is dedicated, in connexion to the CEA, to the study of high energy laser/matter interaction.

The most powerful lasers currently under development will be able to deliver 2 MJ to a target, which is sufficient to heat a target of a few milligrams at the most. The process, in which confinement is achieved only by inertia, involves a series of microscopic explosions. The energy produced by a few mg of fuel gives values in the region of 500 MJ (100 kg of high explosive) which is around the limit of what can be contained on a regular basis in a reaction chamber.

To summarise, the mass burnt is around one mg and the spherical target will therefore have a diameter in the region of 1 mm (the density of DT being 2 g/cm^3 in the solid state).

14.4.2. Target ignition by hot point

- *The heating of a small mass of fuel*

    Without confinement, the target will disperse at a speed equal (in m/s) to $2 \times 10^5 \sqrt{T}$ where T is in keV. When T is between 10 and 20 keV, $V \approx 10^6$ m/s and $\tau \leq \nu s$. The ignition criterion imposes densities above 10^{30} particles/m^3, whereas the density of solid DT is around 5×10^{28} particles/m^3. The number of fusion reactions would be insufficient. As τ cannot be changed, compression is necessary. There would be an advantage in compressing prior to heating.

- *Compression of a solid target while heating the fuel as little as possible until the end*

    The principle consists of extremely rapidly vaporising the external layers of the target through the effects of high-power radiation. In the resulting ablation, the plasma produced expands outwards at a speed of 100 to 1000 km/s. The mass of fuel is subjected by reaction to considerable pressure (tens of Mbar) but which is insufficient to raise the density of the target to many hundred times more than the solid density (100 Gbar would be necessary).

- *The implosion of a hollow sphere was therefore proposed*

    The hollow sphere consists of a shell of solid DT surrounding gaseous DT. Ablation of the outer part under the effect of radiation causes pressurisation and also imparts motion to the shell, causing it to implode. At the end of the implosion, due to the kinetic energy gained, adequate densities are achieved. If, in addition, precautions are taken, relating in particular to the shape of the implosion, so as to heat the gas at the centre at the end of the implosion, the desired temperature conditions are attained at the central hotpoint. Burning of the DT begins at the hotpoint and propagates through the as yet unablated spherical shell. At this moment, the hotpoint is at a temperature exceeding 10 keV. The spherical shell has a temperature of a few hundreds of eV and density of many hundreds of grams per cubic centimetre.

A more sophisticated "quick ignition" process is being studied, featuring the use of extremely intense laser beams with the hope of decoupling compression from heating. This would require less energy from the compression laser beam and assure fuller burning of the target.

14.4.3. Instabilities

The energy source can be a laser system or, possibly, in the future, a particle accelerator (heavy or light ions). Interaction between the energy beam and the target can cause unacceptable disturbance which eliminates the use of some types of lasers (such carbon dioxide lasers). There are also other types of phenomena independent of the energy vector that can have unwanted effects.

The target is compressed by plasma expansion. This is an intrinsically unstable situation (Rayleigh-Taylor instability), with a fluid of low density (the expanding plasma) acting on a fluid of high density (the compressed solid fuel). It is appropriate to establish an interface that is as regular as possible, as any irregularities will necessarily be amplified over time (other effects, such as smoothing, mean that this statement must be qualified – see below).

Sphericality error essentially results from:

- irregularity in illumination of the target,

- geometrical irregularities of the target itself.

Given a target of initial radius R and final radius r (once compressed), with $R/r \sim 30$, for there to be a radius difference between any two points of less than r/4 at the end of implosion, the relative difference between any two points must not exceed 1% anywhere along the compression path, which is a difficult criterion to meet.

The sphere must be accurately calibrated, and sphericality error must be low, requiring polishing to within one-micron limits.

Recent significant improvements have been achieved in modeling Laser Megajoule ignition target behavior, including in the area of DT ice layer characterization. Roughness flaws and grooves are presently included in numerical simulations. It shows (Figure 14.20) that, assuming an evenly distributed flaw on the internal surface of the cryogenic layer, the thermonuclear gain of the nominal target is reduced by 10% for a roughness increase of the ice of 1 μm; of 60% if the increase is of 2 μm [7].

Comparable efforts must be made to ensure uniform illumination of the target. Two approaches are being studied:

- Direct attack, involving illumination of the target as uniformly as possible by increasing the number of beams: with more than 200 beams overlapping in a controlled manner and care being taken to ensure synchronisation and homogeneity of each.

- Indirect attack, with the target placed at the centre of a metal cylinder referred to as a hohlraum (see Figure 14.21). Here, the energy does not impinge directly on the target: as a result of interaction between the beams and the walls of the hohlraum, an intense burst of X-ray is produced. These X-rays then act on the target. The intermediate mechanism reduces the efficiency but helps to combine the incident beams into a single highly-isotropic effect and thus limiting the development of Rayleigh-Taylor instability.

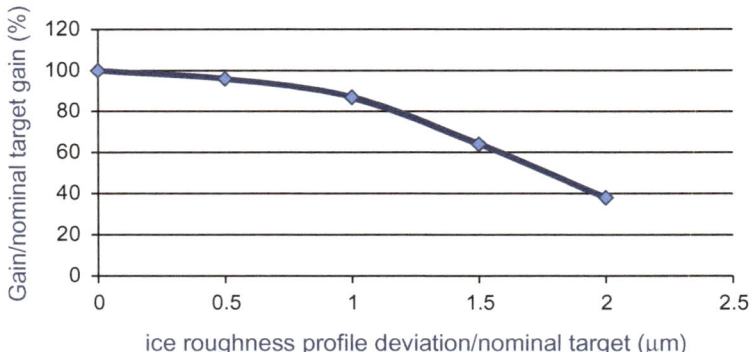

Figure 14.20. Thermonuclear gain of the target (related to the nominal target one) as a function of the ice layer roughness. (from CHOCS – April 2014: [7]).

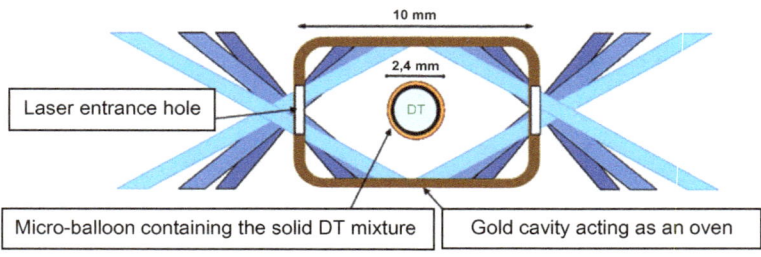

Figure 14.21. indirect attack (CEA).

14.4.4. Findings

- The Centurion-Halite program: experiments carried out as part of weapons testing in the USA have experimentally confirmed that the postulated ablation-compression-combustion is indeed a reality. But although the principle may be established, the experiments involved energy levels higher than those that can be reached in the laboratory, and do not give access to minimum levels necessary for civil use on an industrial scale in the virtually continuous production of energy.

- As concerns the results of experiments with lasers, typical achievements are illustrated by the findings obtained with the Omega laser in the USA in direct attack of a tiny capsule of glass filled with D-T under pressure (5 to 20 bar).

$N \tau T = 10^{19}$ in m^{-3} s kV

when T_{ions} is in the region of 13 keV and $T_{electrons}$ is close to 4 keV.

- The ratio between the energy produced (300 Joules, corresponding to 10^{14} neutrons) and the energy invested in the target is around 1%. Furthermore, in a Japanese experiment (GEKKO XII), densities were raised to more than 600 g/cm^3.

Considerable progress has recently been made in the field of implosion simulation, with:

- 2-D codes combining all the main physical phenomena,

- 3-D codes more fully describing the effects of Rayleigh-Taylor instability,

- codes covering laser-plasma interaction.

These codes have also been the subject of increasing confirmation by experiments.

The NIF Project in the USA and the *Laser Mégajoule* Project in France (Figure 14.22) have capitalised on these advances. In these projects energy levels are being raised from some thirty kJ to values in the region of 2 MJ. This should make it possible to observe the combustion wave and increase the ratio between the energy produced and the laser energy.

Figure 14.22. The Laser MegaJoule (LMJ) Experiment and its reaction chamber (CEA).

14.5. Reactor and associated technology

14.5.1. Reactor principle

Figure 14.23 shows the balance of the plant.

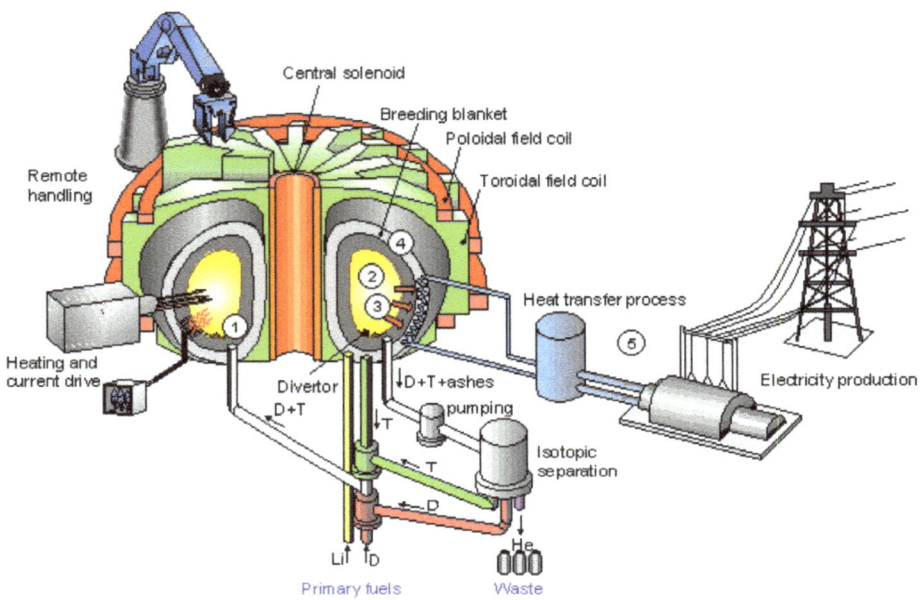

Figure 14.23. The "hot" section of a fusion reactor comprises a first wall, a blanket and a heat removal system within the vacuum chamber.

As concerns the manner in which the fuel is burnt, D-T fusion begins with a 14 MeV neutron reaction with the blanket material in which it loses its kinetic energy (which represents a wasteful use of a monokinetic flux of 14 MeV neutrons, in terms of fission neutronics). This has three results: heat production and removal, neutron shielding and tritium breeding. The heat is removed to a steam generator by a coolant that can be water, a gas (helium), a molten salt or a liquid metal. In the latter case, magnetohydrodynamic (MHD) effects can occur. Two means of limiting these effects have been studied: either the liquid metal is used as a coolant and made to circulate at high speed making it necessary to line the circulation tubes with insulating material, or it circulates more slowly in order to remove the tritium, in which case another fluid is required as a coolant.

14.5.2. Tritium production

Tritium-breeding materials

Figure 14.24 shows schematically the circuitry dedicated to the tritium cycle as it is intertwined with the heat removal systems, starting from the blankets surrounding the plasma.

Figure 14.25 shows a little more of the topology and of detailed integration issues in the specific framework of a project of Test Blanket Modules for ITER.

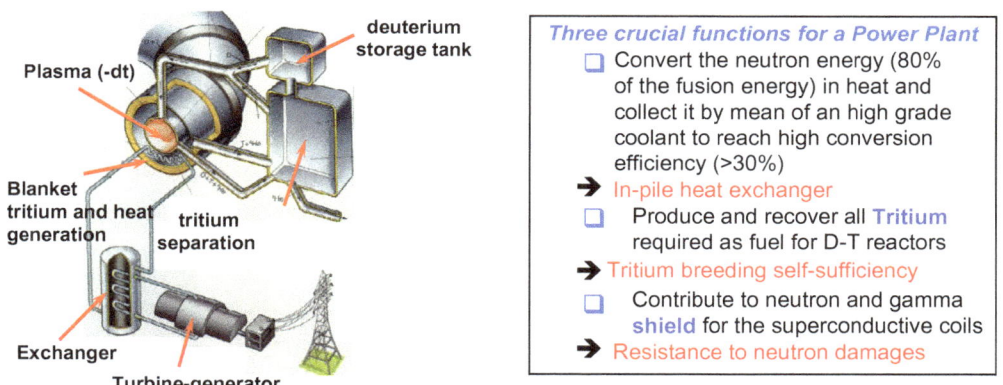

Figure 14.24. The circuitry dedicated to the tritium (breeding) cycle.

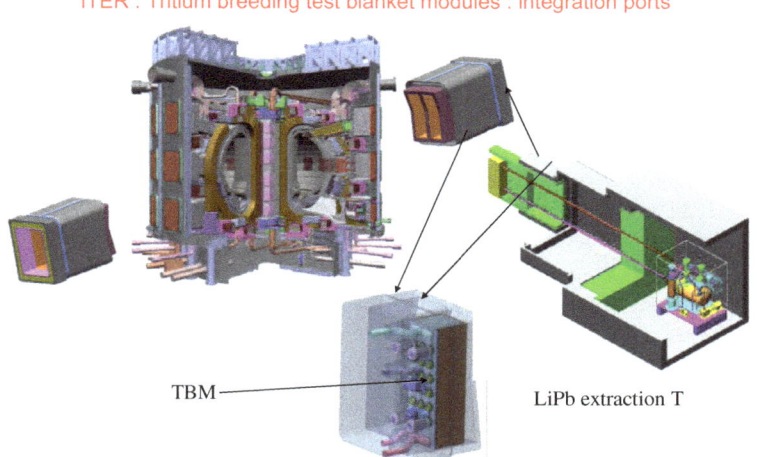

Figure 14.25. ITER: Tritium breeding Test Blanket Modules: integration ports.

The tritium reactions are as follows:

$$n + {}^6Li \rightarrow T + {}^4He + 4.78 \text{ MeV}$$
$$n + {}^7Li \rightarrow T + {}^4He + n - 2.47 \text{ MeV}$$

The second reaction is endothermic and has a threshold. The ^6Li reaction is fostered by using lithium enriched with ^6Li. Overall, these reactions add up to 20% to the energy produced.

The lithium is used in the following forms:

- liquid: metal (liquid), lead-lithium eutectic alloy ($Li_{17}Pb_{83}$), a molten salt Li_2BeF_4 (FLIBE),
- solid: lithium oxide (LiO_2) and various ternary compounds.

Figures 14.26 and 14.27 show a design of Test Module Blankets for ITER, with the associated integration issues and the associated circuits.

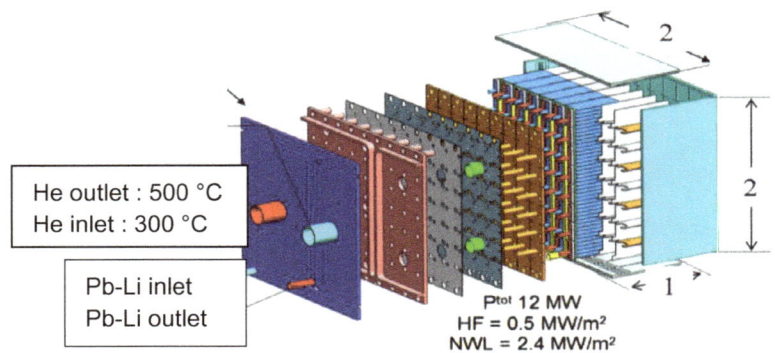

Figure 14.26. Functional/structural integration, HCLL technology (Helium Cooled, Liquid Lithium): for an expected TBR (Tritium Breeding Ratio) of 1.13, the key, tightly coupled issues are related to: thermomechanics; thermal-hydraulics; MHD; integration and systems design and technology; tritium management.

The neutronic balance of tritium breeding has a narrow margin. An amplification by the n,2n reaction is therefore added to the blanket. The lead in the lead-lithium alloy performs this role, as does the beryllium in the FLIBE, and additives in the form of particles, beads or plates can be used as solid tritium-breeding materials. The maximum acceptable rate of tritium loss is less than 10^{-4}, taking into account the existing rules for health protection. This is a formidable challenge, given the necessary circulation of tritium removal through a network of pipes, at high temperature, for rapid recycling.

Moreover, rapid tritium breeding is an absolute requirement for nuclear fusion, with high inventories (tens or even hundreds of kilograms), high fluxes and a short breeding doubling time. This is a more constraining issue than in the fission field, because there is almost no equivalent to the natural uranium playing the role of a "match" for starting up the process.

Architecture

This depends on the choices made in relation to materials, coolants and tritium extraction. It is, however, essential to make sure that there is a useful blanket covering ratio around the plasma to limit neutron loss. This is because if the factor is too low, even with a tritium breeding ratio greater than one in an infinite medium within the blanket, overall tritium breeding cannot be assured. Yet if deficient, it cannot be compensated for by the assistance

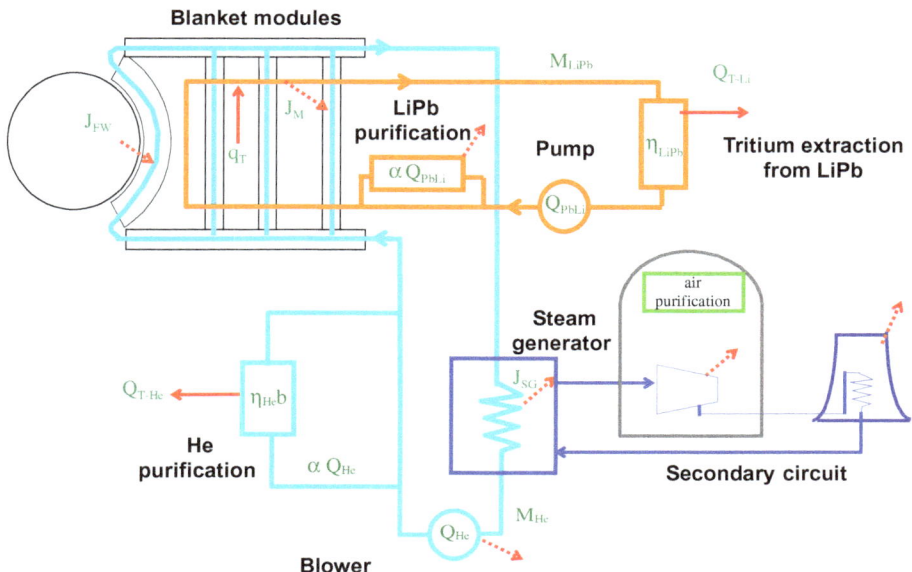

Figure 14.27. LiPb tritium removal systems, HCLL technology. Key related questions concern the need of T-permeation barriers and the extraction and purification efficiency, as well as the He chemistry.

of tritium-producing reactions in auxiliary fission reactors Pu breeding (at the best, one nucleus of T for 200 MeV of fission to complement the fuelling of a few fusion reactions at 17 MeV). Figure 14.28 illustrates how a crowd of densely packed heavy and bulky components is cluttering the space around the blanket.

Overall, the findings of the studies and experiments are reassuring, indicating that it will be possible to self-sustain the T inventory.

Other uses

The value of 14 MeV neutrons is exceptional in fission with $\eta = \nu \sigma_f / \sigma_a > 4$ for every heavy nuclei, including ^{238}U and ^{232}Th. Furthermore, their energy scatter is low, unlike spallation neutrons whose spectrum is of low quality (many neutrons of poor energy levels and high energy component which is of little use and causes considerable shielding problems). Finally, beginning at $Q \approx 2$ and quite perceptibly beyond $Q = 5$ (a point soon accessible, but from a significant unit power), a fusion reactor has an energy balance that is attractive (if not cost-effective) as a source of good-quality neutrons as compared to an accelerator. In the medium to long term, it may be feasible (but probably not required) to use fusion neutrons for applications relating to fission (particularly the net production of fissile nuclei from depleted uranium or thorium) within fusion-fission hybrids acting as amplifiers, under sub-critical conditions as fusion driven systems (FDS).

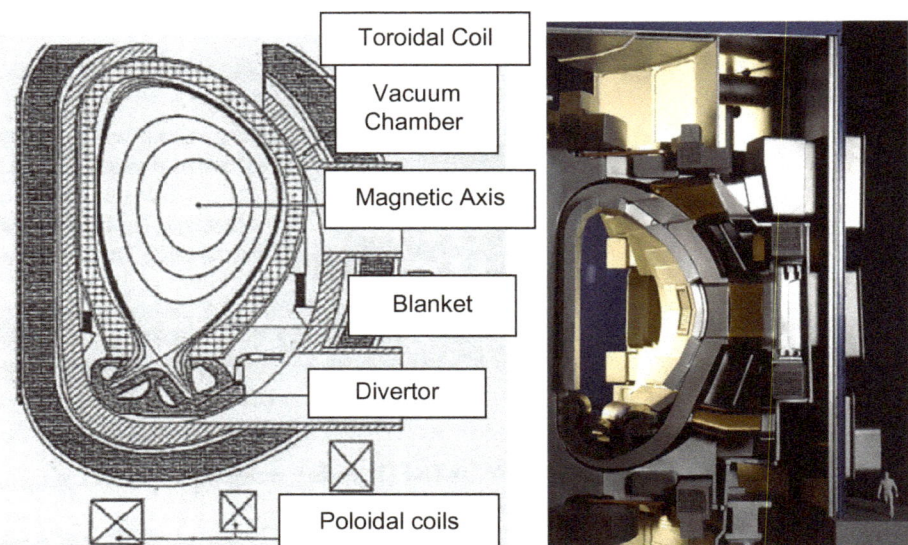

Figure 14.28. Tokamak components: schematic diagram and ITER model.

14.5.3. Materials

The structural materials are subjected to the neutron flux (Figure 14.28), raising problems caused by:

- their possible deterioration during service,
- the desire for minimum activation, to avoid losing the advantage of fusion relating to the radiological waste in the long term.

Three approaches are being investigated with a view to attaining the low-activation goals, ranging from the simplest to the most futuristic, while bearing in mind that in a near-term project use would be made of existing stainless steels with their known shortcomings (such as activation of the nickel of Type 316 austenitic stainless steel). The three are:

- The **adaptation of existing materials**, beginning with the grades of steel whose properties are consistent with the specifications, such as martensitic steels with 8 to 12% chromium. Allied nickel might be replaced. The following substitutions would be necessary in order to reduce the residual activity at 100 years by about two decades. Efforts are being made throughout the world to agree on the required properties of these evolutionary alloys:
    - tungsten and vanadium for molybdenum,
    - vanadium, titanium or tantalum for niobium.
- The **development of new alloys**: the elements least susceptible to activation are first identified then utilised to create new alloys – these are typically vanadium, chromium

(4%) and titanium (4%), these have additional potential qualities and their development represents a long-term effort,

- The **development of new materials**, notably SiC, not in monolithic form as it is too brittle, but as a composite of SiC fibres in a SiC matrix. This SiC/SiC composite is being studied in the aeronautical industry and needs to be tested under irradiation. It is also being considered for fission advanced fuel and structures.

This overview of the current approaches is intended to show the issues, the possible solutions and also the scale of the effort required and of the probable timescale which is of 20 to 30 years, making it one of the main avenues of the current R&D worldwide cooperation.

An advanced and powerful neutron irradiation source (IFMIF) is being considered to assess the properties of structural materials, with a neutron spectrum close to the fusion spectrum (thus "harder" than a fission spectrum). This spectrum leads in particular to a high He/dpa ratio. The best suited reaction looks to be the deuteron stripping on light nuclei (Li, C). IFMIF has a 2 × 125 mA beam of 40 MeV deuterons impinging on a liquid Li target. It can be compared to SPIRAL-2 (GANIL) with a beam of 5 mA of deuterons on a C target. An intermediate power level around 20 mA could represent a good trade-off for developing basic science on materials under irradiation with the help of multi-scale numerical simulation, starting from ab initio calculations. Such a facility would represent an Accelerator Driven Optimised Neutron Irradiation Source (ADONIS, see below, in the "Futuristic Systems" chapter, chapter 15).

14.6. The reactor: magnetic fusion

14.6.1. Energy efficiency

Competitiveness depends in particular on energy efficiency, as the recirculation of excessive amounts of energy for feeding the plasma entails extra costs.

The self-sustainability criterion can be written as $P_\alpha > P_{losses}$.

This criterion is difficult to satisfy and, for convenience, it can be opted to re-inject some power taken from the overall output (recirculated power). This then gives:

$P_\alpha + P_{injected} > P_{losses}$

yet $P_{fusion} = 5 P_\alpha$

and if $P_{fusion}/P_{injected} = Q$

$P_\alpha(1 + 5/Q) = P_{fusion} (0.2+1/Q) > P_{losses}$

Q cannot be reduced too much as the electric power recirculated, $P_{injected}/\eta_{heating}$, ($\eta_{heating}$ being the heating mean, all-inclusive efficiency, from raw power to the right mix of RF heating, of current induction, of neutral particle injection) is costly and should remain below 20% of the total electric available power, which can be written as follows:

$(P_{fusion} + P_{injected}) * \eta_{\theta\delta}$ (thermodynamic efficiency of the plant conversion system)

It is therefore necessary that: $(P_{injected}/\eta_{heating})/[(P_{fusion} + P_{injected}) * \eta_{\theta\delta}] < 0.2$, or equivalently:

$\eta_{heating} * \eta_{\theta\delta} * 0.2 * (Q+1) > 1$ With optimistic but typical efficiencies, it is found that Q should be higher than 20.

Under these conditions, the further potential gain on P_α, hence the value of $n\tau$ to be achieved, remains low. A magnetically-confined fusion reactor can therefore operate just below the ignition point, which can be expressed in a simplified way as $n\tau \sim 10^{20} m^{-3} s$. It works in **amplification mode**. Moreover, active and powerful heating (above a given threshold of heating power, in H-mode – see below in section 14.11.1) can help achieve sustainable and smooth operation during long pulses. In this framework, a pure, self-sustained ignition regime is not an absolute goal any more.

A numerical example avoiding a large extrapolation of the performances beyond ITER can be taken with current data without contradicting the general conclusions. Let us take Q = 10 (similar to ITER), and include $P_{injected}$ as well the neutral and wave heating as the plasma current induction energy tapping (with a typical efficiency of 50% from "raw electricity" to heating and to induction related "active electrical" power). Assuming a net Balance of Plant (from plasma removed heat to available electricity) thermodynamic efficiency of 40% makes it possible to avoid very high temperatures in the systems performing both heat and tritium removal functions with mere liners for limiting tritium escape and potential release.

Taking:

P_{fus}: heat removed by fusion;

$P_{reactor}$: $P_{fus} + P_{inj}$ (+ exothermic conversion reactions)

η_{thd}: thermodynamic efficiency of the fusion reactor conversion system

P_{inj}: "active" power (accelerated charged particles, waves on cyclotronic frequencies, plasma current induction) injected inside the plasma; with Q = 10, $P_{inj} = 0.1 P_{fus}$;

$\eta_{elec \rightarrow active}$: efficiency of "active power" generation from raw electricity. The electricity can be taken from the grid or diverted from plasma heat release and transformation; the second assumption will be adopted.

P_{div}: Power diverted from the plasma-generated electricity.

P_{elec}: power generated from plasma released heat (including P_{fus} and P_{inj}, as well as exothermic nuclear reaction balance bonuses).

P_{grid}: power released to the grid. A main assumption is that the pulse duration is long enough to keep the starting transient phenomena at a low level of power diversion and waste, when compared to the mean values calculated below.

Simplified Global Fusion reactor balance (steady, at "nominal" power):

$P_{reactor} \sim P_{fus} + P_{inj}$; $Q = 10 \rightarrow 1.1 * P_{fus}$ (**thermal**) $\rightarrow P_{elec} = \eta_{thd}(P_{fus} + P_{inj}) = 0.44 P_{fus}$

Back-stream:

$0.44 P_{fus} \rightarrow P_{div}$ (by anticipation: $0.2 P_{fus}$, see below) + Pgrid = $0.24 P_{fus}$

Up-stream:
Power diverted and re-injected: Pdiv = Pinj/$\eta_{elec \rightarrow active}$ =(P_{fus}/Q)/$\eta_{elec \rightarrow active}$: 0.2 Pfus = 0.1 Pfus/0.5
→ "active" injected power: Pinj = 0.1 Pfus = $\eta_{elec \rightarrow act}$*Pdiv = 0.5 × 0.2 = 0.1 Pfus, consistent with Q = 10 (including heating and current induction)

The global efficiency is thus only ~ 25% (even with a thermodynamic efficiency of 40%).

P_{div}: 0.2 P_{fus} → P_{inj}: 0.1 P_{fus} → P_{plas}: 1.1 P_{fus} → P_{elec}: 0.44 P_{fus} → Pgrid: 0.24 P_{fus}
↑———————————————————————←———————————————————————↓
P_{div} = 0.2 Pfus

This approximation holds true during "nominal" operation but the main drawback would be a low availability factor (k_d), caused by the pulsed regime, the maintenance downtime (the maintenance being mainly performed by remote handling on high activity structures) and, even when minimising external disruption events, energetic incidental transients leading to repairs before restart. This has to be compared to the requirements on the availability of current power plants working in a baseload regime.

Still within the context of the Tokamak, two complex yet vital components need to be described: the electromagnet system and the ash removal system, referred to as the divertor.

14.6.2. Superconducting electromagnets

The coils, in which high currents circulate (constant for the toroidal coil and variable for the poloidal coils), would, if of water-cooled copper, consume a considerable amount of energy. Use of superconducting coils would reduce the requirement to adopt cryogenic technology (4K). The particularities of such coils in a fusion reactor are:

- their size. In the context of the ITER project, there would be twenty D-shaped coils for the toroidal magnet, which is 18 metres high and 12 metres wide in order to achieve 6 Tesla inside a disk of 8 m in diameter. The various forces exerted on the coils amount to many hundreds of MegaNewtons. The mass of a coil is around 700 tons, most of which corresponds to the mass of the structural steelwork. Some 100 GJ of energy are stored in the toroidal magnet system, requiring special precautions, particularly to address the case of undesired transition from the superconducting state to the resistive state;

- the use of variable fields imposes additional design and construction constraints as they are sensitive to mechanical and thermal shocks.

14.6.3. Divertor

This device continuously extracts the fusion ash and prevents the reaction from being choked. Figure 14.33 shows an image of the divertor of TORE SUPRA in the WEST project configuration. Beyond a certain surface (the separator), the magnetic surfaces are open and reach the wall at the level of the divertor where pumps evacuate the area, ensuring overall renewal of the plasma. The flow of particles along the two branches of the divertor constitutes a veritable thermal jet, and the energy deposited on the surfaces encountered can reach many tens of megawatts per square metre. Sophisticated impurity management reduces the heat load to less than 5 MW/m². The design of the divertor tiles is a very delicate matter. Experiments are necessary to achieve the full potential of modelling in this area.

Coils are critical components. They are sophisticated and massive. They must be very reliable, as the cost of any downtime and replacement of one or several windings would present a significant drawback.

14.7. The reactor: inertial fusion

14.7.1. The positive energy balance criterion

Analysis of the global energy balance for combustion of a target gives the following criterion for gain G, the total amount of energy produced divided by the energy injected by the energy source (e.g. a laser):

$$G \geq 100$$

Such gain values can only be obtained if the fraction of fuel burnt is $\geq 30\%$.

The following formulation can be used for comparison with magnetic-confinement fusion:

$$n \tau > 2 \times 10^{21} \text{ s/m}^3$$

which is an order of magnitude greater than the ignition condition. Indeed, point ignition is not sufficient, and it is essential to achieve proper propagation of the combustion through a substantial proportion of the mass of the fuel.

The three main parts of the reactor are the energy source, the reaction chamber and the target management system.

14.7.2. Energy source

The efficiency of the lasers currently in use (glass doped with neodymium) is less than 1% and rate of fire of powerful lasers cannot exceed a few a day due to thermal and thermo-mechanical effects. A reactor requires efficiency levels above 10% and rates of fire of around 5 to 10 Hz. Other approaches have therefore been investigated:

- adaptation of lasers: considerably improved laser efficiency can be attained with the possibility of the required 10% being reached, but problems remain (thermal, lifetime, etc.),

- use of particle beams: their power levels can amount to TW and their pulse energy to MJ and they are capable of rates of fire well above the requirements of inertia-confinement fusion. Their (all-inclusive) energy efficiency could in the future exceed 20%; the ion beam penetrates a converter (metal with a high atomic number), causing X-rays to be released. For a given total energy:

- light ions are accelerated to some 10 MeV with high current levels (10 MA),

- heavy ions are accelerated to 10 GeV with lower current levels (10 kA).

Heavy ions are favoured with (cyclopean) linear accelerators, and possibly with storage rings. It is difficult to maintain the quality of the beam (in terms of W/cm^2), as a number of phenomena occur (space charge and instability). Indeed, there is an obvious discrepancy between the current specifications (10 MeV and 10 MA for light ions) and current developments in the field of spallation (ADS): protons: 100 mA at 1 GeV.

14.7.3. Reaction chamber

In the chamber, a series of 600 MJ reactions occur in succession (2/3 neutrons, 1/3 X-photons and charged particles). The depth of the energy deposits is variable, X-rays and charged particles having little penetration power. The first wall suffers ablation accompanied by mechanical loading. It is rapidly vaporised, creating a shockwave that hits the wall structure. Induced stresses can reach many tens of MPa, and occur at a rate of five times a second. This exposes the materials to mechanical and thermal fatigue. In addition, "the final optics", the last mirror (laser beam) or the last quadripolar focusing device (ion beam) is in direct view of the target and is thus exposed to the same effects as the chamber. Therefore, special means of protecting it need to be devised.

Three main chamber concepts are being studied:

- the dry wall type, seeking to avoid evaporation of the first wall which is of refractory material (carbon),

- the wet wall type, with a porous wall through which liquids can percolate (molten lead, molten salt, etc.),

- the liquid wall type, in which the lining is a complex molten salt flow system.

14.7.4. Targets

Fabrication

At a frequency of 5 Hz, the reactor consumes 400,000 targets per day, each of which releases 600 MJ$_{th}$, i.e. 50 kWh$_e$ with an efficiency of 30%. At an estimated production cost of €0.05 per kWh$_e$ from conventional energy sources (coal, nuclear), each target produces an amount of electrical power worth about €2.5. The cost must therefore not exceed around 0.5 to 1 €. Currently, target production rates barely exceed a few a day (for the simplest ones), and the unit cost quoted is in excess of 1 k€. Moreover, the specifications concerning the target ice layer roughness (see above, § 13.3.3) are very impressive.

Injection and aiming

A target would be shot into the chamber by a gas gun every 200 ms. The acceleration exerted on it would be some 100 g and its velocity would be 100 m/s. During its trajectory, the temperature of the frozen target (D-T ice at 19 K) would have to remain constant within 0.5 K in an environment which would be at some 800 K.

Regarding the process of aiming at the target, beam alignment now takes more than one hour. Even when drastically reducing this value, the gun used would not be able to assure proper positioning alone. The beams would thus need to be aligned with a flying target, or is there any alternative?

14.7.5. In summary

The breakthrough cascade required to reach an industrial feasibility, without even trying to address the competitiveness issue, seems to be still beyond the reach of any predictable R&D and qualification path in this current century.

14.8. Nuclear safety

14.8.1. Normal operation: containment of toxic substances

Protection is required against traces of tritium (which is highly mobile and rapidly escapes) and traces of radioactive substances found in the gases and liquids circulating within the reactor. This is because the substances resulting from activation of the structure are rendered mobile by corrosion phenomena and entrained by these fluids.

14.8.2. Accident situations: a few remarks

- Any impurities extinguish the reaction,

- Careful consideration needs to be given to the mass of tritium present (at the multi-kg level),

- Studies of accident scenarios indicate limited consequences, even in the event of the use of water as the coolant which would make the release of hydrogen possible. None of the scenarios seems to necessitate evacuation of the nearby population, which is satisfactory in terms of current and of future criteria.

14.9. Waste

The mass of waste produced is considerable, with low long-term radioactivity.

Fusion with inertial and magnetic confinement represents an energy source of low power per unit volume. For a given power rating, the amounts of concrete, vessel steel etc. are greater than in the case of existing nuclear and conventional power plants. The total mass of active waste (all levels combined) would typically be between 50,000 and 100,000 tons for a power reactor, corresponding to disposal of some 40,000 m^3.

Some of the materials could be recycled. In addition, if a policy of interim storage for less than a century were adopted, deep disposal would not be necessary, particularly if low-activation materials were used (Model 2 as compared to the more conventional Model 1, see Figure 14.29).

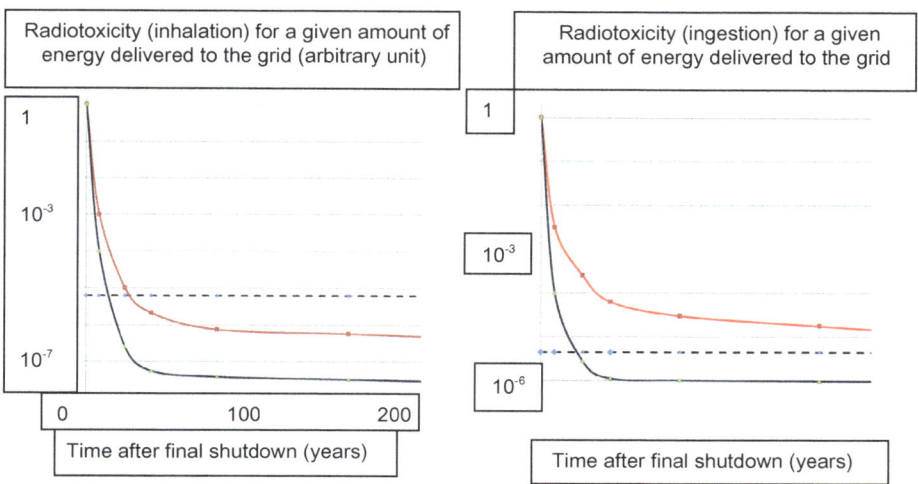

Figure 14.29. comparison of radiotoxicity vs. time (associated respectively to the inhalation and to the ingestion risk) for coal (in dotted line) and for fusion with two different options of materials: Model 1 in red and Model 2 (low activation) in dark blue.

In terms of overall radiotoxicity, for the same amount of energy produced, Figure 14.27 compares fusion using Models 1 and 2 with coal.

14.10. Costs

14.10.1. Composition of costs and orders of magnitude

With a deuterium price of a few k$ per kg (assuming that lithium will be available and affordable in the long term, which seems to be a reasonable expectation given its significant mean concentration in the Earth crust, comparable with the one of the main "intermediate" metals used worldwide at a high – and growing – annual consumption level) and with a consumption rate around 500 g per day, for a 1 GW_e reactor, the cost of the fuel would amount to less than one percent of the total cost per kWh_e. The cost of generation would therefore depend on the capital cost of the massive installation (a low power density reactor section surrounded by heavy components) plus the cost of routine replacement of aging components (such as the first wall).

Rough estimates now place the cost per kWh_e of magnetic confinement fusion power generation, primarily dedicated to baseload power, well above that of conventional production means like coal, oil, gas and fission.

A meaningful capital cost indicator is the steel (and concrete) mass for a given annual power production (or equivalently per "MWe-average").

Obviously, it depends on the specific quality of the materials and, in this case, nuclear facilities specific costs (per ton of steel, see for example the heavy superconducting coils: in ITER – 500 MWth – 18/20 coils of 700 t each) are pretty high.

When included in the following diagram from Per Peterson (UC Berkeley), fusion would certainly be in the upper area when compared with gas, coal and existing nuclear power.

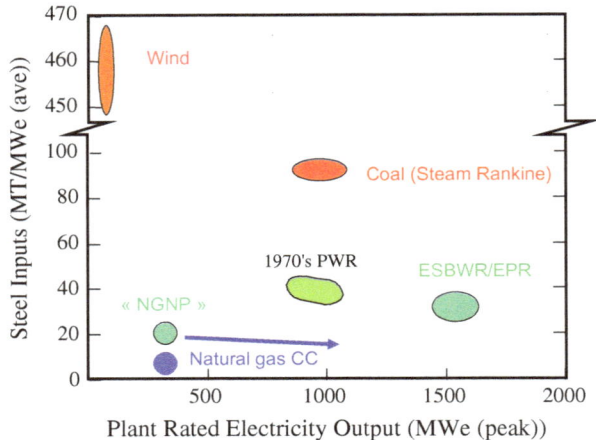

Figure 14.30. Steel and concrete mass per MWeff installed (after P. Peterson / UC Berkeley).

14.10.2. *Ecological impact and external costs*

The ExternE study was conducted in Europe between 1991 and 1998 to assess the external costs of the different existing sources of electrical power. It proposed and substantiated a general method that can be used to predict the life-cycle costs of a fusion reactor (at least a comprehensive assessment in order of magnitude).

The basic finding is that, in terms of the cost per kWh_e generated, the external costs of fusion would be around 1 m€. This can be compared to the following:

- the direct cost of generation per kWh_e is currently around 50 m€,

- the external costs of coal-fired generation are above 10 m€ per kWh_e, (it depends on the "cost" at the CO_2 ton),

- the external costs of photovoltaic generation without storage or support systems exceeds 1 m€ per kWh_e.

Such studies are complex and, in the case of fusion, are very provisional. However, this gives a broad outline of the strong advantages of fusion in terms of quality.

Fission has also a strong position in this field.

14.11. Historical trends, current challenges; R&D ways and needs

14.11.1. Historical trends and current challenges

An intermittent (long pulses) energy amplification operation mode, driven by an active control of current profile and of plasma stability, seems to be the "mainstream" path to magnetic confinement fusion in TOKAMAKS. The active control would make use of the powerful, vectored, sustained and sophisticated heating driven by instrumentation informed real time decision algorithms. Fusion would keep out of the pure ignition regime, with a high Q value around 10 or more, and with an unsteady current drive (consuming some more energy). It would be possible to break the growth of instabilities (evolving along a multi-scale time range to generate chaotic bifurcations and disruptions). The stability would be protected by artificially sustained transport barriers achieving some sort of "quiescent" core plasma zone. This could lead to improvements beyond the present H-mode in terms of plasma performance and of scaling law, while maintaining a sub-ignition regime and an amplification mode of operation.

Figure 14.31 shows a map of some advanced modes of confinement on Tokamaks and of the related plasma pressure distribution through the plasma. It shows the increased gradient inside the plasma in the case of Internal Transport Barrier mode implementation, in addition to the Edge Transport Barrier (ETB) providing a steep gradient between the edge plasma and the internal plasma, characteristic of the H mode.

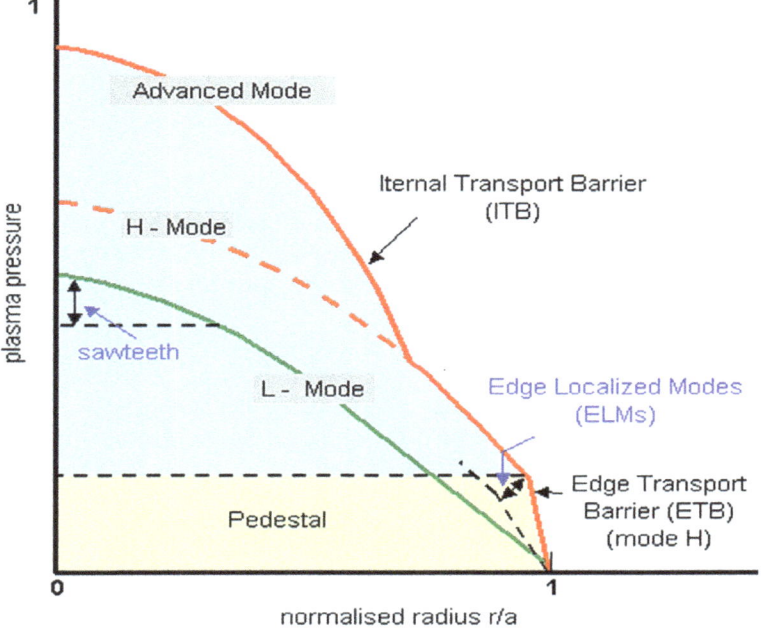

Figure 14.31.

The relation between the edge plasma issue (turbulent 3D transport), the H mode and the construction of transport barriers involving a powerful, vectored heating injection, is a key point.

Separation by the edge plasma of "quiescent" hot plasma core from the separatrix zone colder plasma leads to strong gradients. Increased gradients close to the edge make it possible to reach increased values of plasma density and temperatures in the core plasma region, but the induced (increased) transverse transport (turbulence driven, more convective and ballistic than diffusive/collisional) leads to confinement degradation.

In order to improve the confinement mode (H mode and potentially beyond) in divertor machines, interest lies in studying how artificially forcing a sheared rotation (Figure 14.32) induces a transport barrier: turbulent structures, cells, are disintegrated, shattered and swept away before growing too big.

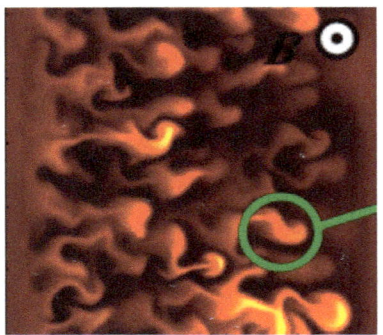 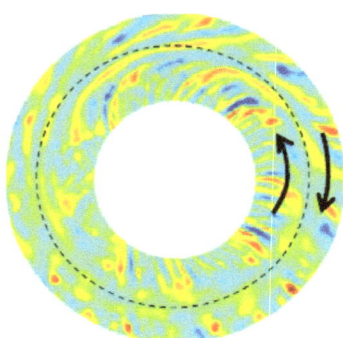

Figure 14.32. A pattern of turbulence related effects and how a sheared rotation could help creating a transport barrier (after Patrick Tamain - CEA).

In the framework of the WEST program on TORE SUPRA (Figure 14.33), shifting to Divertor geometry (W standing for Tungsten – Wolfram – structures facing the plasma), H-mode physics and first wall / plasma interaction will be studied.

The turbulent transport in the edge plasma is a challenge for simulation, involving many phenomena related to the interplay between turbulence and larger scale flows and fluxes, for instance the ballistic propagation of coherent structures (multi-scale phenomenon, avalanche-type).

Could a sustained vectored power injection (from heating?), above a given power threshold, drive and harness the useful phenomena? A global understanding requires a self-consistent description of the transition to operation regimes based on transport barriers as well as on an active control (beyond a threshold power injection). This active control requires a rapid development in the area of the instrumentation and of its on line interpretation.

Another key issue in the perspective of sustained powerful heating is the interaction of the RF antennas with the plasma.

If there is no such breakthrough, with concurrent simulation, instrumentation and control improvements, no technological progress will be able to address by brute force the basic instability weakness because it resembles the Houdini syndrome (after Guy Laval in [6]). Even if this fundamental difficulty is overcome, the cost will be high in terms of the

Separatrix in « limiteur » geometry

Divertor in TORE SUPRA–WEST configuration

 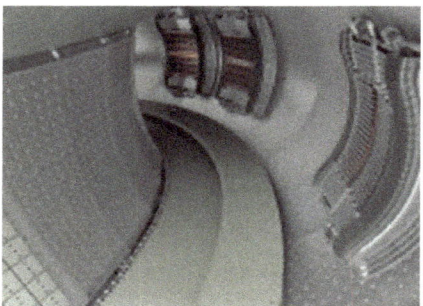

looks better

Figure 14.33. Images of the first wall equipped with rf antennas, as well as of the divertor in the WEST project, on TORE SUPRA.

capital cost per kWe installed (to be released to the power grid, not only re-circulated inside the system to feed heating and current drive). The same holds true for the global efficiency of the plant, even with high Q (higher than 10), which could well be limited to around 25%, whatever the thermodynamic efficiency of the heat to power conversion.

In summary, a low power density nuclear system, with a very large mass of high quality steel per installed MWe (dedicated to the grid), and with a low overall efficiency and a low availability factor, could well be a balanced picture of what could be attained through successful development but without breakthrough. Both tritium issues, breeding and ultra-high efficiency containment, all present difficult challenges.

14.11.2. R&D trends and needs [2]

The first plasma in ITER could be achieved in 2020, and the thermonuclear operation could start around 2027. The other major programs considered at the moment belong to the broader approach of the Japanese – European agreement. It includes a TOKAMAK (JT-60SA) dedicated to some relevant operational regimes at the post-ITER R&D stage, as well as the design and the prototype realisation for the Cyclopean accelerator driven irradiation source dedicated to the simulation of fusion neutron generated damages (dpa, He): IFMIF. These R&D programs are structuring the work to be done during the next 20 years.

In addition, and for Europe alone, the German Stellarator (see Figures 14.7, 14.8) Wendelstein W7-X, is expected to deliver its first plasma around 2015.

As for the TOKAMAKS, two main research topics are:

- plasma optimisation, as well in the hot and dense plasma core where steady fusion will occur during long pulses as at the plasma periphery, in the interaction of the colder plasma with the first wall and the divertor,
- system integration, including (active control) heating, measurements and real-time decision making and control software and associated hardware.

The major related R&D fields and programs are:

- the optimisation of the plasma configuration and control which requires a better understanding of the turbulence in the plasma and of the MHD stability conditions. A modelling approach closer to "ab initio", coupling kinetics and MHD, is necessary. The numerical simulation (numerical/virtual plasma) must be validated by dedicated experiments. It will improve performance by design as well as helping to develop and test plasma control devices (active control).
    - MHD stability: controlling (by active control) the instabilities that could disturb the magnetic structure and lead to disruptions. Furthermore, the ITER operation regime will be new and dominated by the α particles generated by the fusion. They will provide, close to the ignition regime, most of the heating (1/5 of the 500 MWth delivered by the fusion is taken by the α, compared with about 50 MW of external heating in total: antennas plus neutral beams). A major issue will be the study of the instabilities that could be generated in this new, highly non-linear regime. This requires hybrid MHD-kinetics simulation codes.
    - Turbulent transport: a "gyro-kinetics" approach is required to deal with the interactions between the diverse classes of particles. A smart (involving clustering in space, and velocity) multi-scale simulation capability has to be developed to take into account the space scales (from cyclotronic electron trajectory (a few μm) to the reactor scale) as well as the time range for instability development, over numerous decades. Applied mathematics, physics and numerical-software expertise are challenged by this critical issue that makes a dramatic difference when compared to the linear Boltzmann fission case but could beneficially be brought closer to the more conventional case of the multiphase/multi-fluid thermal-hydraulics.
    - Wall-plasma interaction in ITER. Beyond the operational issue, a global understanding would make it possible to set up a self-consistent simulation of potential transition towards transport barrier driven regimes.

The R&D needs in this field include numerical simulation platforms with stringent hardware and software specifications.

- System integration requires powerful tools, as well in the experimental facility area as in the modelling area. The experimental facilities are either of integral or of analytical nature. The integral experiment facilities are first the tokamaks themselves, thus TORE SUPRA (and potentially JET and ASDEX-upgrade (IPP Garching)). Specific test benchs (instrumentation, heating, "flight simulators") must be considered.

Two steps in the system integration will be to firstly come up with an integrated design in which many systems of diverse nature (functional, structural) must cooperate, with numerous interfaces and under hard environmental conditions (temperature, transients, damage sources: radiation, chemistry, thermal shocks); and secondly, to ensure optimised operation towards high plasma energetic performance, under smart control, keeping out of the unstable zones.

The design integration will be based on advanced materials and on innovative material assembly technology, as well as on two critical tools requiring further development: heating and instrumentation.

14.12. Conclusion

Civil research into controlled nuclear fusion truly began in 1958 as a result of cooperation that led to the declassification of the existing knowledge and an openness that was particularly remarkable at that time. Since then, cooperation has never faded away and has resulted in considerable progress: as regards the product $n \tau T$, an improvement factor exceeding several millions has been achieved, and recently, MW and MJ have been produced.

When considered and assessed on the basis of the criteria anticipated for reactors of the future, the concepts that appear to be the most feasible and the best candidates for commercial use in the long term (more than fifty years in the absence of any tightly coupled physical and technical breakthroughs), such as magnetic confinement fusion with a Tokamak, would face the following challenges:

- Competitiveness: This is an immense challenge, but the situation determining the competitiveness of fossil sources will evolve (resources and externalities) so that the competition in the very long run could be presented by sustainable fission and renewables (with adapted high capacity storage tools).

- Nuclear safety: Initial assessments of the nuclear steam supply system are extremely encouraging, but more exhaustive studies are needed with allowance for the entire fuel cycle, and even though the industry may be familiar with it, tritium remains problematic.

- Cleanness: This is a strong point for the long term, however in the short term vigilance and careful containment for several centuries will be required.

- Thriftiness (in terms of – the immensity of the – natural resources): Deuterium will never be scarce and if the D-T cycle is retained it will be lithium whose levels needs to be monitored as a raw material.

- Resistance to proliferation: This depends on both intrinsic properties and on adequate safeguard, while tritium presents a special case, being very remote from easy application and hence without strong related incentive.

- Other utilisations: Before contributing to power generation, fusion has the unique advantage, at least with the D-T cycle with its 14 MeV neutrons, of mutual enhancement of the use of fission, fusion and charged particles; it is a potential source of excellent neutrons (14 MeV) that could complement a tight neutron balance for fission in terms of all the future requirements (breeding of fissile materials and possibly transmutation). However, when any (critical) fission (fast or thermal) breeder – or quasi-breeder with a breeding gain close to 0 succeeds, such a neutron make-up is not needed any more. This means that fission R&D should not be slowed down but should keep a sustained momentum, betting on its own achievements as well as on future synergetic potentialities (see chapter 13).

Reference

[1] R. Aymar, P. Barrabaschi, Y. Shimomura, *The ITER Design*, Plasma Phys. Control. Fusion **44**, 519–565, 2002.
[2] A. Grosman – Private communication about the R&D European and French roadmap; Joseph Weisse, The scientific content of this chapter is partly inspired by an internal CEA document by J. Weisse (around 2000).
[3] J. Wesson, *Tokamaks*, 3rd edition, International Series of Monographs on Physics (118). Oxford Science Publication – Clarendon Press-Oxford, 2004.
[4] J. Lachkar, private communication.
[5] R. Lenain, private communication on Tritium Blanket Modules design and technology for ITER.
[6] G. Laval, *L'énergie bleue – Histoire de la fusion nucléaire*. Odile Jacob, 2007.
[7] L. Bitaud, B. Canaud, *Hydrogen isotopes at the core of inertial confinement fusion – (CEA/DAM) in CHOCS*. Revue scientifique et technique de la Direction des Applications Militaires – N° 43 – April 2014.
[8] A. Ertaud, F. Prevot, *A dialogue about the thermonuclear fusion power production future*, Rev. Gén. Nucl. May – June 1985.

15 Futuristic systems: ADS, Space Nuclear propulsion and power generation, ADNIS

J.-B. Thomas

15.1. Accelerator Driven Systems (ADS)

15.1.1. Introduction

ADS come from time immemorial. Figure 15.1 shows the early systematisation of the synergy between several ways of producing energy and neutrons, primarily for breeding. This synergy has given rise to proposals since the fifties, before the systematisation occurred in the late seventies, in the wake of the INFCEP initiative launched by the US President Jimmy Carter and mainly dedicated to the proliferation issue. Fast breeders were put under pressure, so people looked for alternatives including fusion-fission hybrids and ADS. The main objective was to generate "cheap" neutrons for subcritical fission breeders, feeding a fleet of conventional power reactors. Fusion-fission hybrids and ADS were assessed in many places, including CEA.

The ADS concept was considered again in the early nineties in connection with another issue which also involved a high consumption of available neutrons, beyond keeping the fission chain reaction running (Bowman, for instance). Carlo Rubbia launched around 1993 the "Rubbiatron" which was intended to become the next generation of nuclear power reactor. The ADS had to be assessed anew, either as a functional enabler for transmutation (a dedicated "transmuter"), or as a new and unique type of nuclear system for the worldwide fleet.

In the framework of a closed uranium/plutonium fuel cycle, if an advanced fast reactor is feasible, acceptable and competitive, it can achieve all the objectives because obtaining a fission from each heavy nucleus means simultaneously maximising the amount of energy from natural resources and minimising the long term radiotoxicity. Even in a more limited perspective such as GNEP (Global Nuclear Energy Partnership, a US initiative launched in 2006) in which the fast reactors are only considered as burners (and in so doing escape the constraints of competitiveness), and where there is no *a priori* transmutation option, there is no place for ADS.

The following paragraphs will show how the intrinsic strengths and weaknesses of the ADS concept defined a functional niche as "second strata transmuter", should fast reactors not be able to enter the marketplace.

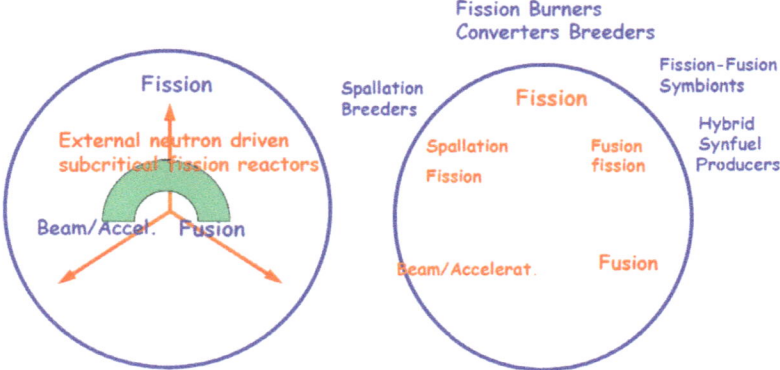

Figure 15.1. Nuclear Energy Synergetics: the nuclear continuum ([2]).

15.1.2. The physics of ADS. Basic principles and first design consequences

The ADS is a fission reactor operated in a subcritical regime, as an amplifier of a neutron source generated by a nuclear reaction between accelerated charged particles and the nuclei of a target, in one or two steps (if intermediate γ are used). Depending on the type (electrons, protons, deuterons) and energy of incident particles, several types of reactions, with diverse efficiencies, advantages and drawbacks, can be used.

As for industrial ADS considered in the field of nuclear energy generation, the energy efficiency of the external neutron source is a significant criterion. The spallation reaction can typically yield up to 30 neutrons per GeV(el) from the beam of charged particles. This is higher than the yield from other reactions, and therefore the spallation reaction is the standard choice, even if this criterion is missing an important point: the peculiar and somehow problematic spectrum of the spallation neutrons. It includes a very high energy part (at a lower than but close to the incident proton energy), peaked forward, and thus requiring a thick shielding, as well as a significant component of neutrons "below the fission spectrum", and thus of moderate value for fission use in a fast spectrum. This specific mix of neutrons, combined with the primary protons (from the beam) and the secondary protons (from the reactions in the target and around) generate a specific blend of damages in the materials.

The basic principles of operation are shown in the Figures 15.2 and 15.3. Starting from the accelerator, one can find:

- the charged particle generation (standard option up to now: protons, accelerated to energies of several hundreds of MeV (typically from 500 MeV to 1 GeV)).

- the interaction of these protons with a target of heavy nuclei (because of a higher efficiency in terms of neutron yield per injected MeV (electric)), in a solid or liquid state (enclosed in a loop). The selected materials can be Pb, W, Pb-Bi, U or Th. The spallation (Figures 15.4 and 15.5) produces nucleons (neutrons, protons (therefore hydrogen in structure materials)), α (therefore helium in structure materials with a

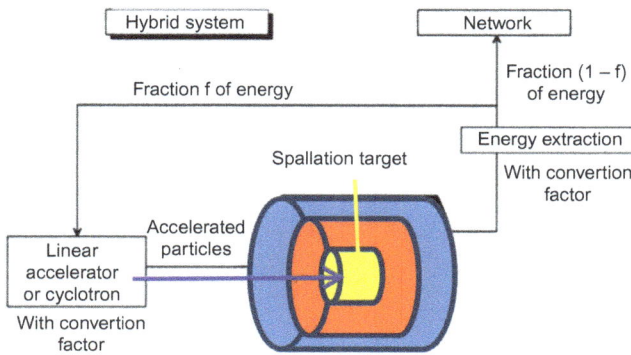

Figure 15.2. Schematic layout of an ADS (CEA).

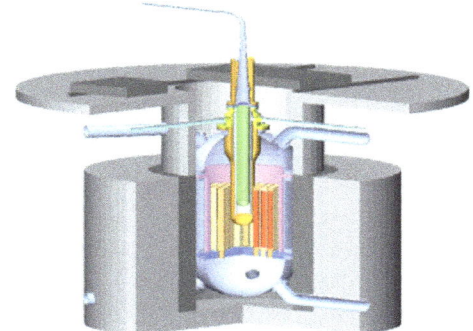

Figure 15.3. An ADS image (CEA) showing the "spallation target module" and the related hull at the centre of the fissile core, the target cooling system, the large shielding, as well as the highly schematic representation of the last portion of the particle beam, downstream of about 400 m of superconducting LINAC. A thin "window" is separating the accelerator vacuum from the spallation target module.

large helium to dpa ratio), and various spallation products, scattered all over the nuclides chart with a high fraction of light nuclei and thus a correlated peak of distribution of heavy elements with atomic masses close to the mass of the target material. Numerous radioisotopes are generated (for instance ^{210}Po, from natural Bi), as well as lighter elements, some of them leading to corrosion or catalysing diverse modes of degradation. A few neutrons are emitted with an energy close to the energy of the incident protons, in a forward peaked high energy neutron beam. These neutrons lead to an unusual shielding issue.

Calculating the order of magnitude of key parameters using approximations

- The amplification of the spallation neutron source in a subcritical fission core can be roughly estimated. The spallation target is generally located in the centre of the core, limiting the leakage and maximising the neutronic importance of the source.

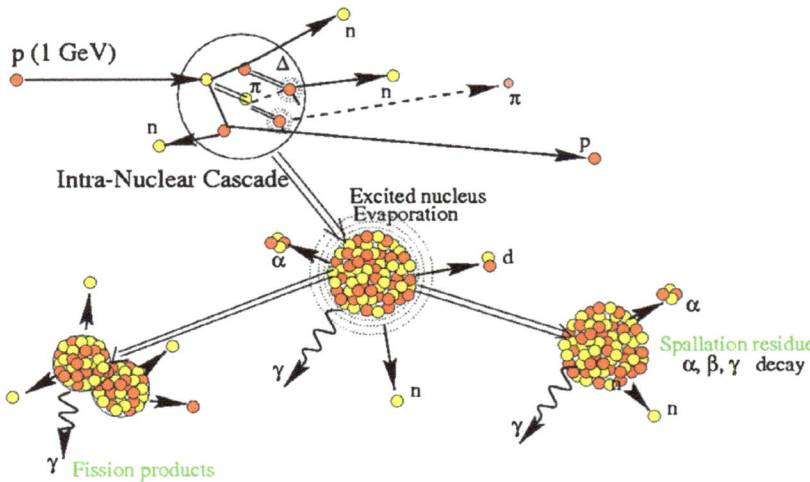

Figure 15.4. Spallation: intra-nuclear cascade, evaporation, fission and decay mechanisms.

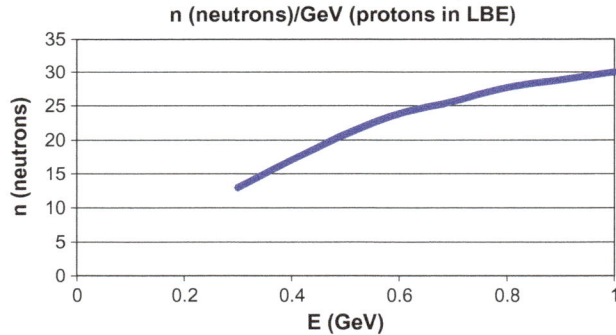

Figure 15.5. Number of neutrons produced by one proton in a lead-bismuth target as a function of beam energy.

By combining several approximations, one can obtain by the sum of the successive neutron generations, from one spallation neutron, a series close to:

$$k_{sp}(1 + k + k^2 + \ldots + k^n + \ldots) = k_{sp}/(1-k)$$

where k is the eigenvalue of the fissile core (in a low subcriticality regime), and k_{sp} is the number of neutrons generated by the spallation neutrons (with the exception of the very high energy tail of the spallation spectrum which escapes from the core). The "mean" spallation neutron is quite different from the neutrons from the subcritical fission chain reaction by its spatial and energetic importance, but the secondary neutrons are closer to the mean fission neutrons.

One can also draw up the balance, neglecting leakage, for one spallation neutron:

S (fission neutron source from spallation neutron source) + νF (further production by fission chain reaction) = A (absorption), which leads, for $k \# 1$, to $A \sim k_{sp}/(1-k)$.

The fraction of spallation neutrons (equal to $1/(k_{sp}/(1-k))$) is therefore close to $(1-k)/k$, i.e. the absolute value of the reactivity in the sub-critical core (because $\rho = (k-1)/k$ is negative). As for orders of magnitude, for a 1% sub-criticality ($k = 0.99$), there are a hundred times more fission neutrons than spallation neutrons entering the core. For a sub-criticality of 5% ($k = 0.95$), the ratio is still twenty. Unless working in a highly sub-critical regime (which is a very expensive option in terms of energy, of accelerating devices and, finally, of cost, as will be seen later on), fission neutrons are largely predominant. They are the "working horse" of the process. They determine the effective spectrum and thus the "nuclear quality" of the interactions between neutrons and nuclei.

The efficiency in terms of transmutation and breeding depends on the design of the fission core, not on the "ADS label". Most of the ADS designs (except the first version of the "Rubbiatron" which takes advantage of the limited degradation of $\eta = \nu\sigma_f/\sigma_a$ of the ^{233}U at epithermal and thermal energy, an advantage which is not specifically linked to the ADS concept), converge on fast neutron cores, apart from some recent concepts generally oriented towards the ^{233}U – ^{232}Th cycle, with an intermediate spectrum. This leads to a competition between the ADS and the "twin critical reactor" which has the same nuclear properties and the same efficiency and which looks, roughly speaking, like the ADS without the accelerator.

- Another specific property of the ADS is its ability to operate as an amplifier, justifying the term "accelerator driven". The fission power delivered by the ADS, as well as the transmutation rate, is proportional to $I_{acc}/(1-k)$.

This is a key property of the ADS. Specifically, it protects from – some – prompt jump energetic reactivity accidents (generating a burst of thermal and disruptive mechanical energy) with potentially severe consequences for prompt neutron lifetimes around 1 μs in fast spectrum. The acceptable reactivity insertion can be significantly higher than β_{eff} and higher than the Doppler feedback (up to the maximum acceptable fuel temperature), provided that it is lower than the ADS subcriticality. This could be used first to protect fast reactor cores loaded with MOX (β_{eff} around 0.3%). In a large, homogeneous conventional sodium cooled fast reactor, the global (intrinsically positive) void coefficient $\Delta\rho_v$ could potentially reach seven to eight times β_{eff} (7 to 8 "\$"). Advanced designs are under study in order to drastically reduce $\Delta\rho_v$ in any situation. On the other hand, a fast neutron gas cooled core $\Delta\rho_v$ can be kept "by design" under β_{eff} (in case of depressurisation).

But the strongest incentive is given by the extreme example of a core loaded with a mix of minor actinides, in which β_{eff} could drop under 0.1%. This case is connected to the "double strata" schema where the first stratum which is the majority stratum is composed of highly competitive "power reactors". This power reactor fleet is exempted from the burden of any minor actinide transmutation. In this case, the transmutaion is concentrated, with a limited amount of MOX content, into the second stratum, and several types of initiators could lead to reactivity insertions of several \$.

Using an ADS can enable to "exclude by design" extreme reactivity induced accidents. By including a set of possible reactivity insertions in the sub-criticality margin, a sub-criticality level of several % (typically 3 to 5%) is required. This leads to a more robust ADS concept from the reactivity viewpoint (even if it does not exclude every type of criticality issue in the case of a molten core), but at the expense of a rather high cost in investment (accelerator and spallation module) and in operation (power for the accelerator, periodic replacement of parts, drop in availability).

Lastly, the amplifier mode operation is an advantage in the case of accidental transient, limiting the power overshoot proportionally to the reactivity insertion, without an exponential jump. However it can be a drawback in normal operation when the accelerator current is fluctuating around nominal intensity in the range of 5 to 10%, or if it turns out to be unstable and/or unreliable.

As for decay heat, which is the primary contributor to the safety degradation in nuclear systems, there is no difference between the ADS and a critical reactor of the same design (coolant, unit power, fuel and lattice, power density, safety systems).

The same equivalence applies to criticality, from the physics viewpoint, as well as for irradiation damage in the core, except for the spallation module and its hull, which is a part of the second barrier and subject to a higher fluence.

As for shielding, the effect of spallation neutrons leads to a very specific issue that will be considered later on.

- Leaving the core, one can follow the path of the energy generated in the system, which is the sum of the fission energy and of the major part of the beam energy. A part of the generated energy must be sent back to the accelerator. In the case of fusion – fission hybrids, one can hope that with modern designs capable of a Q ratio (thermal fusion energy over electric beam energy injected) ≥ 5 (which was out of reach in the late seventies) the neutron source could be self-sufficient in energy (let alone any cost consideration).

As for ADS, an example shows the connection between the following parameters; the fraction of energy that must be tapped to feed the accelerator, the proton energy and the sub-criticality level of the ADS operation. Some assumptions have to be made concerning the thermodynamic efficiency of the ADS as well as the efficiency of the accelerator.

For 1 GeV protons, about thirty neutrons are generated by spallation (there are losses of diverse types and not every spallation neutron is absorbed in the target or even in the fissile core. But every neutron has an energy cost of about 33 MeV (electric, in the beam). It is a fairly good efficiency. By comparison, some reactions involving lower energy charged particles on heavy or light nuclei requiring, in order of magnitude, 1 GeV (beam) per neutron.

Just in order to reach the number to be used in the next section, if the yield of the accelerator (beam power over power injected in the acceleration complex) is supposed to be 0.5, the 33 MeV in the beam require 66 MeV (electric) on the network. If the thermodynamic yield of the electric energy source is 0.33, the thermal primary source required is 200 MeV per spallation neutron (the same result would follow from an efficiency of 33% for the accelerator and of 50% for the electricity generator ...). Obviously, if the global efficiency of the acceleration complex (including the helium refrigeration unit power consumption for superconducting accelerators) remains around 10% or even lower, the industrial and economic consequences would be fundamentally modified.

So, in the specific case under scrutiny, by a strange coincidence, 200 thermal MeV are required for the generation of a spallation neutron.

In a neutron rich, $\eta \geq 2.3$ absorption, a neutron is available for breeding, transmutation and irradiation, beyond perpetuating the chain reaction. In this case, the product is a neutron **plus** 200 MeV (thermal). Spallation cannot compete with an optimised fission process harnessed to fulfil a given function. Moreover, if the fission is short of neutrons, spallation alone is not able to bridge the gap. So, spallation must be connected to the best feasible fission process (and thus, when dealing with nuclear efficiency in a U-Pu cycle, with fast neutron systems). Spallation can then drive the working fission process and result in a functional auxiliary system (ADT; Accelerator Driven Transmuter, or FADT; Fast ADT) which would have a qualitative advantage in the specific case of a low β_{eff} minor actinide intensive core (close to double strata schema): sub-critical amplifier operation, with a reactivity margin enabling the practical exclusion of energetic prompt jump accidents.

This advantage has a cost in energy. Keeping the same data and assumptions (1 GeV proton, 30 neutrons per proton, $\eta_{acc}\eta_{thermodyn} \sim 0.15/0.16$, thus 200 MeV thermal required per spallation neutron), one can assess the sensitivity of the energy amount required by the beam to the sub-criticality level. In this specific case, with the coincidence between the energy required for producing a spallation neutron and the energy generated by a fission (200 MeVth), f = energy sent back to the accelerator/energy generated in the ADS $\sim \nu/|\rho|$. This is related to the ratio of spallation neutrons to the total number of neutrons, and to the fact that a fission (about 200 MeV-th.) delivers ν neutrons.

For instance, with a sub-criticality of 5%, which would prevent most reactivity and criticality accidents, the fraction of energy drained and sent back to the accelerator is about 12.5%, which is a fairly high proportion. External neutrons have a high cost in energy.

This induces an implicit convergence regarding the range of subcriticality for most of the designs, of around a few percent. For a lower acceleration complex yield, 1% subcriticality would be the maximum potential level reasonably at hand.

The simplified estimations performed in the previous section can be compared with numerical simulation results. In terms of accelerator current, they lead to the curves shown in Figure 15.6, which shows, for a 80 MWth ADS of the MYRRHA type, how the current depends on the beam energy (in GeV, from 0.5 Gev to 1 GeV) and on the subcriticality

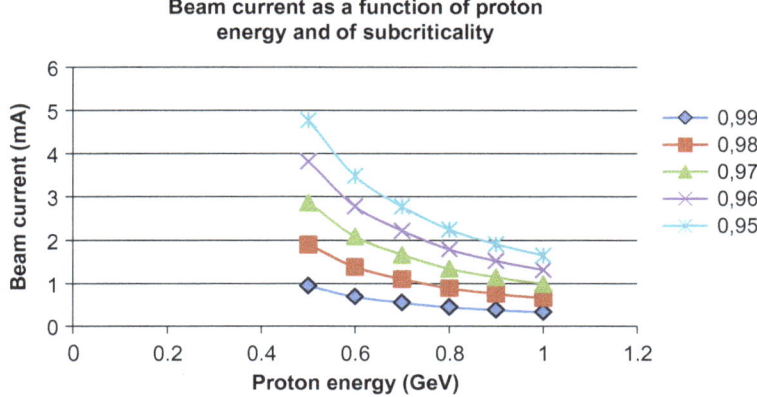

Figure 15.6. For a 80 MWth ADS, beam current as a function of the proton energy and of the subcriticality (simplified estimation).

level (k varying from 0.95 to 0.99). The neutron yield per GeV - of protons - of the spallation reaction is taken from the curve of Figure 15.5 and a single (slight) scalar fit is performed. The results agree (within a few percent) with more detailed calculations, within this range of parameters. As for the required thermal energy to be drained and delivered to the accelerator, the results depend on the thermodynamic efficiency of the power production facility as well as on the accelerator efficiency yield in generating the proton beam.

15.1.3. Technology and design: main specific components, challenges, and key points for feasibility

The functional analysis of the overall process, combined with the review and selection of the available and advanced technologies, leads to the architecture of the ADS and to the assessment of the key challenges and potential default modes, when comparing the ADS to the very close critical reactor fulfilling the same functional requirements.

The accelerator

The main issues are:

- The performances: The proton energy is typically in the range of 0.5 to 1 GeV; the beam intensity is of a few dozen mA for an industrial system, which means a (significant and unusual) beam power of a few dozens MWel. In France, for instance, the IPHI projects deals with the low energy part of the high intensity (and space charge challenge) technology: source (SILHI), injection, up to the superconducting section of the accelerator. A sketch of the whole accelerator (a LINAC which is generally the preferred option at this combined level of intensity and energy, and with a high level of synergy with the scientific applications in high energy physics) is shown in Figure 15.9. The overall length would be about 500 m for 1.3 GeV protons.

Another concept is the cyclotron. A somewhat low intensity but very successful example is the SINQ neutron source of the PSI (Paul Scherrer Institut) shown in Figure 15.7. The proton energy is 590 MeV and the intensity more than 2 mA. One can see the huge shielding blocks surrounding the source.

The challenge of developing such powerful accelerators is made more difficult by the constraints inherent in coupling with a fission system, which are highly demanding in terms of operating quality, reliability, and for safety and competitiveness purposes.

- The specific issue, when comparing to research facilities, is the "ADS quality" of "ADS class" accelerators, related to the stability and reliability needs. Figure 15.8 shows the typical record (in this area) of a 800 MeV LINAC. Obviously, a fast restart is possible in research operations, without coupling to a fission system, leading to very high availability performances. But it would not be eligible as the driver of a fission system.

The required improvements will concern several areas of design and operation, and the space charge effects of high intensity LINAC do not make things easier.

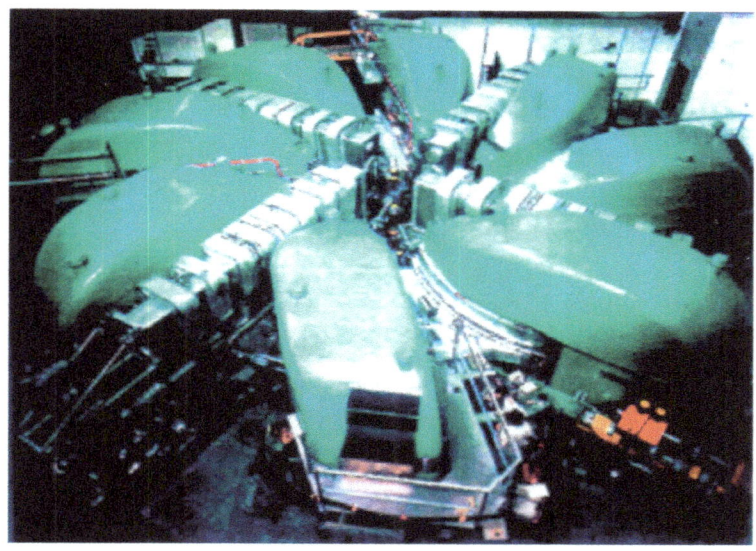
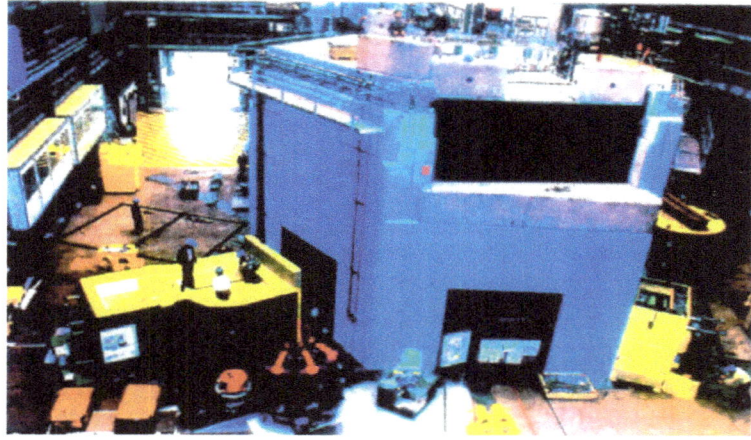

Figure 15.7. The cyclotron driven neutron source SINQ at the PSI. ©PSI.

As for the MYRRHA project, the current reference scheme for the accelerator is a superconducting LINAC of 600 MeV, 2.5 up to 4 mA cw meaning 2.4 MW in the beam at maximum power. For R&D, the "name of the game" is reliability, with benefits for the next generation high intensity projects such as ESS, EURISOL, etc. Figure 15.9 shows a schematic diagram of a LINAC dedicated to an ADS using 1.3 GeV protons.

The proton beam general specifications within EUROTRANS are listed in the Table 15.1.

Basic principles to be implemented include "overdesign", redundancy in some critical areas/devices and fault tolerance (as far as reasonably achievable). The R&D will include the following critical items: management of the space charge (specifically in the injector), continuous regime – high power RFQ, superconducting cavities and accelerator/reactor interface etc.

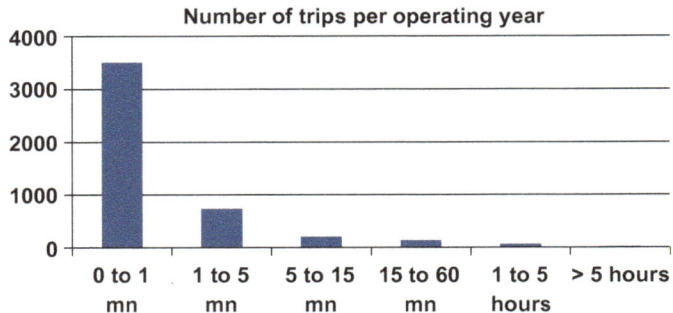

Figure 15.8. Typical beam failure statistics of a LINAC dedicated to physics, with an H$^+$ beam: about 1.6 trip per hour (40 trips per day). Significant improvements have been made. An "ADS quality" LINAC should meet more stringent requirements.

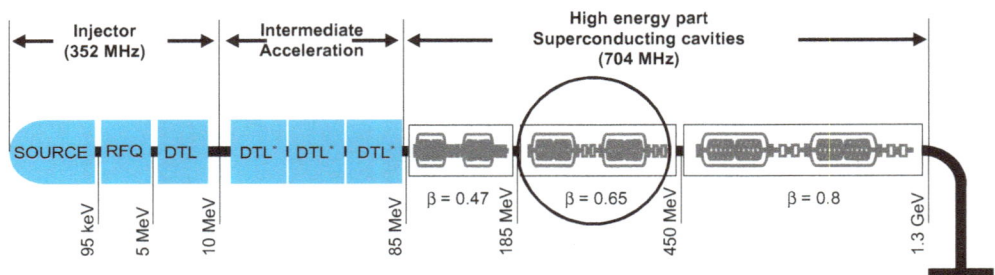

Figure 15.9. Schematic diagram of a new generation LINAC delivering protons at 1.3 GeV (CEA).

Table 15.1. EUROTRANS proton beam general specifications.

	Transmuter demonstrator (XT-ADS/MYRRHA project)	Industrial transmuter (EFIT)
Proton beam current	2.5 mA (up to 4 mA for burn-up compensation)	~ 20 mA
Proton energy	600 MeV	800 MeV
Allowed beam trips(> 3s)	≤ 10 per 3-month operation cycle	≤3 per year
Beam stability on target	Energy: ± 1%; current: ± 2%	idem

The spallation module

In the course of the functional analysis a specific component called "spallation module" emerged. This arises from a global system design choice which is to avoid the direct injection of the beam (through a thin "window", which would be part of the second barrier) into the sub-critical fission core. Figure 15.10 shows a schematic diagram of a modular Pb-Bi spallation module

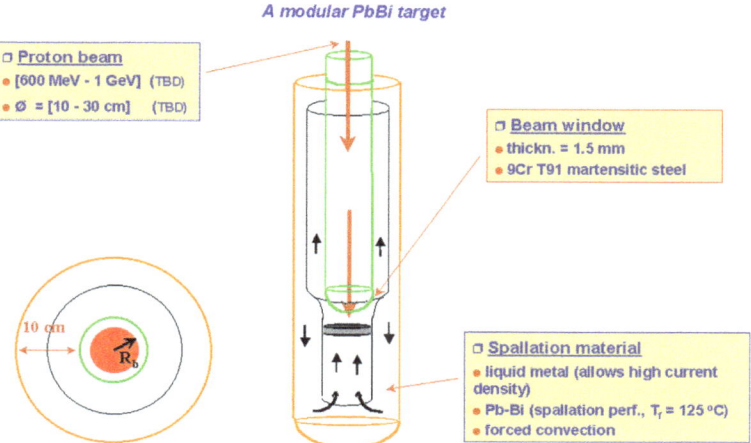

Figure 15.10. A modular Pb-Bi spallation target module. © Cea.

The spallation module design leads to several options:

- A solid target cooled by a liquid (heavy water, liquid metal) or a gas (helium). The axial source density of neutrons can be adjusted (for instance, to be radially flat).

- A liquid heavy metal target (Pb, Pb-Bi eutectic, with a lower fusion point ∼ 125 °C) flowing in a loop, as shown in Figure 15.10. This concept has been adopted for the R&D experiment MEGAPIE. The operational objective was to significantly improve (more than doubling) the neutron per one incident 590 MeV proton of the PSI SINQ target. This improvement was mainly dedicated to solid state structure studies with cold neutrons thermalised in a heavy water tank surrounding the source. This experiment was a success. The lifetime considered for the MEGAPIE spallation module was between a few weeks to a few months. Actually, it has been operated during the second half of 2006, prior to being unloaded. It is the first "MW class" (∼ 1.5 mA × 600 MeV), high performance (with associated incident beam intensity density and duration) experiment in Europe with an up to date design.

Taking into account the diverse constraints associated with the potential default modes, the operational envelope for a typical Pb-Bi target, with a beam of about 1 GeV protons is shown in the Figure 15.11. Beam intensity densities of several tens of $\mu A/cm^2$ seem to be manageable.

Lastly, the spallation module has to be replaced periodically (for instance, every year). The external hull is a part of the second barrier of the fission system. Like magnetic confinement fusion, the ADS is a source driven system. The main functions and the components and systems connected to the risk are respectively tritium breeding on one hand and fission

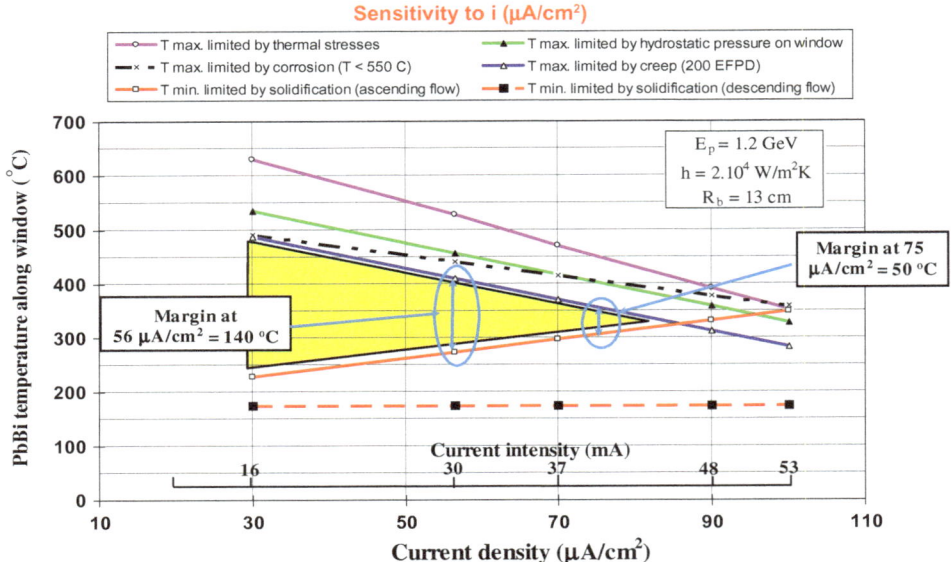

Figure 15.11. A design method illustrated by the target "flight envelope" determination, with a focus on the sensitivity to the beam current density ι on the inlet (in $\mu A/cm^2$) – from CEA/DEN/DM2S/SERMA.

on the other hand. The ADS shares with fusion the drawbacks of this concept. The tritium breeding and heat removal function dedicated components and systems in the fusion case, as well as the fission modules in the ADS case, are only separated from the central source driving the process by a wall, which must be transparent to the radiation flux, must also withstand the damage induced by this flux during a "reasonable" lifetime, and simultaneously be designed to fulfil the mechanical requirements for a second barrier. The spallation module is subjected to the irradiation of the incident beam as well as to the damage caused by the spallation products leading to a degradation of the structural materials properties by the effect of hydrogen (protons), helium (α), and many heavier and chemically active elements which build up under irradiation. A key damage is a result of the fission neutrons. In order to obtain a high "nuclear efficiency" with a U – Pu cycle, most of the ADS are designed with a fast neutron spectrum. The fast neutron fluence (> 1 MeV) is high for a given amount of energy produced by fission. The hull of the spallation module is located downstream from the (necessarily) thin window separating the spallation module from the low pressure accelerator volume. It protects the accelerator from any accidental pollution, which would be very hard to fix. The hull is embedded in the centre of the fission core at the peak location of the fast flux level. In fission reactors, the first barrier (the fuel clad) is subjected to full neutron flux, but the second barrier (the vessel) is protected from the fast fluence by several decades of reduction. This leads to a lifetime of several decades (typically fifty to sixty years; the plant lifetime). In the ADS case, the hull must be frequently replaced (like the first wall in fusion), which will lead to cumbersome and long remote handling operations, when taking into account the topology of the system and the tight fitting of the components,.

Lastly, one has to remember that heavy metals such as lead or the lead-bismuth eutectic, have a dual status as chemically toxic elements, as well as radiotoxic materials after irradiation. Moreover, the presence of significant quantities of α emitters in internal modules and loops (like ^{210}Po from natural bismuth) some of which are highly volatile, is a source of complexity regarding normal and incidental operation.

The sub-critical fission system

In the framework of a sustainable nuclear energy development, and with a U-Pu fuel cycle, if a transmutation role is defined for the ADS in the nuclear fleet as an ADT (AD Transmuter), the ADS will probably be run with a fast neutron spectrum. The main coolant choice is between liquid metals (sodium, lead, lead-bismuth) and gas (helium, potentially CO2 or a blend of helium with a heavier gas). The main criteria are the same as for the critical reactors, along with some specific criteria associated with the coupling of the fission system with the accelerator for example the number of acceptable accelerator related instant shutdowns (mainly for thermo-mechanical reasons) during the ADS lifetime.

The fission system includes the fuel cycle. If the ADS works as a "power reactor" with global recycling of the whole set of actinides, there will be no major difference in the fuel cycle when compared to critical fast reactors generally considered as potential power reactors of the future.

If the ADS works as a dedicated transmuter (ADT, FADT) for the minor actinides (and primarily for Am, generating more Cm as a by-product which in turn needs to be transmuted) with a limited make-up of MOX fuel to adjust the reactivity swing and increase some safety coefficients (including the Doppler feedback), the fuel will be highly loaded with minor actinides. Such a fuel is a challenge and the cost of the fuel cycle would be significantly increased. But under these assumptions, the over-cost would be concentrated in the ADS fleet which represents less than 10% of the nuclear reactor fleet (in terms of installed power) - a fleet mainly composed of conventional LWR and of critical breeder "power reactors".

As for the spectrum, the transmutation objective (from the CAPRA concept of plutonium burning to the more recent trends of global actinide burning) induces a specific situation. In breeders, the plutonium concentration is limited by the criterium of a high conversion ratio as soon as a reasonable burn-up is reached. In a transmuter, a high concentration of transuranics (more than 50%) is considered. With a given coolant and fuel assembly lattice, this increases the spectrum hardness, characterised by the index $r = N_f \sigma_f / \xi \Sigma_s$.

As for the safety assessment, the main issues are very similar to the ones related to the closest critical fission reactor. These concern the fuel and the decay heat removal, etc. The containment issue is worsened particularly as a result of a higher number of complex interfaces (between the fission core and the spallation module and between the spallation module and the accelerator) and the topology and dimensions of the whole system, including the accelerator. The reactivity and criticality issue is eased, but some criticality issues are still present in extreme cases, even with a nominal 5% sub-criticality margin in operation.

As for the shielding issue, specific difficulties arise due to the spallation neutrons. They are illustrated in Figures 15.12 and 15.13.

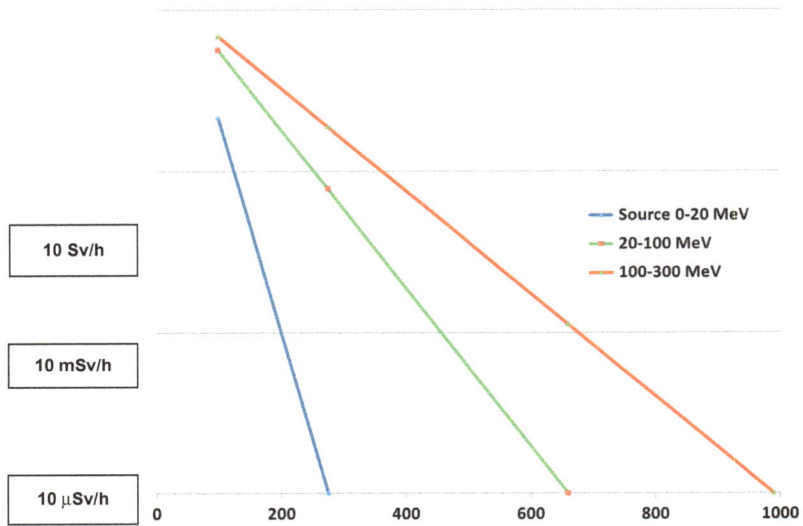

Figure 15.12. Biological shielding design issues: deep penetration of energetic neutrons: equivalent dose rate vs. concrete depth behind the target (in cm). © Cea.

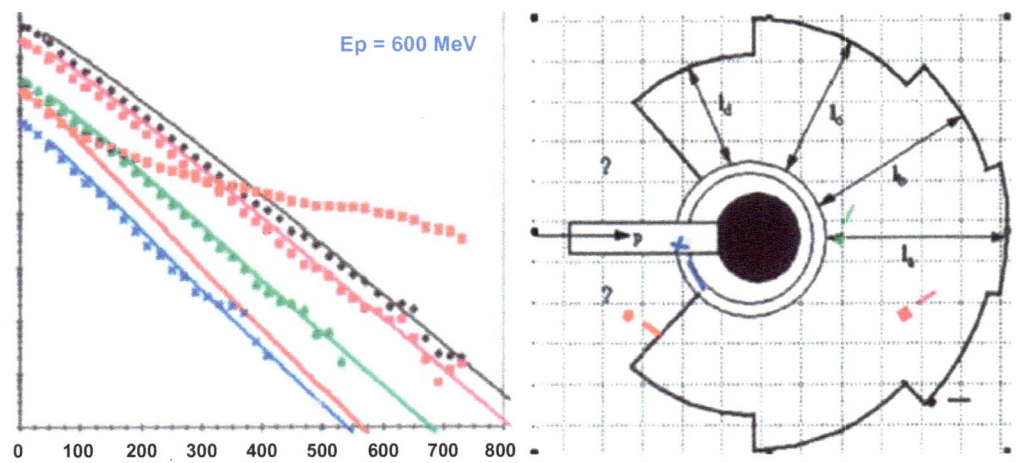

Figure 15.13. Biological shielding issues: "backscattering". © Cea.

The first issue is related to the neutrons ejected with an energy close to that of the incident proton and constituting a highly penetrating, peaked forward "beam". The second is related to a complementary problem; the lower energy neutrons emitted rather isotropically (in the laboratory) by "slow" collective nuclear processes, leading to a kind of "backscattering" emisssion, threatening components located upstream, requiring more shielding in these areas.

15.1.4. Preliminary techno-economic assessment

Several studies have been issued (P. Bacher, OECD/NEA, US roadmap, etc). Figure 15.14 shows a simplified comparison of the following three options:

- the advanced fast reactor (AFR, liquid metal or gas cooled) considered as the basic breeder power reactor of a worldwide nuclear fleet;

- the ADS considered as an alternative system, competing with the AFR, with a closed fuel cycle recycling uranium and the transuranics;

- the fast ADS as a Transmuter (FADT) dedicated to minor actinides transmutation with a make-up of MOX, with a support ratio of one over twenty in the nuclear fleet, meaning that the installed nuclear power of the FADT is about 5% of the global nuclear power.

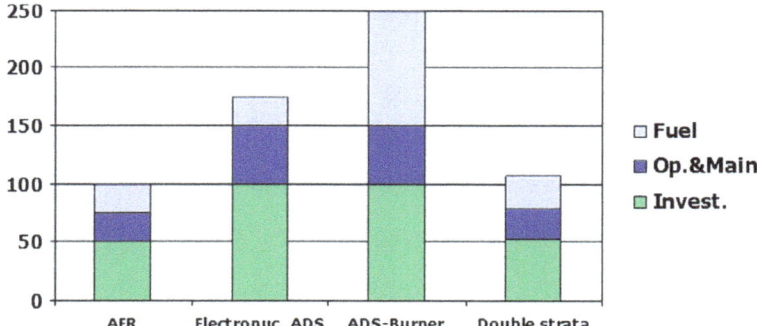

Figure 15.14. A simplified assessment of the over-cost due to ADS integration into the nuclear power generation fleet, dependent on whether it is used as a majority "power reactor" or a second stratum Accelerator Driven Transmuter (ATD).

The AFR (Advanced Fast Reactor of the future) is supposed to be the conventional cost reference of the future (100%).

A "power reactor" ADS seen as the next generation power reactor would lead to a kWhel. cost about 70% higher (50% to 100%), because of extra-costs resulting from:

- investment (accelerator, spallation module, interfaces, shielding),

- operation and maintenance (O&M), for example frequent replacement of parts or components such as accelerator and spallation module,

- decrease in availability due to reliability issues and to maintenance/replacement operations,

- cost of the electricity needed for feeding the accelerator (with a sub-criticality of a few %).

The fuel cycle cost does not differ from the reference.

The kWhel cost from a FADT (Fast Accelerator Driven Transmuter) would be still higher, reaching more than 200% of the reference AFR cost. The difference regarding the "power reactor ADS" cost is partly due to the fuel cycle dedicated to the transmutation of minor actinides, with extremely "hot" fuel and fuel cycle facilities.

But the role of the FADT is to be used as an auxiliary facility dedicated to minor actinides transmutation, thus generating about 5% of the power produced by the nuclear fleet. The over-cost is thus balanced by the support ratio and could be limited to less than 10% for the global fleet. This result depends on a questionable assumption: the feasibility of a fuel and a core only loaded with minor actinides. Adding some plutonium and ^{238}U would lead to an increased ADS fraction of power installed in the fleet and therefore to an increased overcost. The same approach holds, for instance, for the ABR (advanced plutonium fast burner reactor). It could complement the fleet of (Gen-III) LWR operating in once-through regime as the basic "power reactors". The support ratio would probably be about 1/4 and the over-cost would be shared by the power reactors, protecting the first generation of brand new fast reactors from the competition with optimised and "trained" LWR, provided with a fantastic operation feedback.

15.1.5. Defining a role for the ADS in the nuclear fleet: elements for a rationale

The main conclusions are as follows:

- if a given type of AFR (SFR, GFR) is acceptable and competitive, there is no need for developing ADS;

- if not, the ADS could take the burden of transmuting the minor actinides, if it were considered mandatory (which is still an open question).

The ADS is not a straightforward nuclear concept and some potential advantages related to sub-criticality should not hide the many specific and complex issues. Moreover, the extremely "hot" fuel and the related fuel cycle facilities seem to constitute a second area of challenges and, finally, extra-costs, even if they have the advantage of concentrating the problems on a limited fraction of the whole reactor fleet and fuel cycle facilities complex.

15.1.6. The R&D programs

The ADS assessment has generated many interesting R&D programs, focused on the main specific challenges:

- the high intensity protons LINAC, with the SILHI – IPHI program, for instance (CEA-SAC);

- the high (injected) power spallation module with the European MEGAPIE program (PSI);

- the sub-critical, external source driven reactor physics, instrumentation and control, with the MUSE program in MASURCA (CEA-CAD);

- the MYRRHA European project of a demonstrator (SCK/CEN – MOL) (see above).

- etc.

15.1.7. The future in the world, in Europe, in France

At the international level the trend is back to the study of the closed fuel cycle and of dedicated burners or breeders in U-Pu cycle, with a renewed interest in introducing the Th resources and physics into the game. Both concepts involve mostly fast, critical reactors, with some exceptions in the thorium/U-3 centered case.

In Europe, R&D programs related to the ADS are still ongoing at a slower pace, excepting the MYRRHA project.

French organisations are involved in these programs. The ADS concept has been considered as a potential tool which has to be kept in mind when dealing with transmutation perspectives, in the framework of the law passed in 2006 concerning the waste management and the connected R&D.

It seems that the first improvement and demonstration steps needed, are in any case on the common trunk of the closed cycle strategy: developing a robust and competitive (liquid metal, primarily sodium or gas cooled) fast reactor, which can also be used in a critical as well as a sub-critical regime, and developing the fuel cycle processes and facilities able to deal with the demanding requirements (efficiency, cost, resistance to proliferation) of a closed fuel cycle, with a quick processing – fabrication step.

15.2. Nuclear space power and propulsion

This field has always given exceptional opportunities for innovation. Some of them have been tested and qualified, paving the way for major improvements in terrestrial applications and building the slow and winding path of innovation towards the future.

In what follows, propulsion will be predominantly considered, excluding many types of nuclear designs for space power generation:

- ^{238}Pu heat sources from radioactive decay (500 Wth/kg, T1/2 = 80 years) lead to RHU devices (Radioisotope Heat Units), and to RTG for electricity generation by static conversion (thermoelectric or thermo-ionic, with a limited efficiency of 5 to 10% and a poor mass to power ratio index α (kg/kW $\sim$ 200). It is a flight proven technology with long life operation feedback. Some advanced projects aim at improving efficiency beyond 10%.

- The RTG are dedicated to the hundreds of W range. In the kW range, dynamic conversion (closed Brayton or Stirling free piston engines) offers a higher potential efficiency (20 to 25%), with a better but still too high α ratio and large, bulky radiators. The development of renewables offers a new adventure playground to these conversion technologies which are maturing but are probably still deprived of a long life flight operation feedback.

A heat source of ^{238}Pu can be coupled to a compact Stirling engine for a substantial gain in efficiency, and therefore in mass (less plutonium embarked, a potentially improved α, a reduced dependency on ^{238}Pu delivery; the use of other radioisotopes is also considered).

- In the 100 kW to 1 MW range, critical fission reactors, disadvantaged by the critical mass issue as regards the α ratio at low unit power (under a few tens of kW), find a more favourable playing field. The α ratio can drop typically from 50 to 10 in this unit power range. In Europe, the ERATO project was studied from 1982 to 1989 by CNES and CEA).

But the most spectacular projects combining cutting edge innovations in several fields involve propulsion missions including Moon and Mars–bound manned missions and deep space planets scientific missions, for instance the US JIMO concept: Jupiter Icy Moons Orbiter (vs. a very long, one shot, fly-by journey).

As for propulsion, the principle is the one of action and reaction. It follows that the thrust T (N) is given by the flow (kg/s) x the ejection velocity (m/s).

By definition, the specific impulse, which is a merit index as for the efficiency in using the flow of material to be ejected for a given thrust, can be written as:

ISP (s) = $V_{eject.}/g_0$, g_0 being the acceleration of gravity on the – terrestrial – ground.

Typical values of ISP for various implementations, compared with combustion of cryogenic fuels, are:

- cryogenic fuel rocket engine: 450 s,
- nuclear thermal propulsion: 800–1000 s,
- electric propulsion: > 5000 s.

As for nuclear propulsion, two main methods can be used:

- the thermal (nuclear) propulsion, by heating a gas, expanding and ejecting it through a nozzle at critical velocity, which is proportional to $1/\sqrt{M}$, leading to the choice of (cryogenic) hydrogen;

- the nuclear electric propulsion, utilising electric power to ionise propellant and accelerate it to produce thrust, either by electrostatic acceleration in an electric field (successfully used on 1997 Deep-Space-1 Mission to Comet Borelly (2.3 kW Unit)), or by electro-magnetic acceleration via combined electric and magnetic fields (see below a very peculiar magnetic reactor: the VASIMR concept and the dedicated experiment).

The main issue in space is the overall mass balance, so the global optimisation involves three terms:

- the mass to be ejected backwards. The higher the velocity, the lower the required mass (p = mv). On the other hand, the energy grows following the equation $E = 1/2\ mv^2$ (before entering the relativistic range);

- the "engine" mass, including the energy source and the propulsion device (a nozzle in the case of thermal nuclear propulsion, an ion accelerator in the case of ionic propulsion). In the latter case, thermal energy has to be converted into electricity;

- the fuel to feed the energy source.

The mass and dimensions of a radiator, in the case of thermal to electricity conversion, have also to be taken into account.

As for thermal propulsion, the best known project is NERVA (Figure 15.15) with fantastic performances of up to 4.3 GWth unit power (close to EPR) for 125 ton thrust, up to 10 GWth/m^3 power density, about 2200 °C outlet gas temperature and with hydrogen coolant on advanced refractory fuel (with almost no burn up). Another program (PLUTO) was devoted to an air-breathing ramjet for a cruise missile.

Figure 15.15. The NERVA Experiment at Jackass Flats (Nevada).

In CEA, in cooperation with CNES, a smaller (300 MWth, 7.5 ton thrust) thermal propulsion system (MAPS) was studied in the mid-nineties.

In the USSR, advanced fuels enabling very high temperatures and power density were developed, qualified and used.

The most recent studies were devoted to deep space missions, the momentum being generated by US initiatives such as the PROMETHEUS project including the JIMO mission to JUPITER Moons. Unit power up to 250 kWe or even greater was considered, which is a favourable range for nuclear fission application. The strong advantage of a fission critical reactor, beyond the unit power threshold related to the critical mass issue, is the energy density of nuclear fuel, which cannot be matched by any other technology. It enables a long operation at very high specific impulse (or ejection velocity), which is very demanding in energy ($1/2(mv^2)$) but saves on propellant by providing a momentum proportional to mv.

Concerning the nuclear electric propulsion, the last study in CEA led to the OPUS-100 reactor design (Figure 15.16). OPUS is a gas cooled reactor connected to a Brayton cycle. It uses particle based fuel. The mass is about 2400 kg for 100 kWe (the α ratio is 24). In the short term, there seems to be a proper match betwween technologies, the space launcher,

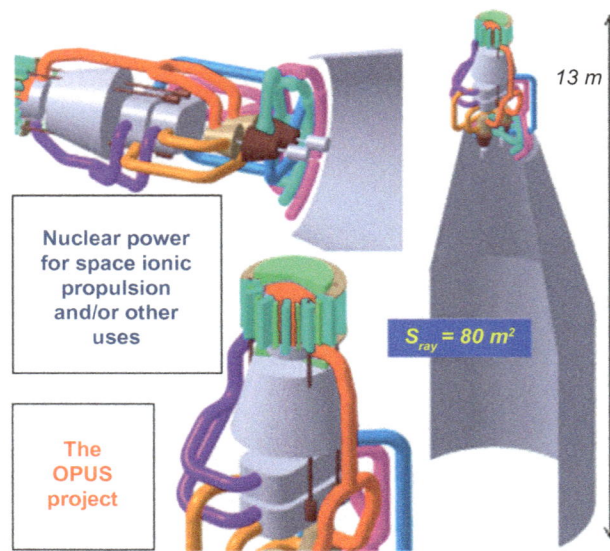

Figure 15.16. Nuclear power generation in the typical power range of 100 to 500 kW: the French OPUS project (CEA)

the reactor and the propulsion balance of plant. In the longer term, another equilibrium point could be reached for long distance/long duration missions in outer space, about 500 kWe and with a reduced α ratio. It could be limited by fuel technology, specifically concerning the fast fluence resistance issue. Innovative fuel technologies could help (see the development around the so called (a little misleading wording) "Accident Tolerant Fuel") as well as an alternative scheme for fission products management.

Very recently, a renewed interest in nuclear potential for space applications emerged in Europe, leading to the MEGAHIT project with a demonstrator around 2022. The global objective is to set up a roadmap towards the implementation of high power electric propulsion systems and to create a technical and scientific community, in cooperation with Russia (see in Figure 15.17 the sketch of a Megawatt class Nuclear Electric Propulsion System from the Keldysh Research Center).

For quite different applications, with a low intended nominal power level of 500 We, a joint team from NASA and LANL has proposed a new reactor concept combining, around a 23 kg enriched uranium core with a single control absorber, eight heat pipes connected to eight Stirling engines for power generation (Figure 15.18). The thermal radiator and the radiation shield are, as usual, bulky and heavy components. The global α is probably not very low and the room occupied by the system is large, but this reactor based concept makes it possible to get rid of a large amount of ^{238}Pu onboard. On the other hand, the ^{235}U inventory is quite large, relating to the low unit power. The system is modular and could be adapted to higher unit power levels. The Demonstration Using Flattop Fission (DUFF) experiment was probably the first one to take place in the USA since 1965. The path from NERVA to DUFF is intriguing but more is expected from nuclear design in the

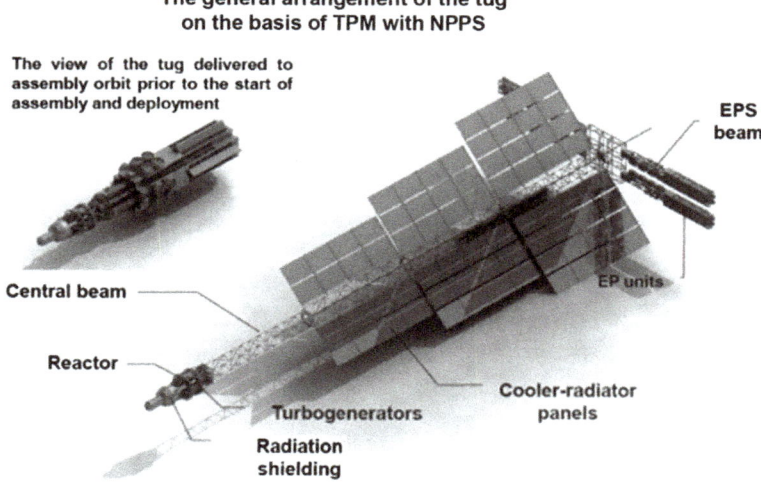

Figure 15.17. Sketch of a Russian design for a Megawatt class Nuclear Electric Propulsion System project. Rights reserved.

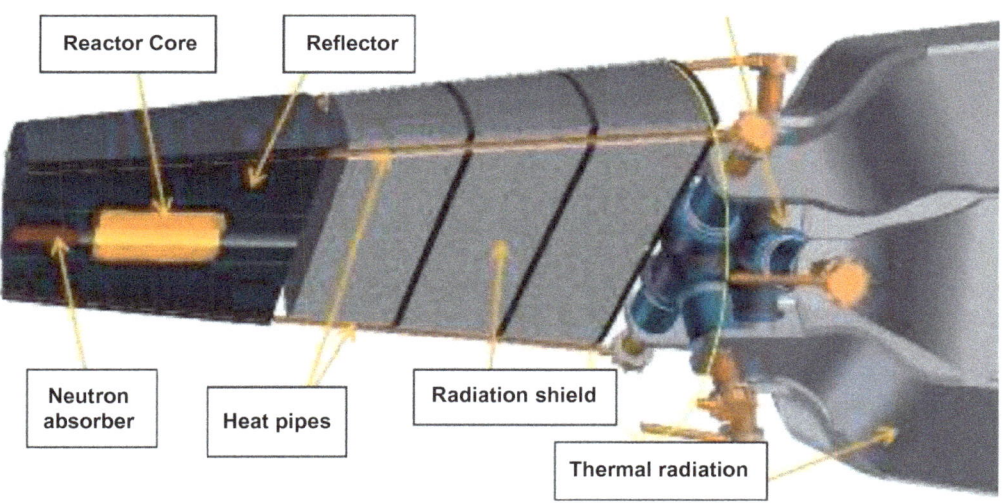

Figure 15.18. "LANL's new reactor concept" (image: LANL) from World Nuclear News – 27 November 2012. © Nasa.

field of adaptation to a broad range of applications, and this is an interesting step showing the versatility of the reactor system technology and design.

Last but not least, advanced nuclear concepts and technologies are finding a way to the future by adapting for use in brand new applications. An example of this is the VASIMR concept (Figure 15.19), inspired by the tandem (mirror) magnetic confinement device

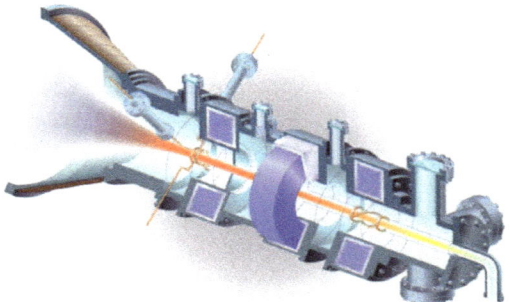 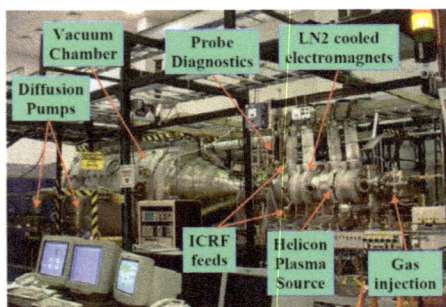

Figure 15.19. VASIMR and the tandem mirror magnetic confinement experiment reminiscence; VX-10 Experiment (NASA – Johnson Space Center Laboratory/ASPL).

developed primarily at the Lawrence Livermore Laboratory and which was considered in the late seventies as a potential source of fusion neutrons in the framework of fusion fission hybrids devoted to fuel breeding for a fleet of (efficient converters) fission power reactors. Actually, even if the potential performances of the mirror concept do not match those of the Tokamaks and Stellarators in terms of confinement and Q (fusion energy produced/energy injected in the plasma), the geometry would have been more convenient for design and for industrial fabrication and operation.

This concept has been transposed to electro-magnetic ion acceleration, with two potential advantages when compared to alternatives (purely electrostatic acceleration).

- In the solenoid, the field lines do not diverge and the geometry is a simple one; on the other hand, the minimal mass of the device probably leads to a high practical threshold in terms of unit power, in order to reach a reasonably low α value.

- Controlling the mirror plugs enables tuning of the plasma temperature and the ejection velocity, thereby providing the capability to optimise the use of the propellant mass ($p = mv$) and the energy ($E \# 1/2\ mv^2$). VASIMR could be used to provide a high thrust during a short period of time and a lower thrust during a longer period. All is needed is a fantastic energy reserve, light, densely packed: i.e. nuclear.

These principles have been tested in an impressive experimental facility, with hydrogen plasma (Figure 15.19).

In summary, the nuclear space power and propulsion field tapped the advanced technologies since the early beginnings of development. It involved the liquid metal cooled fast spectrum reactors, the very high temperature gas cooled reactors, and promoted the use of high performance innovative fuels (high temperature and high power density). Moreover, it can give a second chance to cutting edge technologies and smart concepts when they are stagnated in the course of their development by the competition with "locally" better adapted options.

It is a source of creativity with unusual boundary conditions, and some of the innovations emerging in this field will probably realise their full potential in more conventional applications in the long run.

15.3. Advanced neutron irradiation sources (NIS)

From "New opportunities for "smart" neutron irradiation experiments (ADONIS)" (presented at GANIL in 2004, december) by A. Alberman, A. Barbu, R. Lenain, J.-B. Thomas.

Developing the science of materials under irradiation requires experiments to be performed on samples, with specifications which can differ significantly from those required for the qualification of nuclear fuels. Up to now, both objectives were met in MTR (Material Testing Reactors). In Europe, the JHR (Jules Horowitz Reactor) will take over from SILOE and OSIRIS. In fast neutron spectrum, PHENIX (whose operation ended in 2009) and BN 600 (still in operation but primarily devoted to power production) provided a high level of fast flux (> 1 MeV), thus of integrated fluence (n/cm^2, > 1 MeV) and dpa/year. Table 15.2 shows the main characteristics of a few existing or planned reactors (not including the Russian project of fast neutron irradiator MBIR).

Table 15.2. Some available/projected/shutdown irradiation tools (Reactors).

Name	Power MWth	ϕR*	Volume cm^3	dpa/fpyr	N(efpd)	dpa/yr
OSI RI S	70	2	250	10	220	6
RJH	100	4	n*100	20	260	14
PX (NR)	600	6	n*1000	30	260	21

*ϕR: neutron flux @ E > 1 MeV, in 10^{14} n/cm^2.s

In the future, new trends give an opportunity for developing innovative tools.

The main targets are (for Gen-III and Gen-IV fission reactors, fuel cycle facilities as well as for fusion R&D): the fuel and the core structures (micro/nano-structured, with functional layers), HTR fuel, ODS (oxide dispersion strengthened alloys) and Sic$_f$/Sic. In most cases, the REV (Representative Elementary Volume) is of small size.

Ion irradiation (for instance in JANNUS – CEA-CNRS, a triple beam accelerator providing the combined effects of dpa (with a high build up rate), hydrogen and helium related damage) is a powerful tool and makes it possible to perform on line examination, with a reduced experimental "footprint" (Figure 15.20).

Accelerator driven (sub-critical or without amplification by fission) neutron irradiation sources (ADNIS) exist (LANSCE in the USA) or have been designed, in order to provide several dpa/yr in a reduced volume (several tens to a few hundreds cm^3), with specific or tailored spectra. The project specifications of a few of them are listed in Table 15.3.

In the late nineties, spallation based irradiation systems were designed and the use of a powerful LINAC was considered in order to provide a proton source for several applications (European Spallation Source, radioactive ion beams, an ADS demonstrator, and SPALLAX (Figure 15.21), then Mini-SPALLAX, dedicated to neutron irradiation, in the framework of the CONCERT project (R. Pellat, JF Luciani, JL Laclare et al.).

A further assessment led to the conclusion that for reduced levels of neutron source (around 10^{15} to 10^{16} n/s), other ion beams could provide a better trade-off between cost and performance. This is the case, for instance, for photonuclear production from an electron

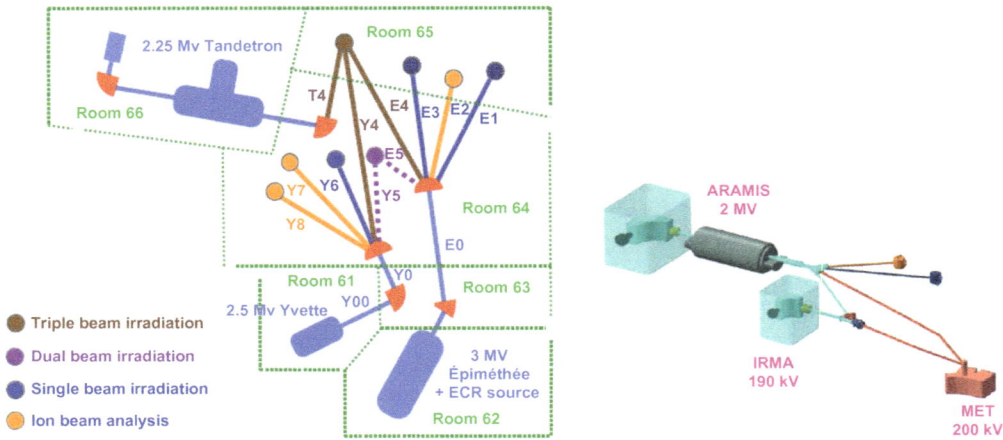

Figure 15.20. General layout of the Saclay's triple beam facility and of the Orsay's dual beam facility coupled with a 200 kV TEM.

Table 15.3. Some projected/conceptual AQDONIS performances.

NAME	SOURCE 10^{15} n/s	ϕR^*	Volume cm³	dpa/fpyr
Mini-SPALLAX in CONCERT (p, 10 mA @ 1.2 GeV)				
	2000	7	n*l	35
IFMIF (d, Li – 250 mA @ 40 MeV on a 20*5 cm² beam spot)				
	100	5	500	25
SPIRAL-2 (d, C – 5 mA @ 40 MeV on a ϕ = 2.83 cm beam spot)				
	1	0.5	14	3
nuclear e⁻ 33 mA @ 150 MeV				
	10	0.25	n*10	1.2
ADONIS (ex.: d, Li – 25 mA @ 40 MeV)				
	10	2.5	~100	13
Alternative: d, C → pushed to the limit?				

*ϕR: neutron flux @ E > 1 MeV, in 10^{14} n/cm².s

beam, and for 40 MeV deuteron stripping on light nuclei (C, for SPIRAL-2 in GANIL, Li for IFMIF dedicated to fusion materials).

An advantage of the deuteron option used in SPIRAL-2 and IFMIF is that the low binding energy of the neutron leads to a forward peaked neutron beam with a very small dispersion. For a given neutron source, it is possible to get a high current entering an experimental tubular volume of a few tens of cm³.

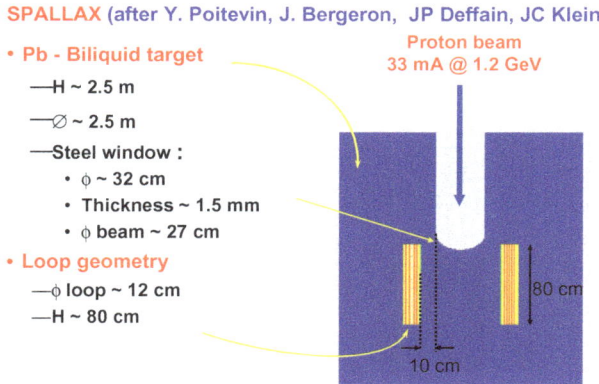

Figure 15.21. SPALLAX in the framework of the CONCERT project: some key parameters.

SPIRAL-2, and beyond: what could be the next steps?

Thanks to the numerical simulation (starting with metals and alloys), to new measurement tools and to innovative instrumentation (smart experiments prepared, interpreted and extrapolated by simulation), even a few dpa in a few mm^3 can be useful, if they are available in a short time for interpretation.

Numerical simulation will bridge the gap by steps: validation, transposition – extrapolation, then moving to the next test, building a shorter path from basic science to design tools and leading to a shorter time to market for innovative nuclear materials and structures.

The dislocation scale is promising (Figure 15.22) as a focal point for observation, measurements and for numerical simulation. The dislocation scale seems well suited to

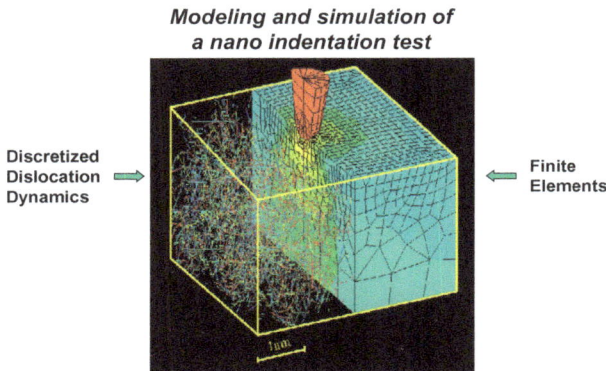

Figure 15.22. Modeling and simulation of a nano indentation test SPIRAL-2 (GANIL) could be the first step in ADNIS facilities, with about one dpa per trimester on small targets for 5 mA of 40 MeV deuterons.

experiments on small samples, to examination of the density and distribution of dislocations and to micro-mechanical measurements (for instance by nano-indentation (500 νm)), with comparison to calculations.

IFMIF looks very ambitious with 10^{17} n/s, 20 dpa/yr in 500 cm^3, for 2 × 125 mA of 40 MeV deuterons.

Updated specifications for the science of nuclear materials under irradiation, with the help of numerical simulation, lead to the consideration of a ADNIS around 20/25 mA of 40 MeV deuterons as a useful next step, dubbed **ADONIS** (O for optimised).

Finally, the technologies (accelerators and targets) considered for ADS transmuters will find their way to the future through other applications (RIBs (Radioactive Ion Beams) for science; nuclear material irradiation), showing how the synergy between fission, fusion and ion beam technologies can be developed.

References

Accelerator Driven Systems (ADS)

[1] M. Boidron, G.L. Fiorini, H. Flocard, J.B. Thomas, *Dossier de motivation pour un démonstrateur hybride*, Joint publication issued by the French R&D organisations involved in the ADS assessment. CEA (ADS Program) and CNRS/IN2P3, January 2001.

[2] A.A. Harms, M. Heindler, *Nuclear Energy Synergetics*. Plenum Press, 1982.

[3] *L'accélérateur de type ADS pour le projet MYRRHA*, presentation by Jean-Luc Biarotte at the Séminaire IPNO, ORSAY – February/07 – 2011, by courtesy of Jean-Luc Biarrotte and Alex Mueller (CNRS).

Nuclear space power and propulsion

[4] Xavier Raepsaet: private communication.

[5] *Nuclear and Stirling engines spur space exploration.* World Nuclear News, 27 November 2012.

16 A few questions fostering further thought on some key issues

J.-B. Thomas

Using the time issue as a starting point; long term problems (typical time range: a century) and short term events and trends (time range: a decade) compete as game-driving forces.

16.1. The designer's carrousel

"It played that same song about fifty years ago when I was a little kid. That's one nice thing about carrousels, they always play the same songs." (Salinger – "The Catcher in the Rye").

In nuclear engineering, designers seem to have been circling the same carrousel for more than fifty years, "trying to grab for the gold ring".

It is not an issue specific to nuclear engineering. For instance, after the "Exposition Universelle of 1900" (a world's fair) in Paris, the expert reporting for the "Mechanics" section summed up the whole thing in a fine-tuned, perfect sentence: "The motor vehicle of the future will obviously have an electric engine, if cheap and light refill batteries are available in the most remote places."

Nevertheless, in both cases, the common feeling is that the revolution will ultimately be driven by the issue of natural resource, that technological breakthroughs will provide the necessary enablers and that a market-pulled, technology-pushed solution will arrive in due time.

In the meantime, in the nuclear field, LWRs have built up an impregnable position similar to that of the piston combustion engine, thanks to a formidable operation feedback and to an incremental but continued adaptation to changing and increasing requirements, specifically regarding safety.

16.2. Entering a new era or circling around a carrousel?

The introduction chapter called for a new era in reactor design and selection, after the highly creative starting phase and after Darwinian selection leading to the LWR supremacy. Taking into account the SWOT analysis of the LWR after a long and successful experience feedback, it seems necessary to set up, internationally, an updated set of specifications and of new design orientations. They should apply to a fleet of acceptable and competitive reactors and fuel cycle facilities for the (near) future, adapted to the expected growth and to the worldwide implementation of nuclear energy.

This approach led to initiatives such as GEN-IV and GIF, INPRO, etc.

By another way, after Fukushima, international harmonisation of nuclear safety requirements and technical specifications has been advancing. It will contribute to update and complement current solutions regarding reactor system design as well as crisis management, including the topics of hardware (an unbreakable, unstoppable "hardened core" implementation, plus mobile devices and "no failure" connections), as well as "software" (organisational and procedural).

On the other hand, the last decade was rich in events that altered the context and the course of the growth of nuclear energy in the world:

- the financial and economic crisis;

- the escalation of power plant capital costs (with some exceptions in the world);

- the buoyant rise of non-conventional gas sources. The cost value currently ranges from a few US$/Mbtu in the USA to typically the double in Europe and four times as much in Asia. But, more importantly, it is rather unstable and sensitive to geopolitics, excepting the case of domestic resources;

- the Fukushima nuclear accident;

- specific and diverging European orientations.

16.3. Main questions to be addressed (combining innovation, design, marketing and acceptance issues)

- Are designers running on a flat circle or climbing up a helix leading, in a finite time, to a safe, competitive and sustainable industrial complex composed of a reactor fleet and of fuel cycle facility centres? Moreover, would this type of solution be acceptable worldwide as well as reasonably proliferation resistant?

- Is there convergence between the Darwinian evolution of the reactor system phylum and the "vertical" axis of the helix heading towards an ultimate jump in value (from the fuel cycle) for increased sustainability? Or are we facing a contradiction forcing a bifurcation and, in this case, how can this revolution be implemented without impeding nuclear competitiveness, primarily against "King Coal", for base-load power production?

- Does the hypothetical helix not resemble the ziggurat-shaped Babel Tower of Brueghel's painting (a feeling supported by listening to discussions between experts on the advantages and drawbacks of their respective preferred options, playing the game according to a "winner takes all" rule rather than a "Nuclear Energy Synergetics" principle)? Is there still a common language, for specifications as well as system assessment (SWOT)?

- Isn't it necessary, from time to time, to copy Daedalus, the architect escaping the prison built after his own plans to satisfy his powerful client? Adding a new dimension to the problem (namely the time dimension), setting a point of the figure outside the conventional boundaries of the drawing frame (optimising a hybrid-symbiotic, multi-strata rather than monolithic reactor fleet) opens new possibilities. Wisdom commands to avert Icarus' fate, the lethal hubris often being a gift from malevolent gods. Higher performances can sometimes be deadly, if the search for them makes simple, safe and cheap solutions out of reach.

- What are the main skills and vision required from the design "dream team" (in the early times, a lonely pioneer) in order to achieve useful steps towards satisfying the needs of the present time and of the future?

16.4. Some answers coming from past and recent history

- Evolution is strongly "market pulled", the word "market" including the conditions for acceptance which, in turn, control costs through regulations. In this context, designers are firing on a mobile target. This requires flexibility and adaptability rather than higher performances in a single area.

- Building up a successful and competitive operation feedback is of the utmost importance. Incremental adaptation is hard to beat, even if some global architecture changes are required to avoid falling into a local optimum design trap. This was the feeling of Eugene Wigner: "Incrementally improved converters may eventually have their day" (even including the sustainability issue, which is still an open question, see below).

- Nuclear design combines functional/structural "genetic chunks" (just as nature does with living beings) in an overall architecture, from fuel and core to the balance of plant. For example, post-Fukushima design requires to strengthen and "harden" the implementation of full defense-in-depth principles, with built-in systems and devices as well as a set of indestructible "slots" for "plug and play" connections to mobile safety devices. Combining existing "genetic chunks" (for instance a passive cooling system connected to an unlimited or refillable, natural and compatible –air or water– heat sink through a "bunkerised" buffer) is a step to be taken, whatever the reactor type. The related pre-conditions and constraints have to be taken into account by the designer. They raise a challenge in terms of architecture and potentially of core coolant and core structure choices.

What can be learned from the diverse visions of the reactor systems issue?

- The plainly descriptive or "kaleidoscopic" vision gives an overview of the "biodiversity" and of the main "genetic chunks" that can be combined. The related SWOT assessment gives more insight into the problems to overcome, the limitations of some basic choices (coolant) and the remedies that design and innovation can offer.

- The "axiomatic" approach starts with the declaration of the foremost priority, the critical issue for the long term sustainability of fission. It is generally related to fuel cycle and natural resources. Then it is shown that a conceptual solution exists and that it can be considered (alone) as a global solution. Finally, a set of specifications and one or a few examples of related designs are proposed. The value of this approach is that it provides a polarisation of the field of nuclear engineering beyond the "business as usual" attitude.

- The "historic" vision brings the knowledge of the mechanisms at work in the process of selecting a reactor type and fuel cycle options. It gives the ability to understand how forces interplay and how the actors' game is played. It leads beyond the consideration of purely technical SWOTs. It opens a door towards resuming the design and marketing of new types, of new options. The historic approach can, in some instances, bring along a key element: the choice of Admiral Rickover for the US submarines was decisive for the course of events. It was well served by a methodical, pragmatic and tenacious determination (see the corrosion issue for instance) as well as a unique concentration of efforts commensurate with the stakes involved.

- A "dynamic" approach gives access to strategic prospective through modelling and simulation of the forces at work, taking into account the main constraints and some of the unanticipated events that can occur. From this viewpoint, nuclear energy has to be embedded into the broader field of energy. This gives some precious clues by comparison to the fate of other energy sources. On top of these considerations, a set of assessment criteria, like those elaborated by the WEC, shows how contradictory requirements must be balanced by decision makers at various scales: world, region, country, then "local" and even individual scale. During the global assessment, techno-economy is a powerful but sometimes puzzling tool. The rules of economic assessment often diverge and can be governed by political choices rather than universal "business" or "public interest" patterns.

Obviously, a global presentation of nuclear systems combines these four main approaches. In the future, the dynamic vision and assessment must be developed and strengthened. It involves the whole set of social sciences as well as philosophy. Nevertheless, the weight of technology and design will remain decisive. The issues of safety harmonisation, risk assessment (including in the case of a severe accident, and with a focus on the effect of low doses delivered at a low dose rate) and proliferation resistance will be the top areas of interest for the future of nuclear energy, in a fierce debate with "the heuristics of fear" (Hans Jonas) and with the "rage against technology" (Heidegger, Ellul, etc.).

The main tools for a successful management of nuclear reactor fleet development

In the commodity business, a "market pull, technology push" strategy is often used. How can it be harnessed to become a relevant tool in the nuclear energy area?

- "Augmented marketing" including the acceptance issue. A set of criteria including the energy "trilemma" has been elaborated by the WEC.

- Technology means incremental improvements, innovation (sometimes, but rarely as a breakthrough) and design.

16.5. Design as a conceptual approach: design wheel and "helix"

- Design runs back and forth to the requirements, through loops involving feasibility and compliance, as well as sizing and trade studies.

- The multi-dimensional "flight envelope" definition is critical. It is tightly connected to the explicit as well as implicit and anticipated requirements, because the lifetime of nuclear systems is of (at least) half a century.

- The first deliverable of design is a comprehensive but simplified drawing with a set of values for critical parameters. It should be traced back to a few simple calculations made by hand, based on basic science as well as databases and expert knowledge bases.

- Fulfilling the requirements means firing on a mobile target. In the long run, the level of the qualities (safety, performances at a given safety level) is increasing. The design "wheel" (Figure 16.1) becomes a helix. People in the field can acknowledge these improvements by comparing successive generations of systems over three to four decades. But sometimes, a poor choice leads to a dead end (at least a durable one for the type of reactor considered).

16.6. Beyond the incremental improvement of LWRs (safety, flexibility, fuel cycle (plutonium), lifetime, availability, uprating), what are the main achievements of recent (in the last three decades) design and operational qualification for power reactors?

The balance is somewhat frustrating. Comparison with the aeronautical field, for instance, gives a striking image (see Figure 16.2):

- the move from the Bell X-15 to the "SpaceShipOne" which successfully reached an altitude of 112 km in 2004 shows the use of technology improvements plus major design improvements: for instance, the shape was adapted for a smoother reentry into the atmosphere;

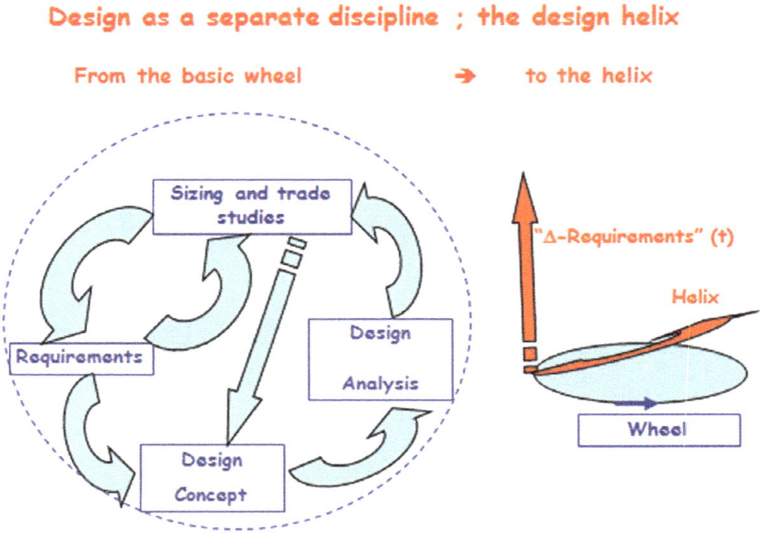

Design as a separate discipline ; the design helix

From the basic wheel → to the helix

Figure 16.1. The design wheel turning into an upward helix.

Figure 16.2. Innovation, design and safety requirements: there are but necessary conditions. X-15 in flight (Air Force photo); SpaceShipOne in flight (SpaceShipOne Press gallery); The NERVA NRX-A3 Experiment at Jackass Flats (Nevada).

- on the other hand, the NERVA project (or the sister-project PLUTO: a nuclear powered cruise missile) shows amazing results: 10 MWth/l power density in the core, the unit power of the EPR, a hydrogen exit temperature well over 2000 °C (for NERVA). Is there a civil spin-off of this (very specific and free of almost any terrestrial safety requirement set) beautiful piece of nuclear engineering? The closest reactor design is the modern HTR concept. Unfortunately it is based on a bright – but of limited efficiency – "total passive" Decay Heat Removal (DHR) implementation. The power density is down to less than 10 kWth/l (depending on the fuel type) and the unit power to about 600 MWth. Some new requirements about extended protection are challenging the "light containment" option. Regaining a decade in power density and an optional increase in unit power could open the way to a gas-cooled fast reactor (GFR from Gen-IV) for the cost of conventional containment and hybrid passive/active safety implementation. On the other hand, the sets of specifications in both cases (NERVA and HTR) are at odds with each other, explaining the major part of the performance gap.

It seemed to be a « no U-turn » road. But natural convection can be used of efficiently (at a given backpressure depending on the decay power level, hence of the time after shutdown) and complemented with innovative Decay Heat Removal systems (DHR-S). A "state driven" type analysis (used for LWRs) - in a simpler temporal logic scheme than for LWRs - during a typical Design Basis Accident (DBA) or a Beyond-DB / "envelope" Accident, would help determine the action planning and the specifications of the DHR5-S in order to get a sequence of "no failure" solutions paving the way to a safe and stable state at low pressure, low temperature and low decay heat power. This would improve the safety demonstration for an intermediate core power density Gas Cooled Reactor, whatever the spectrum. Two examples are the Gen-IV GFR (chapitre 11) in its present design evolution and the GA's EM2 project, as a SMR, HTR and a Gas Cooled Fast Reactor. Such improvements need to be integrated in an advanced overall safety framework for innovative reactors, and specifically for Gas Cooled Reactors. At first a safety framework "for designers", as a tool for a smarter and robust design.

16.7. Other examples

An additional attractive example is given by the most recent BWR concepts: ESBWR + KERENA.

The drawing (of and around the KERENA NSSS, see Figure 16.3) shows a remarkable arrangement of water-steam systems and devices in an annular stack enclosing the NSSS, with the "containment lake" as a huge water reserve and shielding topping the whole cylinder.

- A question at the moment is how to implement the "hardened core" paradigm on existing LWRs (mainly Gen-II).

- The same question holds for a Gen-IV reactor, depending on specific issues concerning potential interactions between the coolant and the ultimate heat sink fluid: water, air.

A critical and generic issue where the design conceptual approach must be revisited and extended is the coolant issue.

Figure 16.3. When the drawing reveals the power of design (KERENA from AREVA).

16.8. The coolant issue: updating some questions

- Using a "specific moderator" is no longer an imperative. On the other hand, heavy water is a soft moderator (giving access to high fuel cycle performance, in a thermal or in a harder spectrum) and a good coolant. Unfortunately, many drawbacks (including the tritium issue) are barring the way. Carbon is also a "soft" moderator and transparent to neutrons (no significant capture). It can be used in structure materials (graphite, carbide) compatible with coolant, for instance in GFRs or HTRs, probably as SiC for additional compatibility reasons.

- Fuel (in the case of solid fuel): in any case, oxide will be a starting point for any reactor type. Then, if and only if it is possible to implement a safe, efficient and competitive reactor/fuel cycle loop (including multi-recycling), exploring improvements with metal, carbide or nitride fuel will be considered. Carbide could be an exception in the case of a very tight neutron balance for some fast neutron systems.

These updates simplify the analysis.

On the other hand, another update will make things worse. It is related to the extended safety assessment taking into account severe accidents and the implementation of a very long term (months, years) closed loop (no feed and bleed) cooling of a corium inside a leak-tight containment (no leaks, no permanent and durable water contamination of hundreds of thousands cubic meters of water to be stored before decontamination). It involves dealing with the potential interaction of the coolant with the fluid of intermediate systems, but also, even if only during short periods of time, with the ultimate natural heat sink composed of air or water. Could the severe accident coolant be different from the normal operation coolant?

16.9. As for the coolant choice, there is no single merit index

A usual viewpoint, stemming from the handicap related to the huge power consumed by pumping in early gas-cooled reactors (UNGG for instance), is to compare (from H. Safa):

- the heat extraction capability: $Q = \dot{m} c_p \Delta T = \rho S v \Delta T$
- the pumping capacity: $P_{pump} = \dot{m} \Delta p/\rho = (\dot{m}/\rho) C \rho L(V^2/2\phi)$

and to define a "Merit Index" as:

Q/P = efficiency of heat extraction (in one-phase–liquid or gas) leading to the merit index F:

$$F = \rho^2 C_p^{2.8} \mu^{-0.2}$$

In this case, and comparing the cooling merit indices of various cores in one-phase flow, one finds (with the Na merit index arbitrarily set to 1) (see Table 16.1):

Table 16.1. A simplified, scalar, coolant merit index.

Coolant	Reactor	Power density	Cooling Merit Index
H_2O	PWR	100	65
Na	FBR	300	1
CO_2	MAGNOX	1	0.0001
He	HTR	6	0.0005

This cooling merit index is useful, but:

- it only takes into account the one-phase core cooling efficiency;
- it is tightly dependent on the core and on the energy conversion system design. For instance, in the case of a natural convection BWR (ESBWR), there is no pumping power at all. So, the cooling merit index is… very high. Moreover, there is no secondary system. So the total pumping power tends to vanish;
- another dimension of cooling performance is the ability to remove high power density generation. In this case, the comparison between the MAGNOX performance (1 MW/m^3) and NERVA (10 GW/m^3) shows that:
    - the safety constraints,
    - and the fuel technology and design,

are of utmost importance. Moving to a fuel (HTR TRISO; NERVA fuel) which is simultaneously "cold" (TRISO particles have a very small diameter and the conduction in the graphite matrix is very good, so $T_{fuel} - T_{coolant}$ is small) and refractory (the fuel can work at high temperatures and still leave large margins against accident heat up), as long as the coolant is chemically inert (He, as long as it is pure enough), makes it possible to drastically reduce the flow required to cool the core by increasing the ΔT. "Flat cores", a larger core section fraction dedicated to coolant flow, moderately larger hydraulic diameters (a possible choice with a coolant that is transparent to neutrons (no significant slowing down, no significant capture, even around 70 bar but the Nusselt coefficient requires careful consideration) can help, too. These reasons show that a higher power density (mandatory for a fast reactor) can be efficiently removed from a gas-cooled core in normal operation. The Decay Heat Removal (DHR) during accidents is setting another demanding framework (see § 16.6) for coolant efficiency assessment, but dedicated solutions exist and must be put into perspective (totally passive solutions at any time, a mix of (temporarily) active and passive solutions, a specific mix of (out of neutron flux) coolants, harnessing the residual power to deliver the coolant flow by "semi-passive" core cooling, etc.);

- one has to enter into the details of comparing one-phase flow and two-phase flow cooling of a core, depending on the situation (normal operation vs. accident and DHR context without pumping). As for water, in both cases two-phase flow is a powerful ally (including feed and bleed situations directly connecting the core with an ultimate/refillable natural heat sink (water)). Sodium is a powerful and efficient one-phase flow coolant at low pressure; the margin to boiling is large in ΔT (with a moderate c_p (1.27 MJ/kg/K)), thus a limited ΔH) but sodium flashing is not a favorable event. Moreover, sodium must be protected from external water and air ingress, both in normal operation and in accident situations, increasing the number of intermediate systems of top quality and resilience, even in extreme situations.

16.10. Main topics involved in the coolant issue

The previous section showed how difficult it is to define a single merit index for the coolant (even for cooling efficiency). Is it possible to identify the main dimensions of the coolant issue, connecting it to almost the whole set of situations, components, physics, etc.?

Putting it first into words would make it possible to set up a table to be used as a framework for representing the main dimensions of the problem and for structuring, then planning the assessment.

a) The main objectives are:

- to efficiently fulfill the power generation functions and the safety functions;
- to avoid the occurrence of any adverse effect on the performances (physics, economy) and on the safety.

b) This has to be satisfied in any type of situation:

- normal operation, including outages, and specifically loading/unloading operations;

accident, up to extreme and durable situations involving non-standard DHR devices and systems including those belonging to the "hardened core" and to the mobile auxiliary safety systems that the "Rapid Action Nuclear Force" team is bringing along.

c) These requirements involve the study of:

- potential interactions with every type of component and material, including in case of breaks, leaks, involving external elements or not, the occurrence being generated by natural or malevolent causes. For instance, a violent interaction between incompatible coolant and fuel could initiate a fast degradation of the situation and lead to a global severe accident;
- every type of interaction involving the whole fields of physics and chemistry (neutronics, mechanics, etc.) as well as multiphysics coupling (neutronics, thermal-hydraulics, fluid-structure interaction: for the SFR, of the ULOF type; for the PWR, of the SLB type);
- severe accidents, managed in the perspective of ultimate resilience of the safety architecture and systems, similar to the "Vauban" design, leading to yet unanticipated situations and "encounters of the third kind" of materials, of fluids.

d) The fields of:

- front end (production (D_2O)/procurement (He), transport);
- back end (decommissioning, decontamination, recycling vs. destruction (sodium), with specific issues: irradiated Pb as double toxic, by several dozen thousand tons; Pb-Bi as a ^{210}Po source);

must give rise to specific studies in order to complement the assessment during the whole "lifecycle" of the coolant. Criteria include economy, independence, safety, health and environmental footprint.

The following diagram (Figure 16.4) tries to summarise the assessment process and logic:

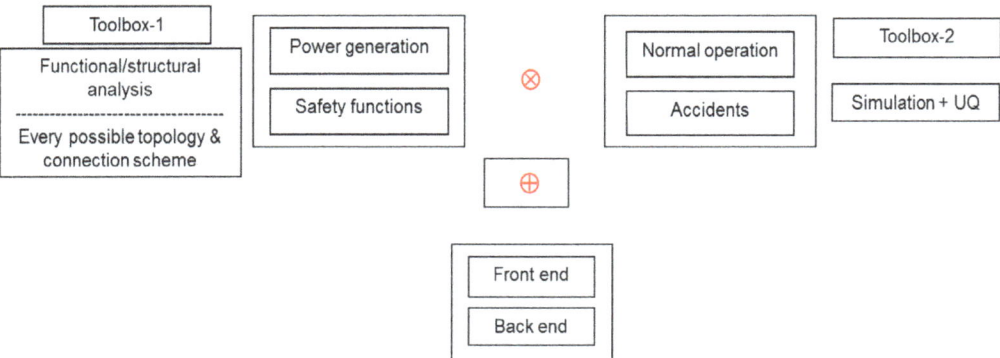

Figure 16.4. A simplified diagram for an "asymptotically comprehensive" coolant assessment. UQ: Uncertainty Quantification.

16.11. Multi-criteria assessment: the representation and computation issue; a tentative representation diagram

Most of the SWOTs related to a given coolant are implicitly (and often unconsciously) dependent on a given context: of spectrum, of fuel composition (U-Pu, Th, U3, a mix), of fuel technology (cold and refractory – HTR type – or conventional rods clad with Zircaloy or with steel; oxide, carbide, metal). They also depend on the way the techno-economic rules of the game (see the various viewpoints about plutonium recycling) are fixed and on the way safety is assessed.

This is the case for the following diagram (Figure 16.5) representing the SWOT analysis of four arbitrarily chosen (and deliberately unnamed) coolants. The diagram would look different if projected on another "leaf" of the same expertise book, related to another set of constraints or to a different weighing of these constraints. Obviously, there could be quantitative assessments and formulae (plus a fuzzy logic engine) hidden behind the points of the diagram. But the resulting values cannot be combined within an optimisation algorithm.

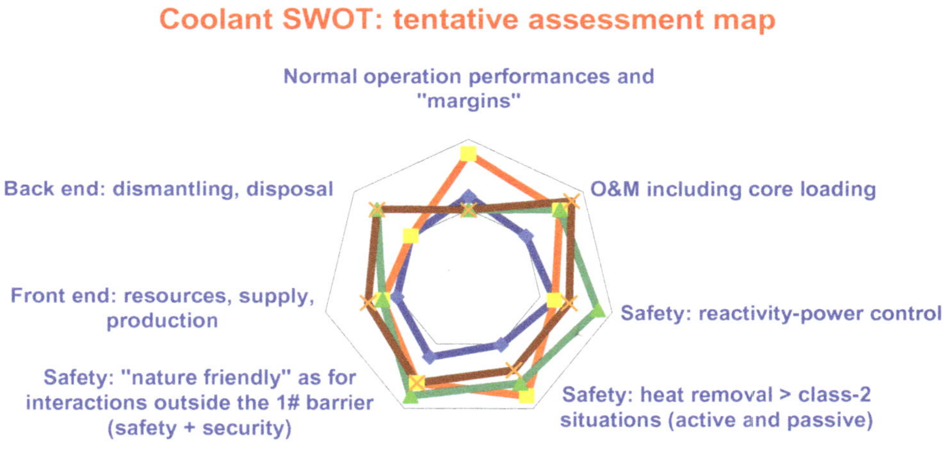

Coolant SWOT: tentative assessment map

Figure 16.5. A tentative, conceptual assessment map.

Last but not least, the "ranking" can be modified in the future by technological breakthroughs, for instance in fuel structure and "cladding" (see for instance R&D on the so called "Accident Tolerant Fuel"), in innovative cooling and DHR devices or in architectural design (combining functional and structural requirements with new potentialities). This is not a "forever" stable mapping and therefore not a very comfortable one for thinking on design options.

16.12. Making a positive contribution to the qualification of Gen-IV "enablers"

Sometimes, some thresholds are exceeded and feasibility is threatened. In this case, a backtrack process is started: the designer has to circumvent the obstacle. The core catcher option can be considered as an example; the open question of a limited unit power (and power density), making it possible to achieve in-vessel corium cooling and retention (IVR), is an alternative. In any case, there is a cost to be paid.

The same holds for a drastic reduction of the positive void coefficient and temperature coefficient in "plainly homogeneous" SFR cores loaded with plutonium. More sophisticated heterogeneous cores, innovative thermo-mechanical arrangement of the structures, optimised fuel composition (mixing various fissile nuclei) can help. Beyond this critical role, the designer must imagine solutions taking into account the worst possible realisation of the uncertainties from basic physics and from numerical simulation. Moreover, the designer has to imagine how benefiting from the bonus given by the first operation feedback will help reduce these uncertainties.

The number of intermediate circuits can be varied in order to exclude by design extremely improbable but potentially dangerous interactions between the coolant and the balance of plant or the outside fluids. There is an overcost but it can also offer some advantages (efficient and flexible energy conversion).

Etc. (other examples)

In some cases, it is harder to find a solution (Pb-Bi releasing volatile ^{210}Po; the enigma of steel embrittlement by liquid metals; extreme corrosion cases where the remedies are difficult to implement and to control at the scale and in the industrial operation context of large power plants (Pb, some molten salts)).

Identifying these issues in due time before launching costly demonstration projects is critical. Once they are detected, fixing them requires inspiration from the "Daedalion" or "Metis": the ability, in problem solving, to combine, with creativity and realism, various symbiotic "assets" in an innovative architecture. It requires a large body of expert knowledge and the availability of light and powerful problem solving methodology and tools (including software "advisers"), as well as advanced optimisation tools (for instance genetic algorithms).

16.13. Knowledge bases and tools

Addressing the assessment issue requires the use of several simulation tools and expert knowledge bases, as well as the rules of the game concerning economy and safety (changing with time, and region-dependent):

- physical and chemical properties: databases, models;

- multi-physics interactions (validated simulation);

- architecture, topology, connections (for the mainstream design of a given reactor type);

- all classes of situations during the whole lifecycle (partially depending on the results of the study and not fixed forever);

- performances/cost: optimisation tools taking into account competitiveness assessment rules as well as acceptance and safety constraints.

16.14. "War" is (or should be) over

For many decades to come, not one reactor system will be able to achieve the whole set of Gen-IV objectives. Initially (as of 2000), Gen-IV was supposed to start around 2030.

But the so-called "Gen-IV" requirements and objectives are not related to a particular reactor system but to a fleet of plants that may include several types of reactors, as well as the fuel cycle facilities necessary to the (sustainable) closed fuel cycle.

The name of the game is: efficient and competitive multi-recycling. The efficiency associates two needs which are physically closely coupled (through the cross sections, starting with the U-Pu example):

- limiting the degradation of plutonium isotopic composition, with a pragmatic criterion based on a comparison of recurrent recycling (*As Fast As Reasonably Achievable*) with a longer out of pile decay without periodic valorisation and "rectifying";

- reducing the additional amount of fissile nuclei (hence of fresh natural uranium) necessary for producing a given amount of energy.

These characteristics translate into the following advantages:

- resuming the progress towards closing the fuel cycle (at first for plutonium and uranium, then possibly shifting to more sophisticated strategies and to blends involving thorium + U3);

- saving natural resources (at first by a factor of two to four, which is highly significant).

As for safety, there will not be any noticeable difference between post-Fukushima Gen-3+ LWR and the best of Gen-IV. The very issue will be to address the new safety challenges in the framework of Gen-IV designs.

As for competitiveness (except in some specific market "niches", see the SMR concept), the (large) LWR will be the only system able to compete with coal and gas on large grids, at least during the next decades. For Gen-IV systems, two causes are slowing down or even limiting the cost reduction:

- intrinsic design features;

- bridging the gap in operation feedback between LWRs and newcomers, whatever the type, starting with the mandatory requirement of high availability.

16.15. Optimisation of a multi-strata nuclear fleet achieving "smart recycling" is the new frontier

The strata structure and the right blend make the Gen-IV fleet.

Some requirements must be fulfilled whatever the reactor type. This is the case for safety. On the other hand, some qualities can be brought by specific "enablers" (transmuters (?), breeders). Competitiveness must be achieved thanks to the "workhorse" of the fleet and with a high support ratio, meaning that more costly enablers support a large number of cheaper "power reactors".

For the current century, the Gen-IV fleet will be composed of **a blend of Gen-3$^+$**, post-Fukushima LWRs, potentially capable of efficient multi-recycling during their lifetime, and of genuine Gen-IV reactors, the leaders of the current competition, namely **SFRs**, while some of their challengers will be developed and qualified.

The fleet composed of different **"strata"** and of their related reactor types, capable of complementary functions through specific features, can be built up incrementally, giving time to the newcomers: time for more operation feedback and credibility (industrial operation and safety); time for optimisation. This process benefits from the formidable operation feedback of LWRs. LWR limits can be pushed further by incremental improvements, even for the fuel cycle. The objective is not to modify the LWR role during the century (power reactors, advanced converters) but to help the fleet as a whole come closer to the objectives (because power reactors form the main part of the fleet), before specialised systems help make the final jump, bridge the gap. That is the only way to value newcomers which can't win the contest on their own.

LWRs should thus be able to reach a multi-recycling regime, as "rectifiers" and "savers", keeping the built-up plutonium in good condition for its future, when it would enter the breeding regime into FBRs, while thorium and ^{233}U could be introduced stepwise and with cross-valorisation into both reactor systems.

Beyond improvements of the reactor systems and of the fuel cycle processes and facilities, optimisation relies on the (slow) evolution of the composition of the fleet and of the related support ratios.

16.16. Qualification (including substantial operation feedback) of all efficient enablers, with an updated design fulfilling the post-Fukushima requirements, must be started ASAP

There is no time left. There is no contradiction between resuming the massive construction of Gen-3$^+$ LWRs and launching ASAP the demonstration, then "NOAK" operation, of SFRs and of promising alternatives. Presently, in the wake of Fukushima, safety improvements, followed by stabilisation, are the top drivers.

Epilogue: J.D. Salinger: the carrousel ("going around and around") and the gold ring.

The playing field of nuclear energy is reminiscent of the final scene in "The Catcher in the Rye" by J.D. Salinger:

On the carrousel, "**All the kids kept trying to grab for the gold ring** and so was old Phoebe, and I was sort of afraid she'd fall off the goddam horse, but I didn't say anything or do anything. The thing with kids is, if they want to grab for the gold ring […] If they fall off, they fall off, but it's bad if you say anything to them."

Whatever the field (design and safety, operation, dismantling, R&D, teaching, etc.), knowledge and understanding of reactor systems will be a valuable resource in trying to grab for the gold ring, reducing the risk of falling off and increasing the pleasure to play as well as the chance to win (contributing to decisive improvements).

Lightning Source UK Ltd.
Milton Keynes UK
UKHW050609070223
416551UK00003B/133